LIPOSOMES FOR FUNCTIONAL FOODS AND NUTRACEUTICALS

From Research to Application

LIPOSOMES FOR FUNCTIONAL FOODS AND NUTRACEUTICALS

From Research to Application

Edited by
Sreerag Gopi, PhD
Preetha Balakrishnan, PhD

AAP APPLE
ACADEMIC
PRESS

First edition published 2022

Apple Academic Press Inc.
1265 Goldenrod Circle, NE,
Palm Bay, FL 32905 USA
4164 Lakeshore Road, Burlington,
ON, L7L 1A4 Canada

CRC Press
6000 Broken Sound Parkway NW,
Suite 300, Boca Raton, FL 33487-2742 USA
4 Park Square, Milton Park,
Abingdon, Oxon, OX14 4RN UK

© 2022 Apple Academic Press, Inc.

Apple Academic Press exclusively co-publishes with CRC Press, an imprint of Taylor & Francis Group, LLC

Library and Archives Canada Cataloguing in Publication

Title: Liposomes for functional foods and nutraceuticals : from research to application / edited by Sreerag Gopi, PhD, Preetha Balakrishnan, PhD.
Names: Gopi, Sreerag, editor. | Balakrishnan, Preetha, editor.
Description: First edition. | Includes bibliographical references and index.
Identifiers: Canadiana (print) 20210381736 | Canadiana (ebook) 20210381795 | ISBN 9781774637548 (hardcover) | ISBN 9781774637555 (softcover) | ISBN 9781003277361 (ebook)
Subjects: LCSH: Liposomes. | LCSH: Functional foods.
Classification: LCC QH602 .L59 2022 | DDC 571.6/55—dc23

Library of Congress Cataloging-in-Publication Data

CIP data on file with US Library of Congress

ISBN: 978-1-77463-754-8 (hbk)
ISBN: 978-1-77463-755-5 (pbk)
ISBN: 978-1-00327-736-1 (ebk)

About the Editors

Sreerag Gopi, PhD
Chief Scientific Officer, ADSO Naturals, India;
Vice President, CureSupport, The Netherlands,
Mobile: +91-8594023331,
E-mail: Sreeraggopi@gmail.com

Sreerag Gopi, PhD, is a Chief Scientific Officer at ADSO Naturals, India, and Vice President at CureSupport, The Netherlands. He graduated with a degree in Chemistry from Calicut University, Kerala, India, and an advanced degree from Madras Christian College, Chennai, India. He is a recipient of a prestigious Erasmus Mundus Fellowship from the European Union during his PhD period. He is a materials chemist and nanomaterials scientist and has expertise in nanomaterial synthesis, characterization, biocomposites for natural products, and biomedical and water purification experiments. He has published over 20 peer-reviewed international papers and several book chapters and has book projects in the works with several publishers, including the Royal Society of Chemistry, Springer, and Wiley. He was selected as an Associate Member of the Royal Society of Chemistry in 2018, and he is a chartered member of the Royal Australian Chemical Institute.

Preetha Balakrishnan, PhD
Principal Scientist, QA/QC, ADSO Naturals, India,
and CureSupport, The Netherlands,
Mobile: +91-7025921175,
E-mails: preetha@adsonaturals.com;
P.Balakrishnan@curesupport.com;
b.preethabalakrishnan@gmail.com

Preetha Balakrishnan, PhD, is the Principal Scientist, QA/QC at ADSO Naturals, India, and at CureSupport, The Netherlands. She graduated in Chemistry from Calicut University, Kerala, India, and earned her postgraduate degree at Mahatma Gandhi University, Kerala, with a gold medal and first rank. She is a recipient of the prestigious INSPIRE Fellowship from the Government of India. She was a postdoctoral researcher in the research group of Professor Sabu Thomas, Vice Chancellor, a renowned scientist in this area who has sustained international acclaim for his work in polymer science and engineering, polymer nanocomposites, elastomers, polymer blends, interpenetrating polymer networks, polymer membranes, green composites and nanocomposites, nanomedicine, and green nanotechnology. She completed her PhD in Chemistry at Mahatma Gandhi University under Dr. Thomas's guidance. She has visited many foreign universities as a part of her research activities and has published over 15 research articles and over 20 book chapters. She has edited 10 books with leading publishers, including Elsevier, Springer, Wiley, and the Royal Society of Chemistry. Dr. Balakrishnan has received a number of national and international presentation awards. She also worked as a guest lecturer in chemistry at the Department of Chemistry, Morning Star Home Science College, Angamaly, Kerala, India.

Contents

Contributors

Chirom Aarti
Research Department of Plant Biology and Biotechnology, Loyola College, Chennai – 600034, Tamil Nadu, India

Hossein Adelnia
Australian Institute for Bioengineering and Nanotechnology, University of Queensland, Brisbane, QLD, Australia; School of Pharmacy, University of Queensland, Woolloongabba, QLD, Australia

Khalid Saad Alharbi
Department of Pharmacology, College of Pharmacy Jouf University, Aljouf, KSA

Nasser Hadal Alotaibi
Department of Clinical Pharmacy, College of Pharmacy Jouf University, Aljouf, KSA

Nabil K. Alruwaili
Department of Pharmaceutics, College of Pharmacy, Jouf University, Aljouf, KSA

Sultan Alshehri
Department of Pharmaceutics, College of Pharmacy, King Saud University, Riyadh – 11451, Saudi Arabia

Chettupalli Ananda
Department of Pharmaceutics, Center for Nanomedicine, School of Pharmacy, Anurag University, Venkatapur, Ghatkesar, Hyderabad – 500088, Telangana, India

Sophia G. Antimisiaris
Laboratory of Pharmaceutical Technology, Dept. of Pharmacy, School of Health Sciences, University of Patras, Rio 26504, Greece
Foundation for Research and Technology Hellas, Institute of Chemical Engineering Sciences, FORTH/ICE-HT, Rio 26504, Greece

Preetha Balakrishnan
ADSO Naturals, India, and CureSupport, The Netherlands, Mobile: +91-7025921175, E-mails: preetha@adsonaturals.com; P.Balakrishnan@curesupport.com; b.preethabalakrishnan@gmail.com

Farid Abedin Dorkoosh
Department of Pharmaceutics, Faculty of Pharmacy, Tehran University of Medical Sciences, Tehran, Iran; Postal address: No. 1462, Kargar Ave, Tehran – 1439804448, Iran, Telephone/Fax: +982188009440, E-mail: dorkoosh@tums.ac.ir

Maryam Edalat
Pharmaceutical Sciences Research Center, Islamic Azad University, Tehran Medical Sciences University (IAUTMU), Tehran, Iran

Reza Baradaran Eftekhari
Department of Pharmaceutics, Faculty of Pharmacy, Tehran University of Medical Sciences, Tehran, Iran

Chunli Fan
School of Food Science and Technology, Collaborative Innovation Center of Food Safety and Quality Control in Jiangsu Province, Jiangnan University, Lihu Road 1800, Wuxi, Jiangsu – 214122, People's Republic of China

Valker Feitosa
PhD Degree in Biochemical and Pharmaceutical Technology, University of Sao Paulo, São Paulo, SP, Brazil

Sadaf Jamal Gilani
Department of Basic Health Sciences, Preparatory Year, Princess Nourah Bint Abdulrahman University, Riyadh, Saudi Arabia

Sreerag Gopi
ADSO Naturals, India; Vice President, CureSupport, The Netherlands,
Mobile: +91-8594023331, E-mail: Sreeraggopi@gmail.com

Dipak Kumar Gupta
Department of Pharmaceutics, School of Pharmaceutical Education and Research, Jamia Hamdard, New Delhi, India

Mohammad Hajimolaali
Laboratory of Pharmaceutical Technology, Department of Pharmacy, University of Patras, Rio – 2504, Greece

M. Akiful Haque
Department of Pharmaceutical Analysis, Anurag University, Venkatapur, Ghatkesar, Hyderabad – 500088, Telangana, India

Thais Aragão Horoiwa
Postgraduate in Industrial Process Masters by Coursework-Institute for Technological Research of Sao Paulo State, Cidade Universitária, SP, Brazil, E-mail: thaisaragao.h@gmail.com

Syed Sarim Imam
Associate Professor, Department of Pharmaceutics, College of Pharmacy, King Saud University, P.O. Box 2457, Riyadh – 11451, Saudi Arabia, E-mails: sarimimam@gmail.com; simam@ksu.edu.sa

Parinaz Inanloo
Islamic Azad University, Tehran Medical Branch, Tehran, Iran

Sheida Iranpour
Department of Pharmaceutics, Faculty of Pharmacy, Tehran University of Medical Sciences, Tehran, Iran

Mohammed Jafar
Department of Pharmaceutics, College of Clinical Pharmacy, Imam Abdulrahman Bin Faisal University, Dammam, KSA

Mohammed Asadullah Jahangir
Department of Pharmaceutics, Nibha Institute of Pharmaceutical Sciences, Rajgir, Nalanda – 803116, Bihar, India

Chandra Kala
Department of Pharmacology, Faculty of Pharmacy, Maulana Azad University, Jodhpur – 342802, Rajasthan, India

Parsa Khoshkhat
International Campus, School of Pharmacy, Tehran University of Medical Sciences, Tehran, Iran

Ameer Khusro
Research Department of Plant Biology and Biotechnology, Loyola College, Chennai – 600034, Tamil Nadu, India, E-mail: armankhan0301@gmail.com

Niloufar Maghsoudnia
Department of Pharmaceutics, Faculty of Pharmacy, Tehran University of Medical Sciences, Tehran, Iran

Abdul Muheem
Department of Pharmaceutics, School of Pharmaceutical Education and Research, Jamia Hamdard, New Delhi – 110062, India

Farheen Fatima Qizilbash
Department of Pharmaceutics, School of Pharmaceutical Education and Research, Jamia Hamdard, New Delhi, India

Mohamad Taleuzzaman
Professor, Department of Pharmaceutical Chemistry, Faculty of Pharmacy, Maulana Azad University, Jodhpur – 342802, Rajasthan, India, Mobile: + 91-7251892850, E-mail: zzaman007@gmail.com

Chen Tan
School of Food and Health, Beijing Technology and Business University, Beijing – 100048, China; Department of Food Science, Cornell University, Stocking Hall, Ithaca NY – 14853, USA

Xingwei Wang
School of Food Science and Technology, Collaborative Innovation Center of Food Safety and Quality Control in Jiangsu Province, Jiangnan University, Lihu Road 1800, Wuxi, Jiangsu – 214122, People's Republic of China

Shuqin Xia
School of Food Science and Technology, Collaborative Innovation Center of Food Safety and Quality Control in Jiangsu Province, Jiangnan University, Lihu Road 1800, Wuxi, Jiangsu – 214122, People's Republic of China, E-mail: sqxia@jiangnan.edu.cn

Ameeduzzafar Zafar
Assistant Professor, Department of Pharmaceutics, College of Pharmacy, Jouf University, Aljouf, KSA, Mobile: +966562789137, E-mail: zzafarpharmacian@gmail.com

Abbreviations

AFM	atomic force microscopy
AI	anemia of inflammation
AIDS	acquired immunodeficiency syndrome
ALA	alpha-linolenic acid
APL	alkyl phospholipids
ATP	adenosine triphosphate
AUC	area under curve
Ca	calcium
CAGR	compound annual growth rate
CARPA	complement activation-related pseudoallergy
CaS	calcium stearate
CAT	catalase
CEO	cardamom essential oil
Chol	cholesterol
CL	conventional liposomes
CLL	CL alone
CM	casein micelles
CoQ_{10}	co-enzyme Q_{10}
CUR	curcumin
DCP	dicetylphosphate
DDPC	didecanoyl phosphatidylcholine
DE	double emulsion
DHA	docosahexaenoic acid
DLPC	dilauryl phosphatidylcholine
DMPA	dimyristoyl phosphatidic acid
DMPC	dimyristoyl phosphatidylcholine
DMPG	dimyristoyl phosphatidylglycerol
DOBAD	dioctadecyl diammonium bromide
DOGS	dioctadecyl amido glycylspermine
DOPC	dioleolyl phosphatidylcholine
DPPA	dipalmitoyl phosphatidic acid
DPPC	dipalmitoyl phosphatidylcholine
DPPG	dipalmitoyl phosphatidylglycerol
DSPA	distearoyl phosphatidic acid
DSPC	distearoyl phosphatidylcholine
DSPG	distearoyl phosphatidylglycerol

EE	encapsulation efficiency
EO	essential oils
EPA	eicosapentaenoic acid
EPC	egg phosphatidylcholine
EPG	egg phosphatidylglycerol
EPR	enhanced permeability and retention
EYPC	egg-yolk phosphatidylcholine
FAC	ferric ammonium citrate
FATMLV	frozen and thawed multilamellar vesicle
FDA	Food and Drug Administration
Fe	iron
FL	ferrous sulfate liposome
FMO	frankincense and myrrh oil
GPx	glutathione peroxidase
GSH	glutathione
GUV	giant unilamellar vesicle
HCl	hydrochloric acid
HDI	human development index
HEME-LIP	heme liposomes
HIV	human immunodeficiency virus
HM	heating method
HPLC	high-performance liquid chromatography
HPM	high-pressure homogenization
HPMH	high-pressure microfluidic homogenization
HSPC	hydrogenated phosphatidylcholine
IL-2	interleukin-2
Lc	crystalline lamellar
LMV	large multilamellar vesicle
LUV	large unilamellar vesicles
MC	methylcellulose
MCFAs	medium-chain fatty acids
MDR-TB	multidrug-resistant TB
MLV-REV	multilamellar vesicle made by reverse-phase evaporation vesicle
MLVs	multilamellar vesicles
MM	Mozafari method
MPS	mononuclear phagocytic system
MRP	multidrug resistance protein
MS	modified starch
MUV	medium unilamellar vesicle
MVV	multivesicular vesicles
NDDS	novel drug delivery system
NK	natural killer

NLCs	nanostructured lipids carriers
OLV	oligolamellar vesicle
O-SAP	O-steroyl amylopectin
PA	phosphatidic acid
PBS	phosphate buffer solution
PC	phosphatidylcholine
PCBs	polychlorinated biphenyls
PDI	polydispersion index
PE	phosphatidylethanolamine
PEG	polyethylene glycol
PEMT	phosphatidylethanolamine N-methyltransferase
PG	phosphatidylglycerol
PI	phosphatidylinositol
POPC	palmitoyl-2-oleoyl phosphatidylcholine
PPE	phosphatidylethanolamine
PS	phosphatidylserine
PUFA	polyunsaturated fatty acids
PVP	poly(vinyl pyrrolidone)
QC	quality control
QUE	quercetin
RES	reticuloendothelial system
RESS	rapid expansion of supercritical solutions
RESV	resveratrol
REV	reverse-phase evaporation vesicles
ROS	reactive oxygen species
SA	sodium alginate
SC-CO$_2$	supercritical carbon dioxide
SCRPE	supercritical reverse-phase evaporation
SDS	sodium dodecyl sulfate
SEC	size-exclusion chromatography
siRNA	short interfering RNA
siTGF-β1	short interfering transforming growth factor-β1
SLN	solid lipid nanoparticles
SNEDDS	self-nano-emulsified delivery system
SO	salvia oil
SOD	superoxide dismutase
SPC	soybean phosphatidylcholine
SPLV	stable plurilamellar vesicle
SUV	small unilamellar vehicle
TB	tuberculosis
TBA	thiobarbituric acid
Tc	transition temperature

TEER	transepithelial resistance
TFDH	thin-film dehydration-hydration
TFH	thin-film hydration
TMC	N-trimethyl chitosan chloride
ULV	unilamellar vesicles
UV	unilamellar vesicle
V_C	vitamin C
V_E	vitamin E
Vit.	vitamins
Vit.C	vitamin C
Vit.D2	vitamin D2
Vit.D3	vitamin D3
Vit.E	vitamin E
W/O	water-in-oil
WP	whey protein
XDR-TB	extensively drug-resistant TB
γ-CD	gamma-cyclodextrin
ω-3 FAs	omega-3- fatty acids

Preface

Nanomaterials are keystones of nanoscience and nanotechnology. Nano-structure science and technology is a broad and interdisciplinary area of research and development activity that has been growing explosively worldwide in the past few years. In this book, we are focusing on the advanced trends and applications of liposomes in the nutraceuticals industry. They have become a rapidly developing material technology due to their numerous advantages and tremendous potential in medical, nutraceutical, and energy applications.

Liposome technology has found a number of very successful applications in the pharmaceutical and cosmetic industries, but there has been only limited development of this technology in the food industry. The book begins by discussing the processes and protocols of the formation of liposomes and the structures of liposomes produced by different methods. It then reviews their physicochemical properties and the science of encapsulation of bioactive compounds using liposomes. Liposomes are delivery vehicles for transporting substances into the body effectively via facilitating absorption directly in the mouth or by preventing breakdown by stomach acid. Since the 1970s, liposomes have been investigated as potential drug delivery systems because of their biocompatibility and ability to incorporate both hydrophilic and hydrophobic therapeutic agents. Despite early promise, decades later, in the late 1990s to the present, liposome technologies could create successful commercial products. The book split into eight chapters that are exclusively deal with liposomes in nutraceuticals: from basics to advanced applications of liposomes in nutraceuticals.

CHAPTER 1

Liposomes: Methods and Protocols

THAIS ARAGÃO HOROIWA,[1] VALKER FEITOSA,[2] and
HOSSEIN ADELNIA[3,4]

[1]*Postgraduate in Industrial Process Masters by Coursework-Institute
for Technological Research of Sao Paulo State, Cidade Universitária,
SP, Brazil, E-mail: thaisaragao.h@gmail.com*

[2]*PhD Degree in Biochemical and Pharmaceutical Technology,
University of Sao Paulo, São Paulo, SP, Brazil*

[3]*Australian Institute for Bioengineering and Nanotechnology,
University of Queensland, Brisbane, QLD, Australia*

[4]*School of Pharmacy, University of Queensland, Woolloongabba,
QLD, Australia*

ABSTRACT

Encapsulation of biocompounds into liposomes has been attracting the
food industry as its manufacturing features evolved in the last decade,
increasing scale-up production and reducing process stress preventing the
degradation of these sensitive molecules. Liposomes provide a controlled
release and increase stability/bioavailability of these compounds, thus
might improve the nutritional value in dietary and employed in functional
food. These technologies might improve the production of enriched food
or nutraceutical products, and then contribute to the better health of the
world population. This chapter provides a discussion regarding the liposomal encapsulation of food-related molecules, including the conventional
methods and protocols up to recently developed techniques.

1.1   INTRODUCTION

Liposomes are core-shell structured vesicles consisting of one or more concentric phospholipid bilayers containing an aqueous inner core [1, 2]. These structures are very similar to human living cells, with the bilayer protecting the intracellular content from the external environment while still allowing an exchange of molecules across the membrane [3]. Hence, the liposomes have been applied as a structured model to understand the intracellular uptake mechanisms regarding close resemblance to biological membranes [4].

Liposomes are also often employed as drug delivery systems in order to control and delay the release of bioactive molecules due to their unique structures and properties [1, 5, 6]. As they are biologically compatible and non-toxic [7], several technologies have been developed allowing incorporation of almost every substance ranging from hydrophilic, hydrophobic even amphiphilic molecules, thus a wide range of bioactive compounds such as drugs, proteins/enzymes, and RNA/DNA resulting in variable yield (i.e., encapsulation efficiency, EE%) [8].

The releasing mechanism of the entrapped compound into liposome relies on the osmotic principle, based on the concentration gradient. In other words, the compound diffuses out through the bilayer as its concentration in the external phase is low [1, 3]. The diffusion continues until the concentration eventually is the same everywhere, and equilibrium in concentration change is achieved. Parameters affecting the liposome bilayer and its stabilization, such as pH and temperature, would play a key role in the release of the carried content [1].

The main advantage of liposomal encapsulation is their increased stability in food products with typically high-water content [9]. Furthermore, as liposomes usually are prepared from naturally occurring phospholipids, regulatory barriers preventing their application in the food industry are potentially reduced or avoid; thus, novel formulations could be easily developed [10, 11]. Therefore, liposomal structures have been employed in the last decades for encapsulating several substances in food formulation, such as proteins/enzymes [12], antimicrobial agents [13–15], vitamins (Vit.) [16, 17], minerals [18, 19], antioxidants [20, 21] and functional peptides [17, 22]. Herein liposomes are generally employed both to improve nutrient/bioactive solubilization and to enhance its absorption/uptake [1]. Furthermore, liposomes are excellent candidates for nutraceuticals delivery purposes, due to their well-known preparation and tunable physicochemical properties such as size, charge, and number of lamellae [23, 24].

The traditional approaches to produce liposomes are: (i) thin-lipid film dehydration-rehydration, also known as Bangham method; and (ii) solvent injection method; might be followed by size reduction techniques (e.g., sonication, extrusion, or high-pressure homogenization: HPM), as well as, postprocessing procedures, such as freeze-drying, to increase shelf-life stability. Nevertheless, the use of organic solvents [8], the high cost of commonly employed phospholipids, and scaling-up challenges [11, 25] have considered being the main disadvantage of those conventional methods for liposomal encapsulation.

To overcome these mentioned limitations, novel technologies such as microfluidic technique [26], crossflow solvent injection [27, 28], and heating method (HM) [12, 29] have been recently evolved, in addition to the recent studies focused on the production of liposomes from non-purified lipids to reduce the final costs [11, 30].

In this context, this chapter will provide a discussion about the composition and production of liposomes, including a description of the conventional as well as recently developed techniques employed for liposomal encapsulation (see Table 1.1 for a summary of the protocols included herein). These technologies may enhance the production of enriched food or nutraceutical products with improved physicochemical stability and bioavailability, thereby boosting nutritional value.

1.2 LIPOSOMES STRUCTURES AND COMPOSITION

Structurally, liposomes are phospholipid-based vesicular structures composed of one or more lipid bilayer membranes that surround an aqueous core [2, 31]. Phospholipids are versatile, biocompatible, and biodegradable ingredients, generally used in oral formulations as wetting, emulsifying, solubilizing, and matrix-forming agents. Naturally occurring phospholipids, particularly those extracted from renewable sources, are regarded to be environmentally friendly excipients, and thus are preferred as a substitute for the synthetic counterparts and non-phospholipid analogs [7]. Within the clinically approved liposome preparations, the most common source of the bilayer-lipidic compounds such as phosphatidylcholines (PC), cholesterol (Chol), and phosphatidylglycerol (PC) is

TABLE 1.1 A-List Summarizing the Methods and Protocols Employed to Produce Liposome Loading Food-Related Compounds

Nutraceutical	Production Method	Postprocessing Technique	Lipids/Surfactants	Solvent/ Continuous Media	Hydration Solution	References
β-carotene, lycopene, lutein, or canthaxanthin	Thin-film dehydration-rehydration	Sonication	EPC:Tween-80	Chloroform	PBS (pH 7.4)	[6]
PUFAs DHA and EPA (Omega-3)	Microfluidic	NA	Tween-80, SDS, lecithin, or quillaja saponin	Water	NA	[76]
PUFAs DHA and EPA (Omega-3)	Mozafari (evolved heating method)	Sonication	Soy lecithin	NA	Water:glycerol (2% v/v)	[93]
Calcitriol (Vit. D3) + Peptide	Thin-film dehydration-rehydration	Sonication	EPC:EPG	Chloroform/ ethanol (9:1)	HEPES (pH 7.4)	[17]
Calcitriol (Vit. D3) + Peptide	Microfluidic	NA	EPC:EPG	Methanol	HEPES	[17]
Scavenging protein (Superoxide Dismutase)	Solvent crossflow injection	NA	DPPC	Ethanol	PBS	[22]
α-Tocopherol (Vit. E)	Thin-film dehydration-rehydration	Extrusion	Marine lipid	Chloroform/ methanol (2:1)	HEPES (pH 7.4)	[99]

TABLE 1.1 *(Continued)*

Nutraceutical	Production Method	Postprocessing Technique	Lipids/Surfactants	Solvent/ Continuous Media	Hydration Solution	References
α-Tocopherol (Vit. E)	Thin-film dehydration-rehydration	NA	PC, PC:CaS, or PC:SA	Ethanol	Acetic acid	[16]
α-Tocopherol (Vit. E)	Solvent injection	Freeze-drying	Marine lipid	Chloroform/ methanol (2:1)	HEPES (pH 7.4)	[98]
Resveratrol	Thin-film dehydration-rehydration	Sonication	DPPC:DC-Chol or DPPC: Chol	Chloroform/ ethanol	Deionized water	[23]
Resveratrol	Thin-film dehydration-rehydration	Extrusion	Soy PC:Chol	Ethanol	Deionized water	[69]
Resveratrol	Thin-film dehydration-rehydration	Extrusion+Sonication	P90G: Chol (2.3:1 w/w)	Ethanol	Distilled water	[10]
Resveratrol	Thin-film dehydration-rehydration	Sonication	Soy PC: Chol	Chloroform/ methanol (3:1)	PP or PBS (pH 7.4)	[58]
Tea polyphenol	Solvent injection	Microfluidic	PC: Chol: Tween-80 (8:1.2:2)	Ethanol	PBS (pH 6.0)	[97]
Quercetin	Thin-film dehydration-rehydration	Sonication	Soy PC: Chol	Chloroform/ methanol (3:1)	PP or PBS (pH 7.4)	[58]

TABLE 1.1 *(Continued)*

Nutraceutical	Production Method	Postprocessing Technique	Lipids/Surfactants	Solvent/ Continuous Media	Hydration Solution	References
Phytochemicals (Phenolics, Triterpenes, Meroterpenoids)	Thin-film dehydration-rehydration	NA	PC: Chol	Chloroform	Distilled water	[61]
Flavonoids (quercetin, kaempferol or luteolin)	Thin-film dehydration-rehydration	Sonication	EYPC: Tween-80	Ethanol	PBS (pH 7.4)	[21]
Curcumin	Thin-film dehydration-rehydration	Extrusion	PC: Chol:D-α-tocopheryl polyethylene glycol 1000 succinate	Ethanol	PBS (pH 6.6)	[68]
Curcumin	Microfluidic	NA	Soy lecithin	Ethanol:Water	lecithin water suspension	[30]
Curcumin	Solvent injection	Sonication	EPC: Chol (2:1)	Ethanol	Deionized water	[82]
Acid ascorbic (Vit. C)	Thin-film dehydration-rehydration	Microfluidic	PC, PC:CaS, or PC:SA	Ethanol	Acetic acid	[16]
Acid ascorbic (Vit. C) + (Vit. E)	Thin-film dehydration-rehydration	Microfluidic	Soy PC: Chol	Ethanol	Distilled water	[75]

TABLE 1.1 *(Continued)*

Nutraceutical	Production Method	Postprocessing Technique	Lipids/Surfactants	Solvent/ Continuous Media	Hydration Solution	References
Iron	Microfluidic	NA	Emultop (soy lecithin + lysophospholipids)	Water	NA	[18]
Flavorenzyme	Heating method	NA	Lecithin: Chol	Glycerol (3% w/v)	Tris buffer	[12]

Abbreviations: Chol: (cholesterol); DC-Chol: (3β-[N-(N9,N9-dimethylaminoethane)-carbamoyl]cholesterol); DHA: (docosahexaenoic acid); DPPC: (1,2-dipalmitoyl-sn-glycero-3-phosphocholine); EPA: (eicosapentaenoic acid); EPC: (egg phosphatidylcholine); SDS: (sodium dodecyl sulfate); P90G: Phospholipon 90G; PP: (potassium phosphate buffer); PBS: (phosphate buffered saline); PUFA: (polyunsaturated fatty acids); CaS: (calcium stearate); SA: (stearic acid).

from soy and egg. However, there has been recently growing interest in employing milk-derived phospholipids as alternative ingredients [32, 33].

When phospholipids have exposed to an aqueous solution, their molecules self-aggregate or self-assemble into vesicles due to the interaction between their hydrophilic polar head groups and the external aqueous environment. On the other hand, the tail groups, which are hydrophobic fatty acid chains, form Van der Waals' attraction (e.g., hydrophobic-hydrophobic interaction), leading to the spontaneous formation of closed bilayers [2, 34]. Thus, the unfavorable interactions between the fatty acid chains and water molecules are minimized when the edges fold together to form closed sealed vesicles [1, 24].

Overall, one can conclude that the main characteristic of these bilayer-forming phospholipids is amphiphilicity as a result of the presence of a polar head group covalently attached to one or two hydrophobic hydro-carbon tails [2]. These unique features of liposomes allow encapsulation of hydrophilic and hydrophobic compounds. Herein, the hydrophilic may become entrapped inside the liposome, at the aqueous compartment, whilst hydrophobic material might be incorporated between phospholipid bilayers, as depicted in Figure 1.1 [2, 31] Regarding the latter, entrapment occurs in the lipid bilayers due to their hydrophobicity. Thus, challenges such as the loss of an entrapped compound in long-term storage are rarely created [35]. In addition, other molecules such as proteins might adsorb onto the surface of the liposome membrane through electrostatic interactions [33].

Lipid e.g.Phosphatidylcholine (PC)
$(C_{15}H_{31}CO_2CH_2)_2\text{-}CH_2OPOO\text{-}(CH_2)_3N^+(CH_3)_3$

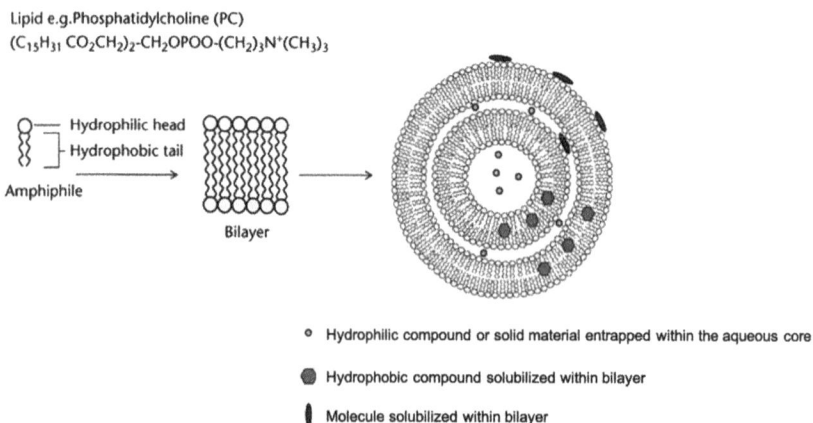

° Hydrophilic compound or solid material entrapped within the aqueous core

● Hydrophobic compound solubilized within bilayer

❙ Molecule solubilized within bilayer

FIGURE 1.1   Liposome structure and its self-assemble formation.
Source: Adapted with permission from: Ref. [33]. ©2020 John Wiley & Sons Books.

A characteristic of amphiphilic lipids that enables them to arrange at lamellar structures, and lately to form enclosed vesicles, is in regard to the ability of transition between three different lamellar-phases: crystal-line lamellar (Lc), lamellar gel (Lβ), or lamellar liquid-crystalline (Lα). This transition is depending on temperature and, when the lipid membrane passes from a tightly ordered gel (Lβ) to a liquid-crystalline lamellar phase (Lα), described as the lipid transition temperature (Tc). This temperature is intrinsic for each lipid (see Table 1.2 for consultation), been related to their molecular structure, chain length, and saturation degree in the alky chains. For the reason that it involves disruption of the crystalline phase (Lc), due to a dramatic increase of molecular movement with the heating input, until it reaches a freedom state observed at the fluidic state (Lα) [33, 36].

TABLE 1.2   Some Properties of Mainly Phospholipids Applied for Liposomes Production

Lipid	Abbreviation	Lipid Number	Transition Temperature (Tc), °C	Charge at pH 7.4
Dilauryloylphosphatidylcholine	DLPC	12:00	−1	0
Dimyristoylphosphatidylcholine	DMPC	14:00	23	0
Dipalmitoylphosphatidylcholine	DPPC	16:00	41	0
Distearoylphosphatidylcholine	DSPC	18:00	55	0
Dioleoylphosphatidylcholine	DOPC	18:01	−20	0
Dimyristoyl phosphatidylethanolamine	DMPE	14:00	50	0
Dipalmitoyl phosphatidylethanolamine	DPPE	16:00	63	0
Dioleoyl-sn-glycero-3-phosphoethanolamine	DOPE	18:01	−16	0
Dimyristoyl phosphate	DMPA·Na	14:00	50	−1.3
Dipalmitoyl phosphate	DPPA·Na	16:00	67	−1.3
Dioleoyl-sn-glycero-3-phosphate	DOPA·Na	18:01	−8	−1.3
Dioleoyl-sn-glycero-3-phosphoethanolamine	DMPG·Na	14:00	23	−1
Dimyristoyl-sn-glycero-3-phospho-(1'-rac-glycerol)	DPPG·Na	16:00	41	−1
Dioleoyl-sn-glycero-3-phospho-(1'-rac-glycerol)	DOPG·Na	18:01	−18	−1

TABLE 1.2   *(Continued)*

Lipid	Abbreviation	Lipid Number	Transition Temperature (Tc), °C	Charge at pH 7.4
Dimyristoyl-sn-glycero-3-phospho-L-serine	DMPS•Na	14:00	35	−1
Dipalmitoyl-sn-glycero-3-phospho-L-serine	DPPS•Na	16:00	54	−1
Dioleoyl-sn-glycero-3-phospho-L-serine	DOPS•Na	18:01	−11	−1
Dioleoyl-sn-glycero-3-phosphoethanolamine-N-(glutaryl)	DOPE-Glutaryl•(Na)$_2$	18:01	−10	−2
Dimyristoyl-sn-glycero-3-phospho]-glycerol	Tetramyristoyl Cardiolipin • (Na)$_2$	14:00	59	−2
Distearoyl-sn-glycero-3-phosphoethanolamine-N-[methoxy(polyethylene glycol)-2000]	DSPE-mPEG-2000•Na	18:00	N/A	−1
Distearoyl-sn-glycero-3-phosphoethanolamine-N-[methoxy(polyethylene glycol)-5000]	DSPE-mPEG-5000•Na	18:00	N/A	−1
Distearoyl-sn-glycero-3-phosphoethanolamine-N-[maleimide(polyethylene glycol)-2000]	DSPE-Maleimide PEG-2000•Na	18:00	N/A	−1
Dioleoyl-3-trimethylammonium-propane	DOTAP•Cl	18:01	0	1

Source: Adapted from: Avanti Polar Lipids. Website: https://avantilipids.com/tech-support/liposome-preparation/lipids-for-liposome-formation [50].

It is worth mentioning that some phospholipids are unable to form liposomes by self-assembling in an independent way without the presence of other lipids. However, they are employed to modify the properties of other liposomes. As an example, Chol up to 50% of total lipids (m/m) is usually incorporated into liposomes to reduce the bilayer permeability as it strongly influences the freedom of the phospholipid alkyl chains [33].

Another important aspect of phospholipids is the type of functional moiety in the lipid head group which can potentially influence the surface charge of the liposomes. For example, phosphatidylserine (PS), phospha-tidylglycerol (PG), phosphatidylinositol (PI), and phosphatidic acid (PA), prevent liposome aggregation or fusion by electrostatic repulsion because of their negatively charged phosphate head group, resulting in higher colloidal stability for long-term storage purposes. Moreover, the inclusion of charged phospholipids increases the distance between concentric membranes in multilamellar vesicles (MLVs), increasing entrapped volume [1].

Therefore, it could be concluded that EE% of molecules by liposomes is dependent upon the bilayer composition [35, 37]; since the selection of compounds used within the formulation will influence the characteristics of the liposomes, including their surface charge, their bilayer fluidity, permeability, and molecules retention [33]. Figure 1.2 illustrates the effects of the molecular shape of the amphiphilic species on the self-assembled structure. As an example of a cylindrical-shaped phospholipid, PC has an approximately equal size of hydrophilic and hydrophobic moieties, which are assembled into flat bilayers with the thickness of 4–5 nm [38]. On the other hand, cone-like molecules, such as phosphatidylethanolamine (PE), usually pack into structures with high radii of curvature, which forms reverse micelles. However, inverse-cone-like molecules, such as surfactants or lysophospholipids, establish structure analogous with small micelles (i.e., large negative curvature). A combination of the cone- and inverse-cone-like molecules can culminate in a flat bilayer structure which could be unstable and become converted into either regular or inverse-micellar structure [39].

1.3 PRODUCTION OF LIPOSOMES

Since the 1960s, several methods have been developed for the production of nano/micro-sized liposomes, such as thin-film dehydration-hydration (TFDH) [40–43], solvent injection methods [44, 45], microfluidic technologies [18, 26, 31], and heating-techniques [25, 46].

Regardless of the manufacturing methodology, liposome formation relies on the spontaneous self-assembly of the lipid molecules. As mentioned above, hydrophilic/hydrophobic interactions between lipid-lipid and lipid-water molecules result in the spontaneous arrangement of the lipid molecules in the form of liquid-crystalline bilayer vesicles [24].

The liposomes tend to fold together then form large multilamellar vesicles (LMVs) in order to minimize the free energy of the system and to reach a thermodynamically stable condition [1, 47]. Nonetheless, to produce unilamellar systems, which have higher free energy and larger surface area/volume ratio values, an energy input (e.g., sonication, homogenization, shaking, and/or heating) must be applied to overcome the energy barrier and to achieve a thermodynamic equilibrium in the system, as depicted in Figure 1.3 [24, 47].

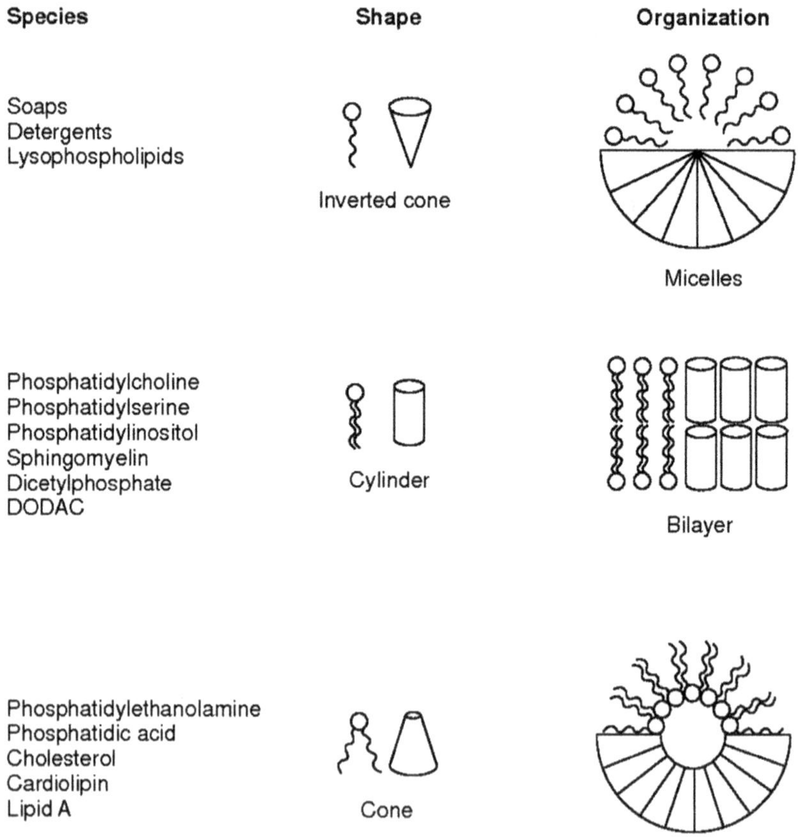

Species	Shape	Organization
Soaps Detergents Lysophospholipids	Inverted cone	Micelles
Phosphatidylcholine Phosphatidylserine Phosphatidylinositol Sphingomyelin Dicetylphosphate DODAC	Cylinder	Bilayer
Phosphatidylethanolamine Phosphatidic acid Cholesterol Cardiolipin Lipid A	Cone	

FIGURE 1.2 Effects of the molecular structure (shape) of amphiphilic molecules (species) on their aggregate state (organization).
Source: Adapted with permission from: Ref. [39]. ©2020 Elsevier.

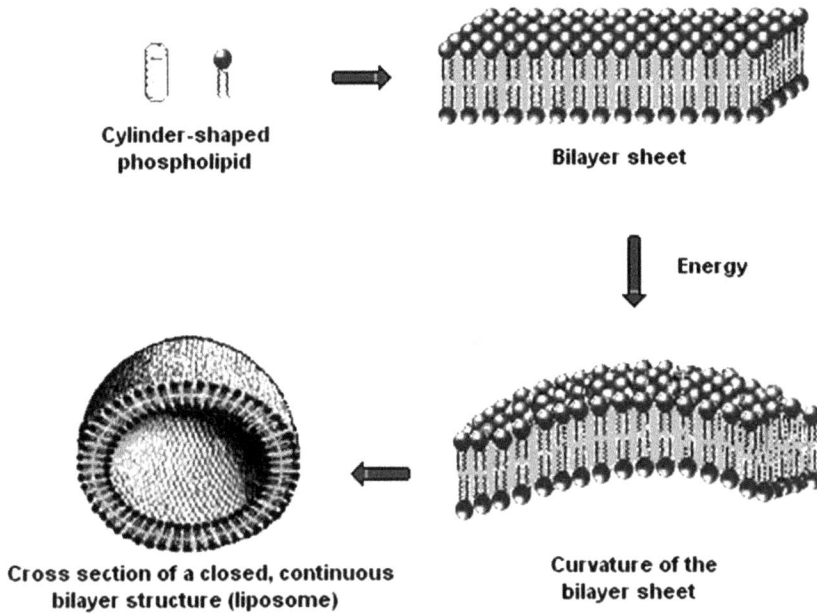

Cylinder-shaped phospholipid

Bilayer sheet

Energy

Cross section of a closed, continuous bilayer structure (liposome)

Curvature of the bilayer sheet

FIGURE 1.3   Self-assembly of phospholipids in liposomes.
Source: Reproduced with permission from: Ref. [34]. ©2005 Trafford Publishing.

1.3.1   THIN-FILM DEHYDRATION-REHYDRATION (TFDH) METHODS

One of the most well-known methods for liposomes preparation is thin-film dehydration-rehydration (TFDH). The first step in this method is creating a thin film of lipids in which lipophilic compounds could be entrapped [35, 48].

These thin films could be obtained by dissolution of lipids in an appropriate organic solvent (e.g., ethanol, chloroform, and/or methanol) followed by its evaporation under reduced pressure using a dry nitrogen or argon stream in a fume hood (for small volume) or in a rotary evaporator (for large scale), until a lipid thin-film has formed [36, 49]. Then, this film is hydrated using a proper aqueous solution forming MLVs, as illustrated in Figure 1.4 [36].

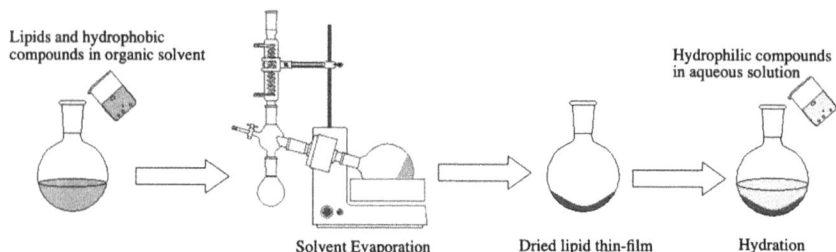

FIGURE 1.4 Liposome production by lipid thin-film dehydration-hydration method.
Source: Adapted from: Ref. [36] with permission via the Creative Commons Attribution
3.0 License. ©2013 Lopes et al.; Licensee Intech.

In the hydration step, before adding the aqueous media on dried lipid film, it is crucial to heat it above the temperature required to reach the liquid-crystalline transition phase of the lipid with the highest Tc in the lipidic mixture. For instance, the Tc values of the most widely used phospholipids to produce liposomes are shown in Table 1.2 along with their net charge value when are dissolved in a neutral media (pH 7.4), and the corresponding lipid number, which is related to their number of carbon atoms and the number of double bonds at the fatty acid [50, 51]. It is also worth mentioning that, though it may differ depending on the lipid characteristics, a hydration time of one hour under severe mixing has been generally recommended [49].

The application of liposomes is a determining factor in choosing the hydration solution. The most widely used hydration media include but are not limited to distilled water, saline, buffer (e.g., PBS) as well as nonelectrolytes (e.g., sugar) solutions [49]. Therefore, hydrophilic compounds (e.g., ascorbic acid [16, 52], functional proteins [17, 27], or even solid minerals such as iron [18, 19]) could be incorporated at the liposomal vesicle in the rehydration step then been protected against degradation.

Furthermore, some lipophilic biocompounds such as carotenoids [6], vitamin D [17, 53, 54], resveratrol (RESV) [10, 23], that are highly prone to degradation through oxidation, can also be benefited by liposomal encapsulation preventing their degradation [55, 56].

Despite its simplicity, the application of this method is restricted due to low EE%, the non-uniform size distribution of vesicles, as well as scaling-up challenges [36, 57]. However, the TFDH method has a significant contribution in explaining the mechanisms of thin-film liposome production, and their benefits in the food industry. For example, Cadena et al. [58] revealed that antioxidant capabilities of typical flavonoids such as quercetin (QUE) and RESV could be increased by loading into liposomes [58]. Their liposomes formed small unilamellar vesicles (SUVs) with a mean diameter of 149 nm, polydispersion index (PDI) of 0.3, slightly negative zeta potential (−13 mV) and, almost 97% of QUE and RESV EE into liposomes [58]. The physicochemical stability assessment at 4°C for 45 days showed low variation in the particle mean size and PDI, as well as fewer pH variations during the storage time, suggesting that lower lipid peroxidation occurred. Furthermore, the *in vitro* release assay revealed that the QUE and RESV have a burst released profile at the first hour (56% and 53%, respectively), then established a sustained release until reaching their maximum level at 57 h (98%) and 54.5 h (96%), respectively [58]. Therefore, the obtained results present the liposomal encapsulation as a promising approach to prolong the flavonoids' antioxidant effects as a food ingredient or supplement.

Moreover, several other studies [10, 23, 59] have also been conducted to stabilize and protect RESV, in an attempt both to improve its aqueous solubility and to achieve targeted and sustained release [60]. Caddeo et al. [59] study, for example, demonstrated that the liposomes reduce in a cell proliferation the cytotoxicity of RESV at high concentrations (100 μM) by preventing its immediate and massive intracellular distribution. Also, liposomes have been found to be highly efficient in enhancing the viability of the cells to survive under stress conditions caused by UV-B radiation, and in improving the RESV performance in cell stimulation and proliferation [59].

A further example of the TFDH application has been provided at Saber et al. [61] work, whose investigated liposomal encapsulation of several phytochemicals from two *Psidium* species fruits (*P. guajava* and *P. cattleianum* extracts), employing PC from soybean, Chol, and Tween-80 as the vesicle compounds. It was demonstrated that the liposomal systems loaded with the *Psidium* extracts reduced the paracetamol-induced hepatotoxicity in rats [61].

Marsanasco et al. [16] also studied the association of hydrophilic and hydrophobic compounds by the same liposomal carrier where vitamin E (Vit.E) and vitamin C (Vit.C) were incorporated into PC liposomes in the presence and absence of stearic acid or calcium stearate (CaS). They employed the followed dehydration-rehydration steps: (i) lipid thin-film formation by the dissolution of lipids in an ethanolic solution of Vit.E at 22.4 mM; (ii) solvent drying in a rotary evaporator at 37°C, purged with nitrogen; (iii) rehydration with an acetic solution of Vit.C at 5 mM for peroxidation assays or 90 mM concentration for other assays. It was shown that the PC-liposomes have a protective effect on the antioxidant activity of vitamins before and after pasteurization, and the formulation containing stearic acid had the most efficient encapsulating rate with 38%. Moreover, the combination of vitamins-loaded liposomes with orange juice showed microbiological stability after pasteurization and storage at 4°C for 37 days, and the juice organoleptic characteristics were not comprised [16].

1.3.2   SIZE REDUCTION AND POSTPROCESSING TECHNIQUES

The vesicles obtained by the phospholipid self-assembling, employing the TFDH method, are spontaneously arranged in LMVs, ranging from 200 nm to 3 um in diameter size [62], which leads to lipids concentration fluctuations and consequently a heterogeneous system [31]. Hence, to reduce the size and to narrow the size distribution of the expected liposomal system, energy input is required, which could be obtained by one or a combination of the following postprocessing procedures, including sonication, extrusion, and HPM (see representation in Figure 1.5) [31, 63, 64].

1.3.2.1   SONICATION

Several techniques have thus far been established to reduce the size and to improve the size uniformity of vesicles. Sonication is considered an efficient method for the size reduction of the vesicles. In sonication, high-intensity ultrasound energy is applied to the liposome dispersion [10, 65]. The applied energy is, in fact, in the form of cavitation either directly or indirectly with a tip or a bath sonicator, respectively (see Figure 1.6) [57]. Through the application of sonication, the sound waves are propagated throughout the liposome dispersion, leading to the formation and collapse of tiny cavities (i.e., bubbles), which creates high pressure and, as a result, a turbulent flow condition [66]. The latter circumstances create a high shear force on LMVs, disrupting their structures and rearrangement to form SUVs, as a consequence [10, 11].

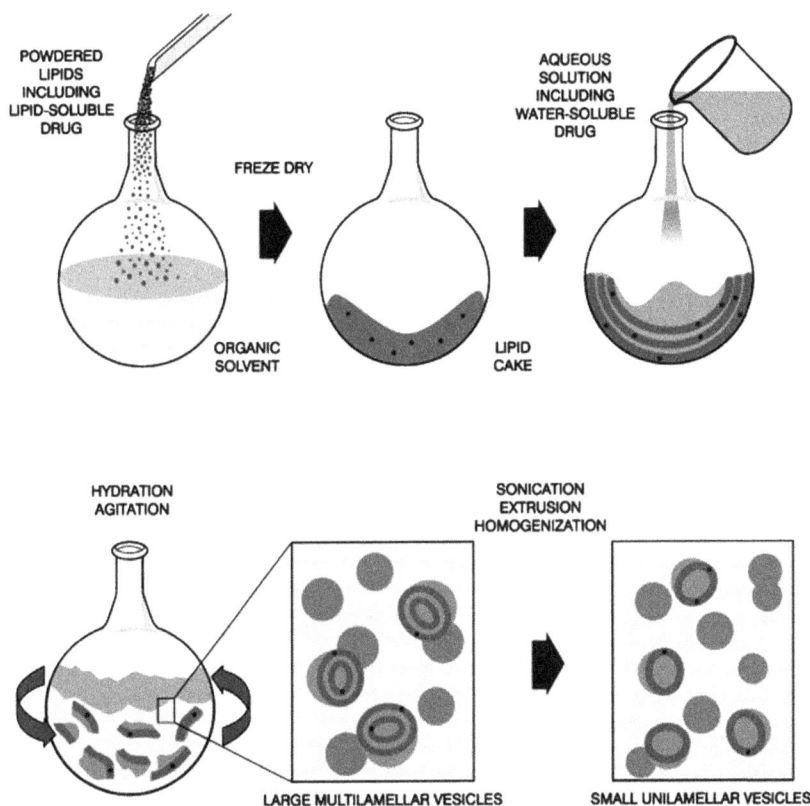

FIGURE 1.5 The process applied for size reduction of liposomes prepared by lipid thin-film dehydration-rehydration (TFDH).
Source: Reproduced from: Ref. [49] with permission via the Creative Commons Attribution 4.0 License.

Sonication has been utilized in many studies for the formation of well-defined liposomes. As an example, Huang et al. [21] employed probe sonication for 10 min at 300 W to produce flavonoids-loaded liposomes, using egg-PC and Tween-80, by the TFDH method. It was shown that sonication-assisted protocol leads to uniformly dispersed liposomes vesicles with the size ranging from 180 to 250 nm, depending on each flavonoid. The obtained vesicles containing QUE showed higher stability and better performance in preventing lipid peroxidation compared with vesicles carrying kaempferol or luteolin [21].

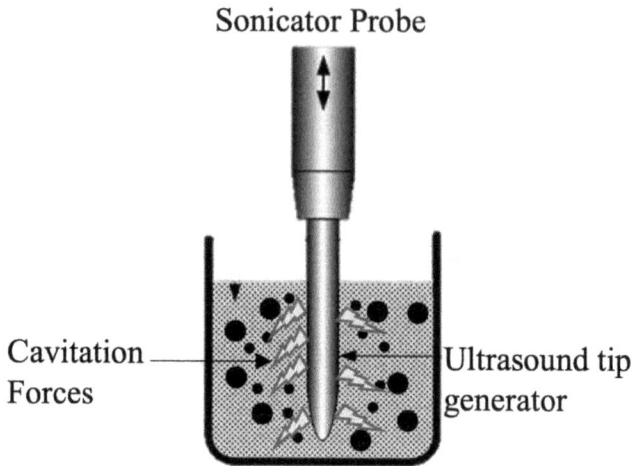

FIGURE 1.6 Liposome size reduction by ultrasound.
Source: Created by Aragao, H. T., using Microsoft® PowerPoint.

Tan et al. [6] likewise prepared carotenoids-loaded LMVs, with egg-PC and Tween-80, by TFDH method and sonication postprocessing in an ice bath for 10 min at 240 W. They were able to fabricate nano liposome with vesicle size ranging from 66 to 195 nm, depending on flavonoids/phospholipid concentration ratio.

The lutein- and β-carotene-loaded liposomes showed a sustained release profile, while lycopene- and canthaxanthin-loaded ones had a fast release. Therefore, one can conclude that the different carotenoids affect the membrane dynamic and structure differently and, as a result, the particle integrity and stability as depicted in Figure 1.7 [6].

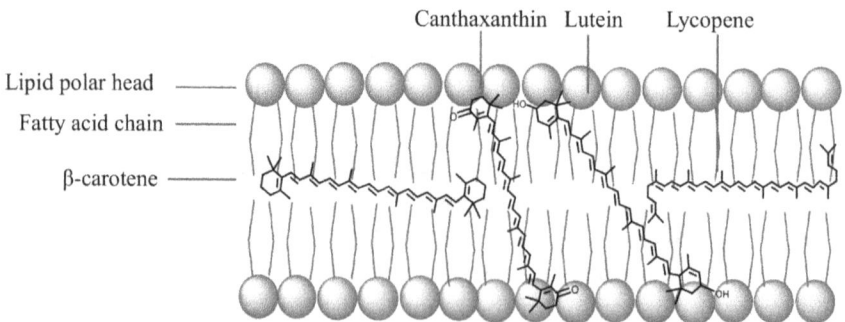

FIGURE 1.7 Carotenoids arrangement into the phospholipidic liposome bilayer.
Source: Reproduced from: Ref. [6] with permission. ©2020 Royal Society of Chemistry.

Therefore, it is crucial to consider the nature and physicochemical characteristics of the molecule desired to be encapsulated, as well as evaluate the nuances of each available downsizing process that might affect the EE and stability of the formulation compounds.

Although sonication is a widely used and effective method for size reduction, its application is limited as it increases the temperature of dispersion and may lead to degradation of the employed compounds (mainly the thermo-sensitive compounds) due to high energy input [57, 66]. Additionally, it is not suitable for large-scale production due to the difficulty to obtain a standard size distribution of the vesicles when high volumes are processed, also because a high energy input is required to promote the propagation of cavitation throughout the whole batch [67], then it is costly to be implemented at an industrial level thus it will be hardly employed in the food industry [57].

1.3.2.2 EXTRUSION

The extrusion technique is described as a well-defined method for size reduction in postprocessing procedures in which liposomes are passed through membrane filters, having a homogenous pore size distribution [57]. An extrusion device includes two syringes that are connected to a membrane holder, allowing the back-and-front passage of the sample (see Figure 1.8). The sample is passed through the membrane from one syringe to the other for multiple times. These passages decrease the vesicle size and narrow the size distribution [10].

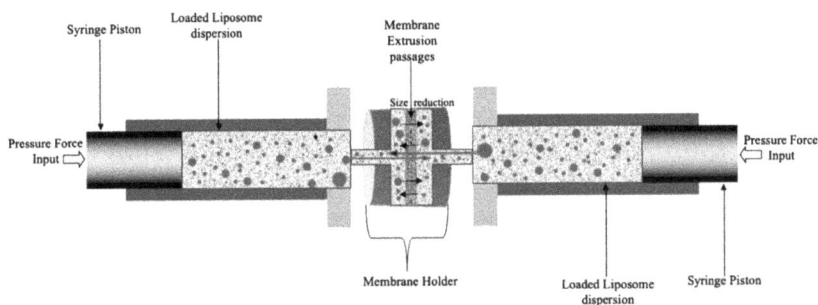

FIGURE 1.8 Liposome processing through membrane extrusion apparatus.
Source: Created by Aragao, H. T., using Microsoft® PowerPoint.

The main advantage of the extrusion method is the fact that a heating device can be coupled to the system to conduct the process above Tc. Also, this can be employed either for saturated or unsaturated lipids and leads to a high degree of reproducibility [57].

Chen et al. [68] employed extrusion processing (0.8 μm pore size polycarbonate membranes) to obtain a homogeneous particle size dispersion, which is suitable to be coated with N-trimethyl chitosan chloride (TMC). A thin-film pre-processed liposome composed of soybean-PC, Chol, D-α-tocopheryl polyethylene glycol 1000 succinate, and curcumin (CUR) was prepared and extruded.

They reported enhanced bioavailability for the TMC-coated CUR liposomes compared to uncoated liposomes and free-CUR suspension so, despite a similar *in vitro* release profile of CUR between the uncoated and coated liposomes was demonstrated, Chen et al. [68] findings present a promising strategy to improve oral delivery of this poorly water-soluble compound.

Zu et al. [69] also demonstrated that trans-RESV aqueous solubility and physicochemical stability are improved when encapsulated into liposomes. The RESV-loaded liposomes were prepared using soy-PC and Chol by TFDH method followed by a membrane extrusion procedure [69].

Though efficient in terms of size reduction and improving polydispersity index, the main disadvantage of this method is that it can just be performed in a prior produced liposomes by a beforehand technique, which is time-costly thus unfavorable for food industry application. Furthermore, one may encounter a high amount of product loss, particularly if clogging of the extrusion membranes occurs through the procedure, which limits its use for large-scale liposome production of expensive compounds [57].

1.3.2.3   *HIGH-PRESSURE MICROFLUIDIC HOMOGENIZATION (HPMH)*

The homogenization technique is another effective method for reducing the size of large MLVs and for achieving a homogeneous size distribution. This is carried out through a laboratory or industrial equipment in which a beam of dispersion is passed under high pressure through an orifice (see Figure 1.9) [70]. Under high-pressure passage and continuous pumping, the dispersion (and so LMVs) is exposed to severe and multiple collisions with the propeller and wall of the homogenizer, leading to a remarkable size reduction of LMVs. Size reduction can also be attributed to turbulence, cavitation, and shear forces generated by high pressure [57, 71].

Microfluidic methods, in general, rely on these shear force-based technologies that enable precise control and manipulation of fluids in the micrometer scale. They have been successfully applied to reduce the size of dispersed systems and overcome the aforementioned scale-up problems faced by the conventional techniques (e.g., sonication or extrusion) because it is possible to conjugate several microfluidic devices to work parallelly, thus increases the volume of a batch that might be processed at the same time [30, 31, 68]. These technologies cause size reduction by the collision of two high-velocity immiscible fluid streams, pressurized into microchannels of the microfluidizer chamber (as represented in Figure 1.10), where a well-defined interface is established by laminar flow, as opposed to the typically turbulent flow in a conventional high-pressure homogenizer [31, 57, 71]. It has been reported that microfluidizers have high efficiency in the production of liposomes with small sizes compared to comparable high-pressure homogenizers [11].

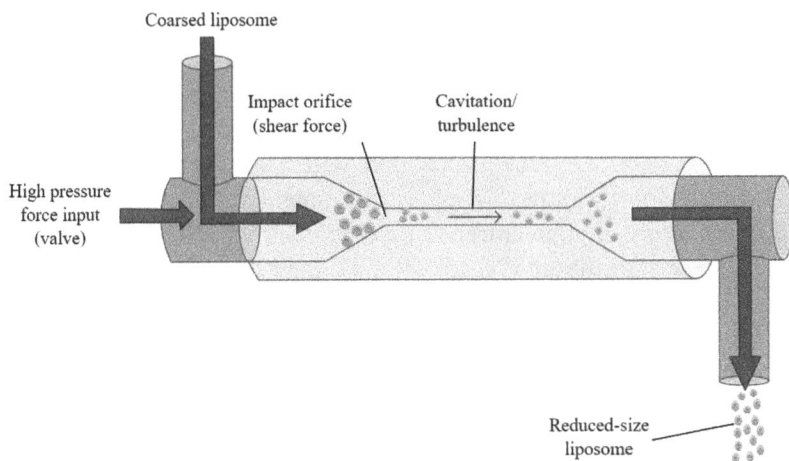

FIGURE 1.9 Liposome size reduction by high-pressure homogenization.
Source: Adapted from: Ref. [72] with permission via Creative Commons Attribution 3.0 license. ©2012 Seok-Cheol et al.

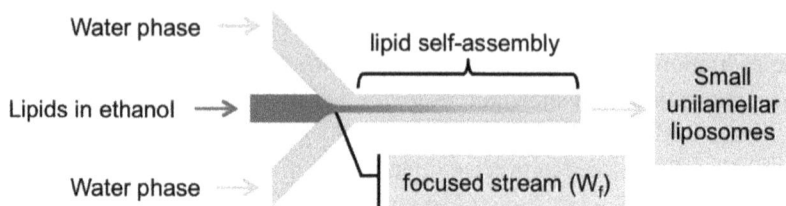

FIGURE 1.10   Liposome production into microfluidizer device.
Source: Reproduced from: Ref. [26] with permission via the Creative Commons Attribution 4.0 International License. ©2016 Springer Nature.

The rapid and tunable mixing provided by the microfluidic method is suitable, not only for size reduction of the preformed emulsions but also for controlling the size of the self-assembled liposomes [73, 74]. For example, Li et al. [75] employed HPMH to reduce the size of soy-PC liposomes prepared by double-emulsion (DE) technique which allows loading of both Vit.C and medium-chain fatty acids (MCFAs) (containing 8 and 10 carbons). In the DE method, the lipid solution containing MCFAs (soy-PC:Chol:Vit.E; 100:25:4 wt.%) in ethanol was injected into water under vigorous stirring at 50°C. Then, the solvent was completely removed under reduced pressure at 50°C in a rotary evaporator. After that, the aqueous solution of surfactant (Tween-80) and Vit.C was added, and, finally, the resulting emulsion was passed through a microfluidizer three times under 120 MPa. This method led to MCFAs- and Vit.C-loaded liposomes with a size around 90 nm. The prepared liposomes showed higher entrapment efficiency of MCFAs (48%), relatively higher entrapment efficiency of Vit. C (64%), and better storage stability at 4°C for 90 days than those prepared only by double emulsion (DE) [75].

Kosaraju et al. [18] also prepared a liposome, with a size ranging from 150 to 200 nm, by the microfluidization technology for loading of ferrous sulfate (iron), achieving 58% EE. They demonstrated the reduced oxidation of the iron-loaded liposomes in long-term storage stability assessment [18].

Uluata et al. [76] also investigated the benefits of nanoemulsions prepared by microfluidizers in terms of decelerating fish oil oxidation. They compared the effectiveness of the two synthetic (Tween-80 or sodium dodecyl sulfate: SDS) and two natural (lecithin or quillaja saponin) surfactants in fish oil nanoemulsification and found that Tween-80, and lecithin are less stable in thermal treatments. Furthermore, *Quillaja saponins* provided the highest stability in terms of the photosensitizer promoted oxidative stability due to its ability to scavenge free radicals [76].

Galea et al. [17] also reported the preparation of liposomes loaded with the combination of peptides and Vit.D3 (Pep/D3-liposomes). The liposomes were prepared by a microfluidic system where flow ratios of the two inlet streams, contained egg-PC and egg-PG dissolved in methanol and HEPES buffer, were controlled with syringe pumps. No-entrapped compounds were removed by dialysis and ultrafiltration [17]. The resulting Pep/D3-loaded liposomes were shown to reduce the severity of rheumatoid arthritis and Goodpasture's vasculitis models with suppression of antigen-specific memory T cell differentiation and function. Thus, such liposomes may have the potential to be utilized as a nutraceutical enhancer for the treatment of inflammatory diseases.

In an interesting study, Takahashi et al. [30] employed lecithin to prepare a large volume (about 10 L) of liposome loading CUR by mechanic/chemical method, using homogenizer and microfluidizer to obtain SUVs around 114 nm. The protocol included emulsification of a mixture of an aqueous suspension of CUR (previously obtained by extraction and lyophilization of an aqueous-ethanolic solution of *Curcuma longa*) with an aqueous suspension of soy lecithin, by a homogenizer at 4,000 rpm for 15 min. Then, the ensuing dispersion was once passed through the microfluidizer device at 100 MPa [30].

This study is of utmost importance as it demonstrates the preparation of nano-liposomes at large scales (i.e., 10 L) with high entrapment efficiency (80% for CUR) employing inexpensive sources of phospholipids from soybean lecithin.

Overall, in light of the abovementioned, it can be concluded that compared with ultrasonication and extrusion, HPMH can produce smaller droplets, better emulsion homogeneity, and higher productivity [77]. Moreover, a high lipid phase ratio, as well as the potential for large-scale production, are the main advantages of HPMH [78]. However, the increased temperature, as a consequence of the shear, turbulence, compression during isostatic pressure, and severe mixing during the fluid discharged through the homogenization valve, is a disadvantage for encapsulation of sensitive products such as several nutraceuticals [79].

1.3.2.4   FREEZE-DRYING POSTPROCESSING TECHNIQUE

Freeze-drying is an alternative technique for dehydration of heat-sensitive materials and is usually employed to improve the shelf-life and long-term storage of liposomes, particularly those containing thermosensitive bioactive compounds [8, 80]. This method is performed on preformed 'empty' SUVs that have been subjected to dehydration-rehydration cycles. The rehydration above the Tc results in a fusion of small, preformed vesicles to produce MLVs [11].

The freeze-drying process comprises four distinct stages: (i) freezing the preformed SUVs; (ii) sublimation; (iii) desorption; and (iv) storage. The mixture is firstly flash-frozen for immobilization of the continuous and discontinuous phases, then sublimation occurs under vacuum until the moisture content is below 5% [66].

For example, Moeller et al. [53] used the freeze-dried technique to prepare casein micelles (CM) loaded with Ergocalciferol (Vit.D2) through the following steps: (i) water hydration of pre-filtered and dried casein obtained from raw cow's milk at a 3.5% dried matter; (ii) agitation in a double-walled tank at 2°C; (iii) pH adjustment to 5.5 with 10% HCl (hydrochloric acid), and incubation overnight; (iv) addition of Vit. D2 from ethanolic stock solution within 10–30 min, using a circulating peristaltic pump and stirring under light exclusion, preventing Vit.D2 from photodegradation, for 15 min; (v) pH readjustment, and incubation for 1 h at 10°C while stirring for re-association of CM and incorporation Vit.D2; (vi) freezing suspension at –20°C; and (vii) drying under vacuum for 76 h [53].

The obtained particle had a size around 100 nm, and the recovery rate for Vit.D2 was measured to be 88% in weight compared to the added concentration in the initial formulation, which remained nearly constant for long-term storage (up to 4 months). These results indicate that the prepared CM is a promising carrier as carrier for Vit.D2 with high storage stability.

Despite the usefulness of the freeze-drying process, particularly for thermosensitive compounds, this method is a highly time-consuming and energy-intensive method, which limits its applications in the food industry [11, 64, 66].

1.3.3   SOLVENT INJECTION TECHNIQUES

The solvent injection methods are based on the replacement of the organic solvent by an aqueous solution. It usually involves the gentle injection of the organic solution carrying the phospholipid into a warm aqueous phase containing a compound for encapsulation, as illustrated in Figure 1.11 [57, 81]. The injection pressure is usually sufficient to reach a uniform mixing so that the solvent (e.g., ethanol) quickly dilutes in the aqueous solution, and thus phospholipid molecules disperse in the medium homogenously [45].

Depending on the final application of the formulation, solvent removal can be carried out by one of the following methods: (i) dialysis; (ii) simple mechanical stirring-forced evaporation at room temperature; (iii) inert air-assisted evaporation; or (iv) heating under reduced pressure using an inert gas stream [45, 49, 82].

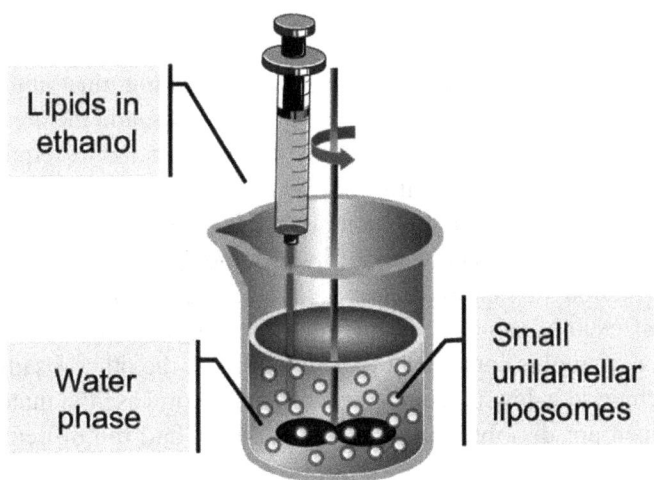

FIGURE 1.11   Liposome production by solvent injection method.
Source: Adapted and reproduced from: Ref. [26] with permission via the Creative Commons Attribution 4.0 International License. ©2016 Springer Nature.

Batzri and Korn [44] were the first who reported this technique in the early 1970s for the preparation of SUVs without sonication [64]. By the addition of the ethanol solution to the aqueous phase, the phospholipid molecules form bilayer-planar fragments and self-assembled them into liposomes, thereby encapsulating the inner aqueous phase [57, 83].

For example, Shin et al. [82] prepared liposomes of EPC and Chol (2:1 molar ratio) for CUR encapsulation. The ethanolic solution containing phospholipids and CUR was injected into the water, followed by ethanol evaporation. Afterward, sonication was applied for 15 min to reduce vesicles size, and purification was performed by centrifugation in order to separate CUR-liposomes from the un-encapsulated CUR as aggregates or pellets. Finally, the liposomes have coated by a dropwise addition of a chitosan solution to the dispersion of the former. The resulting chitosan-coated CUR-liposomes reached 55% EE, which was significantly higher compared to those prepared by the TFDH method (43% EE). Furthermore, chitosan coating improved the EE up to 65%. The chitosan-coated CUR-liposomes exhibited stronger adsorption to mucin compared to the non-coated ones, suggesting that the chitosan coating can improve the residence time of the carriers at the target site, thereby prolonging the absorption in the gastrointestinal tract as a result of mucoadhesion [82].

This study further verifies the fact that the solvent injection method is a viable alternative to the TFDH method when heating must not be used to prevent degradation of thermosensitive compounds, and the presence of harmful solvents must be avoided at the final formulation as required, for example, in the majority of nutraceuticals formulation.

Even though the ethanol injection method is simple, it still suffers from some drawbacks, such as poor solubility of some lipids in ethanol, leading to the formation of heterogeneous liposomes [84]. Also, the presence of residual ethanol can still be regarded as a concern, unless complete removal of ethanol is carried out [45]. In addition, the other disadvantage of this technique is low percentage encapsulation, in case the materials to be entrapped are dissolved in an aqueous phase, and the difficulties for scale-up.

1.3.3.1 *CROSSFLOW INJECTION TECHNIQUE*

The aforementioned challenges of the firstly developed solvent injection technique motivated Wagner et al. [28] to develop a scalable and sterile solvent injection process [27]. This improved method was called the crossflow injection technique, and it comprises the following units process (see Figure 1.12): (1) crossflow injection module; (2 and 3) vessels for the polar phase; (4) vessels for ethanol/lipid solution; (5) a nitrogen pressure device; and (6) peristaltic pump [27, 57]. Herein, the crossflow injection module might be particularly designed for its purpose, thus enable to have higher control of the injection streams, as well as customize it to being suitable for manipulation of any kind of phase mixture. This technique might manufacture dispersion at great industrial scales, being limited only by the volume of the disposable vessel available [57]. In addition, the residual solvent and non-entrapped molecules might be removed by coupling, at the outlet of the injection module, a tangential flow filtration system [57] where the produced liposomal dispersion flows parallel to a filter, rather than forced through it, thus it does not suffer from the clogging difficulty as observed, for example, in the extrusion technique.

The crossflow injection technique also has the advantage of controlling liposome size by the lipid solution concentration, the injection hole diameter, the injection pressure, and the aqueous phase flow rate [57]. Furthermore, these well-defined parameters are responsible for highly reproducible results in terms of vesicle size and EE% [85].

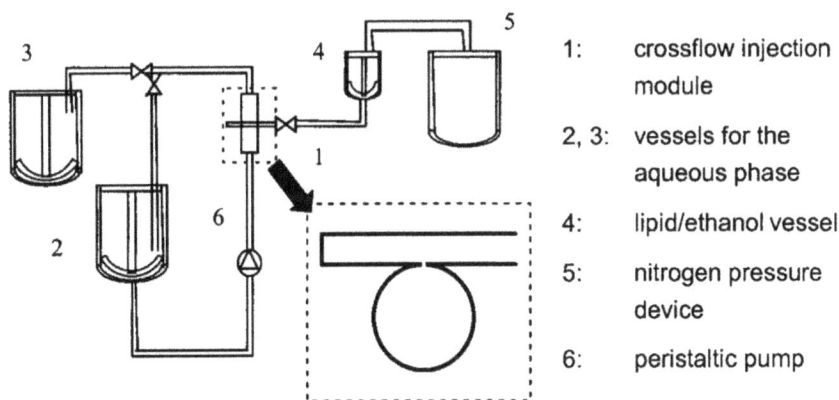

FIGURE 1.12   Liposome production by crossflow injection technique.
Source: Reproduced from: Ref. [27] with permission via Copyright Clearance Center. ©2020 Taylor & Francis.

Another advantage of this method is the suitability of the entrapment of different molecules such as large hydrophilic protein by passive encapsulation, [27] as well as small amphiphilic compounds by a one-step remote loading technique [86]. Moreover, all raw materials such as buffer solutions, lipid ethanol solution, and even nitrogen for applying the injection pressure might be transferred into a sanitized and sterilized system through a 0.2 µm filter, as the process vessels can be sterilized, either by steam or autoclavation, to guarantee an aseptic production [22].

Wagner et al. [27] applied the crossflow injection technique for encapsulation of a radical scavenging protein (rh-Cu/Zn-SOD: recombinant human superoxide dismutase) into liposomes at a scalable ratio (300 mL). Herein, firstly an ethanolic lipid mixture containing dipalmitoyl-phosphatidyl-choline (DPPC) was prepared in a lipid vessel, and the polar phase, composed of PBS-buffer and the rh-Cu/Zn-SOD solution, was prepared in a dedicated aqueous vessel (both vessels tempered at 55°C); then, lipid mixture was pumped through the crossflow injection module and injected into the polar phase, while the protein solution was pumped from aqueous vessel to process vessel (in continuous mode) or back to aqueous vessel in the batch mode [27]. Therefore, liposomes loading rh-Cu/Zn-SOD-l within a narrow size distribution (100 to 200 nm) and 25–30% EE were obtained and assessed regarding storage stability, which demonstrated these liposomal carriers were stable for at least one. Furthermore, it was demonstrated that the characteristics of the liposomes remained reproducible even after scaling up the process to 2.4 L and, interestingly, it provided a more homogeneous liposomal dispersion, without prejudice to the vesicle-size distribution, EE, or stability [27]. Overall, it can be concluded that this technology has the potential for industrial scale-up production of liposomal systems loading large molecules such as proteins.

1.3.4   HEATING-METHOD

Although the conventional methods were shown to be efficient for liposome preparation, their application in the food industry is limited due to the toxicity of the used organic solvents (e.g., chloroform and/or methanol) and surfactants (e.g., SDS). Apart from toxicity, they may adversely affect the stability of entrapped compounds. In addition, these methods are not energy-efficient enough, and therefore presenting the scaling-up challenge [24, 34, 47, 87].

Moreover, as mentioned, the conventional techniques usually require a postprocessing step to achieve the preferred vesicle size. It generally involves mechanical stresses, which may be harmful to some sensitive nutraceutical compounds in terms of the structural disintegration as well as degradation process [29].

Focus on finding solutions for these challenges, and a new method was recently introduced by Mozafari et al. [46, 88] for fast production of liposomes in the absence of any hazardous chemical or process. This is referred to as a simple HM and involves hydration of the liposome components in an aqueous medium followed by heating of the compounds to up to 120°C, in the presence of glycerol (3% v/v) [88–90].

Glycerol, as a water-soluble and physiologically acceptable chemical, is not required to be removed from the final product and is employed to improve the stability of liposomes dispersed in the aqueous medium at the hydration step. Moreover, exposing the compounds to high temperatures eliminates the need to carry out any further sterilization procedure, thereby reducing the time and cost of liposome production by this method [24].

Later on, Mozafari [29] improved HM to eliminate the film hydration step, which enables one to produce liposomes in only one step, in less than an hour, and employing a single apparatus. The improved method was named after his inventor, Mozafari method (MM), and involves heating and stirring (<1000 rpm) of the ingredients, employing or not Chol, in the presence of a polyol, at 40–120°C, based on the properties of the compounds and type of material to entrapped [25].

It was reported that liposomes prepared by the MM are non-toxic to the cultured epithelial cells, while those prepared by a conventional method using organic solvents showed cytotoxicity [25, 91]. Therefore, the MM was introduced [29] to be an efficient, simple, and economical technique for manufacturing bioactive carriers, in the absence of toxic compounds with monodisperse size distribution and high storage stability [25, 88, 92].

Hence, HM, and MM has been the subject of many studies for encapsulation of nutraceuticals and food ingredients as an approach to improve their nutritional value and to decelerate degradation processes such as oxidation of sensitive compounds. Jahadi et al. [12] employed the HM to incorporate flavoenzyme into liposomes. The enzyme-loaded liposomes were mixed with cow milk, which resulted in the formation of white-brined cheese. The product was evaluated in terms of proteolysis degree, chemical constituents of whey plus curd, and yield. The ~25% EE was achieved for the enzyme, while apparent changes in the sensory characteristics of the cheese were not detected. Furthermore, the chemical composition of both curd and whey remained constant when compared to the control sample statistically [12].

Rasti et al. [93] loaded docosahexaenoic acid (DHA) and eicosapen-taenoic acid (EPA) as omega polyunsaturated fatty acids (PUFA) into liposome and evaluated its oxidative stability against a stressful condition of the organic solvent presence, as well as, long-time (up to 10 months) at cold storage (4°C). The liposomes produced by MM, when compared to liposomes prepared by the TFDH method, showed improved oxidative stability in terms of dienes and cyclic peroxides resultant after mixing with absolute ethanol. It was correlated to the employing of an aqueous system to hydrate lipids in the one step of MM, which might protect them from the attack of oxidative molecules. On the other hand, both TFDH or MM-produced liposomes showed significant differences regarding size variation of the liposomal system during cold storage [93].

The same group [94] also studied the effect of several parameters including shear rate (600–1000 rpm), mixing time (30–60 min), and soni-cation time (10–20 min) on the MM in order to prepare liposomes with a maximum EE of PUFAs at the smallest particle size. It was revealed that the highest mixing time (60 min) leads to high PUFAs entrapment (100% EE). On the other hand, the EE% value increased up to a certain level of shear rate, which was found that 795 rpm performed better, and, regarding sonication time, the shorter (10 min) was better to improve EE%. Therefore, PUFAs-loaded liposomes produced by MM under the optimized conditions (795 rpm; 60 min mixing; and 10 min sonication) had an average diameter around 81 nm and 100% EE [94].

Later on, Rasti et al. [95] evaluated the applicability, stability, and sensory effects of the prior optimized PUFAs-loaded liposomes incor-porated in bread and milk. It was compared against likely enriched food with unencapsulated PUFAs in the form of fish oil and microencapsulated PUFAs. Significantly higher PUFAs recovery (EPA-95% and DHA-96% in freshly cooked bread; EPA-94% and DHA-95% in freshly pasteurized milk) and lower peroxide (0.14 meq/kg in both enriched products) and anisidine (0.91 meq/Kg in freshly cooked bread; 2.1 meq/Kg in fresh pasteurized milk) values were observed in the liposomal PUFAs enriched samples in comparison with others samples. Interestingly, there was no significant variation of PUFAs recovering from enriched bread after 7 days of storage. From the sensory evaluation, milk and bread containing liposomal PUFAs were similar to the control, while samples enriched with unencapsulated or microencapsulated PUFAs showed significant fishy flavor [95].

In conclusion, liposomal systems produced by HH and MM are a favorable alternative to produce protective carriers for sensitive chemical nutrients which and to increase food nutritional value. However, in spite of all the advantages of HM and MM mentioned above, these techniques yet required relatively high temperature, which is a drawback for encapsulation of some nutraceuticals as it may potentially lead to degradation and loss of the desirable properties of thermosensitive compounds.

1.3.5 COMBINED PROTOCOLS

A combination of the previously mentioned technologies is also a common approach to prepare liposomes in the required scale with homogeneous size. For example, Chen et al. [96] prepared CUR-loaded liposomes employing the TFDH technique followed by dynamic high-pressure microfluidization. The prepared liposomes had the 57% EE, size around 68 nm, polydispersity index of 0.246, and zeta potential of -3.16 mV. The vesicles also showed appropriate sustained release properties and improved CUR solubility and stability under alkaline pH conditions, also in the presence of oxidative-promoter ion (e.g., Fe^{3+}, Al^{3+}, Fe^{2+}, Cu^{2+}, K^+, Li^+), which contributed to a longer shelf life at 4°C. The cellular antioxidant assay demonstrated the encapsulated CUR has equal activity as the free CUR, which was attributed to its lower cellular uptake [96].

Zou et al. [97] also employed the combination of ethanol injection and high-pressure microfluidization methods to produce tea polyphenol liposomes. They reported enhanced stability of encapsulated tea polyphenol under alkaline media and a relatively good sustained release profile as compared to the tea polyphenol solution. It was shown that around 30% of the loading is released *in vitro* up to 24 h.

Likewise, Bonnechi et al. [23] prepared polyphenol-loaded liposomes by the TFDH as a first step to obtain MUVs of encapsulated RESV. Chloroform and ethanol were used for the dissolution of lipids (DPPC, DC, Chol) RESV, respectively. Next, the following three techniques were carried out to obtain uniform liposomes with reduced or no lamellarity: (i) homogenization by freeze-thaw in nitrogen and water bath at 50°C; (ii) sonication with a tip sonicator; and (iii) extrusion through 100 nm polycarbonate membranes. It was shown that the cell viability was not affected by the administration of RESV-loaded liposomes [23].

Nacka et al. [98] also freeze-dried the α-tocopherol-loaded liposomes previously prepared by a simplified TFDH method, where the solvent (chloroform/methanol) of the marine lipid's solution was first removed under a nitrogen stream. Afterward, HEPES buffer was used for rehydration of the lyophilized lipid film, which was then extruded through polycarbonate membranes. The method resulted in spherical MLVs with a size around 4 μm. These studied α-tocopherol-loaded liposomes improved PUFAs absorption in rats (98%) compared to fish oil (73%); in addition, DHA proportion in lymph was higher after liposome ingestion (78%) than after fish oil ingestion (47%) [98].

1.4 CONCLUSIONS

The liposomal carrier technology has been growing very fast throughout the years, mainly focused on medicinal applications. The knowledge about its production features has been broadly discussed in the literature, including the physicochemical aspects involved in the process. However, it is factual for application on medicinal therapies, when it is regarded to nutrition field, less information is found, in part because of inherent challenges of implementing safety, solvent-free, as well as, low-cost processes suitable for producing liposomes at large scale; however, the interest of the food industry in this nanotechnology is gradually increasing. Despite the mentioned difficulties, the advantages provide by liposomal encapsulation in increasing stability and bioavailability of nutrients, and bioactive compounds have been continuality considered to improve the nutritional value in dietary or even advance to the therapeutic application of nutraceuticals. Thus, it has recently motivated the search for improved technologies, which overcome the issues of conventional methods (e.g., TFDH, or solvent injection) for manufacturing liposomal carriers with consistent characteristics and on a large scale required by the food industry.

Microfluidic, crossflow solvent injection, heating, or Mozafari techniques are the most recently developed technologies that evolved as alternative methodologies to the conventional protocols. It is because, by employing these methods, liposomes might enable to be prepared at large scales without or with less harmful solvents. Also, these methods are less aggressive to the sensitive compounds.

Nevertheless, it is still valuable to consider the physicochemical characteristics of the desired ingredient to be protected. In other words, based on the application of the final formulation, the most suitable liposomal encapsulation protocol needs to be chosen. Also, the association of several techniques should increase the production of liposomes with nutraceuticals and food supplements.

Finally, this chapter summarized several methods employed for encapsulation of different food-related compounds into liposomes which could be useful as a guide for the development of novel enriched food or nutraceutical supplement formulation.

KEYWORDS

- **biocompounds**
- **food supplements**
- **functional food**
- **liposomes manufacturing**
- **nutraceuticals formulation**
- **nutrients encapsulation**

REFERENCES

1. Singh, M. H., Thompson, P. A., & Corredig, C. M., (2012). Liposomes as food ingredients and nutraceutical delivery systems. In: Garti, N., & McClements, D. J., (eds.), *Encapsulation Technologies and Delivery Systems for Food Ingredients and Nutraceuticals.* Woodhead Publishing.
2. Frézard, F., (1999). Liposomes: From biophysics to the design of peptide vaccines. *Brazilian J. Med. Biol. Res., 32*(2), 181–189. https://doi.org/10.1590/S0100-879X1999000200006.
3. Gibbs, F., Selim, K., & Inteaz, A. B., (1999). Encapsulation in the food industry: A review. *Int. J. Food Sci. Nutr., 50*(3), 213–224. https://doi.org/10.1080/096374899101256.
4. Gregoriadis, G., & Neerunjun, D. E., (1974). Control of the rate of hepatic uptake and catabolism of liposome-entrapped proteins injected into rats. Possible therapeutic applications. *Eur. J. Biochem., 47*(1), 179–185. https://doi.org/10.1111/j.1432-1033.1974.tb03681.x.
5. Torchilin, V., (2009). Multifunctional and stimuli-sensitive pharmaceutical nanocarriers. *Eur. J. Pharm. Biopharm., 71*(3), 431–444. https://doi.org/10.1016/j.ejpb.2008.09.026.

6. Tan, C., Xue, J., Lou, X., Abbas, S., Guan, Y., Feng, B., Zhang, X., & Xia, S., (2014). Liposomes as delivery systems for carotenoids: Comparative studies of loading ability, storage stability and *in vitro* release. *Food Funct., 5*(6), 1232. https://doi.org/10.1039/c3fo60498e.

7. Van, H. P., (2017). An update on the use of oral phospholipid excipients. *Eur. J. Pharm. Sci., 108*, 1–12. https://doi.org/10.1016/j.ejps.2017.07.008.

8. Yokota, D., Moraes, M., & Pinho, S. C., (2012). Characterization of lyophilized liposomes produced with non-purified soy lecithin: A case study of casein hydrolysate microencapsulation. *Brazilian J. Chem. Eng., 29*(2), 325–335. https://doi.org/10.1590/S0104–66322012000200013.

9. Desai, K. G. H., & Park, H. J., (2005). Recent developments in microencapsulation of food ingredients. *Drying Technology* (pp. 1361–1394). Taylor and Francis Ltd. https://doi.org/10.1081/DRT-200063478.

10. Isailović, B. D., Kostić, I. T., Zvonar, A., Đorđević, V. B., Gašperlin, M., Nedović, V. A., & Bugarski, B. M., (2013). Resveratrol-loaded liposomes produced by different techniques. *Innov. Food Sci. Emerg. Technol., 19*, 181–189. https://doi.org/10.1016/j.ifset.2013.03.006.

11. Taylor, T. M., Weiss, J., Davidson, P. M., & Bruce, B. D., (2005). Liposomal nanocapsules in food science and agriculture. *Crit. Rev. Food Sci. Nutr., 45*(7, 8), 587–605. https://doi.org/10.1080/10408390591001135.

12. Jahadi, M., Khosravi-Darani, K., Ehsani, M. R., Mozafari, M. R., Saboury, A. A., & Pourhosseini, P. S., (2015). The encapsulation of flavorzyme in nanoliposome by heating method. *J. Food Sci. Technol., 52*(4), 2063–2072. https://doi.org/10.1007/s13197-013-1243-0.

13. Da Silva, M. P., Daroit, D. J., & Brandelli, A., (2010). Food applications of liposome-encapsulated antimicrobial peptides. *Trends in Food Science and Technology* (pp. 284–292). Elsevier. https://doi.org/10.1016/j.tifs.2010.03.003.

14. Were, L. M., Bruce, B., Davidson, P. M., & Weiss, J., (2004). Encapsulation of nisin and lysozyme in liposomes enhances efficacy against Listeria monocytogenes. *J. Food Prot.* https://doi.org/10.4315/0362-028X-67.5.922.

15. Taylor, T. M., Gaysinsky, S., Davidson, P. M., Bruce, B. D., & Weiss, J., (2007). Characterization of antimicrobial-bearing liposomes by ζ-potential, vesicle size, and encapsulation efficiency. *Food Biophys.* https://doi.org/10.1007/s11483-007-9023-x.

16. Marsanasco, M., Márquez, A. L., Wagner, J. R., Del, V. A. S., & Chiaramoni, N. S., (2011). Liposomes as vehicles for vitamins E and C: An alternative to fortify orange juice and offer vitamin C protection after heat treatment. *Food Res. Int., 44*(9), 3039–3046. https://doi.org/10.1016/j.foodres.2011.07.025.

17. Galea, R., Nel, H. J., Talekar, M., Liu, X., Ooi, J. D., Huynh, M., Hadjigol, S., et al., (2019). PDI-L1- and calcitriol-dependent liposomal antigen-specific regulation of systemic inflammatory autoimmune disease. *JCI Insight, 4*(18). https://doi.org/10.1172/jci.insight.126025.

18. Kosaraju, S. L., Tran, C., & Lawrence, A., (2006). Liposomal delivery systems for encapsulation of ferrous sulfate: Preparation and characterization. *J. Liposome Res., 16*(4), 347–358. https://doi.org/10.1080/08982100600992351.

19. Xia, S., & Xu, S., (2005). Ferrous sulfate liposomes: Preparation, stability and application in fluid milk. *Food Res. Int., 38*(3), 289–296. https://doi.org/10.1016/j.foodres.2004.04.010.

20. Kostecka-Gugała, A., Latowski, D., & Strzałka, K., (2003). Thermotropic phase behavior of α-dipalmitoyl phosphatidylcholine multibilayers is influenced to various extents by carotenoids containing different structural features- evidence from differential scanning calorimetry. *Biochim. Biophys. Acta - Biomembr., 1609*(2), 193–202. https://doi.org/10.1016/S0005-2736(02)00688-0.

21. Huang, M., Su, E., Zheng, F., & Tan, C., (2017). Encapsulation of flavonoids in liposomal delivery systems: The case of quercetin, kaempferol and luteolin. *Food Funct., 8*(9), 3198–3208. https://doi.org/10.1039/C7FO00508C.

22. Wagner, A., Vorauer-Uhl, K., & Katinger, H., (2002). Liposomes produced in a pilot-scale: Production, purification and efficiency aspects. *Eur. J. Pharm. Biopharm., 54*(2), 213–219. https://doi.org/10.1016/S0939-6411(02)00062-0.

23. Bonechi, C., Martini, S., Ciani, L., Lamponi, S., Rebmann, H., Rossi, C., & Ristori, S., (2012). Using liposomes as carriers for polyphenolic compounds: The case of trans-resveratrol. *PLoS One, 7*(8). https://doi.org/10.1371/journal.pone.0041438.

24. Mozafari, M. R., (2005). Liposomes: An overview of manufacturing techniques. *Cell. Mol. Biol. Lett., 10*(4), 711–719.

25. Mozafari, M. R., Khosravi-Darani, K., Borazan, G. G., Cui, J., Pardakhty, A., & Yurdugul, S., (2008). Encapsulation of food ingredients using nanoliposome technology. *Int. J. Food Prop., 11*(4), 833–844. https://doi.org/10.1080/10942910701648115.

26. Carugo, D., Bottaro, E., Owen, J., Stride, E., & Nastruzzi, C., (2016). Liposome production by microfluidics: Potential and limiting factors. *Sci. Rep., 6*(1), 25876. https://doi.org/10.1038/srep25876.

27. Wagner, A., Vorauer-Uhl, K., Kreismayr, G., & Katinger, H., (2002). Enhanced protein loading into liposomes by the multiple crossflow injection technique. *J. Liposome Res., 12*(3), 271–283. https://doi.org/10.1081/LPR-120014762.

28. Wagner, A., Vorauer-Uhl, K., Kreismayr, G., & Katinger, H., (2002). The crossflow injection technique: An improvement of the ethanol injection method. *J. Liposome Res., 12*(3), 259–270. https://doi.org/10.1081/LPR-120014761.

29. Mozafari, M. R., (2005). *Method and Apparatus for Producing Carrier Complexes.* WO2005084641A1.

30. Takahashi, M., Inafuku, K., Miyagi, T., Oku, H., Wada, K., Imura, T., & Kitamoto, D., (2007). Efficient preparation of liposomes encapsulating food materials using lecithins by a mechanochemical method. *J. Oleo Sci., 56*(1), 35–42. https://doi.org/10.5650/jos.56.35.

31. Yu, B., Lee, R. J., & Lee, L. J., (2009). microfluidic methods for production of liposomes. *Methods Enzymol., 465*(C), 129–141. https://doi.org/10.1016/S0076-6879(09)65007-2.

32. Spitsberg, V. L., (2005). Bovine milk fat globule membrane as a potential nutraceutical. *Journal of Dairy Science* (pp. 2289–2294). American Dairy Science Association. https://doi.org/10.3168/jds.S0022-0302(05)72906-4.

33. Gregoriadis, G., & Perrie, Y., (2010). Liposomes. In: *Encyclopedia of Life Sciences.* John Wiley & Sons, Ltd: Chichester, UK. https://doi.org/10.1002/9780470015902.a0002656.pub2.

34. Mozafari, R. M., & Mortazavi, S. M., (2005). *Nanoliposomes: From Fundamentals to Recent Developments*. Trafford Publishing Ltd: Oxford.

35. Mufamadi, M. S., Pillay, V., Choonara, Y. E., Du Toit, L. C., Modi, G., Naidoo, D., & Ndesendo, V. M. K., (2011). A review on composite liposomal technologies for specialized drug delivery. *J. Drug Deliv., 2011*, 1–19. https://doi.org/10.1155/2011/939851.

36. Arajo, L. S. C. D., Santos, G. C. D., Ribeiro, T. G., Santos, F. D. D., Amaral, L. E., & Cristina, M., (2013). Liposomes as carriers of anticancer drugs. In: *Cancer Treatment - Conventional and Innovative Approaches*. InTech. https://doi.org/10.5772/55290.

37. Johnsson, M., & Edwards, K., (2003). Liposomes, disks, and spherical micelles: Aggregate structure in mixtures of gel phase phosphatidylcholines and poly(ethylene glycol)-phospholipids. *Biophysical. J., 85*, 3839–3847.

38. Watwe, R. M., & Bellare, J. R., (1995). Manufacture of liposomes: A review. *Curr. Sci., 68*(7), 715–724.

39. Lasic, D., (1998). Novel applications of liposomes. *Trends Biotechnol., 16*(7), 307–321. https://doi.org/10.1016/S0167-7799(98)01220-7.

40. Amselem, S., Gabizon, A., & Barenholz, Y., (1990). Optimization and upscaling of doxorubicin-containing liposomes for clinical use. *J. Pharm. Sci.* https://doi.org/10.1002/jps.2600791202.

41. Tseng, L. P., Liang, H. J., Chung, T. W., Huang, Y. Y., & Liu, D. Z., (2007). Liposomes incorporated with cholesterol for drug release triggered by magnetic field. *J. Med. Biol. Eng.*

42. Lu, Q., Li, D. C., & Jiang, J. G., (2011). Preparation of a tea polyphenol nanoliposome system and its physicochemical properties. *J. Agric. Food Chem., 59*(24), 13004–13011. https://doi.org/10.1021/jf203194w.

43. Alexopoulou, E., Georgopoulos, A., Kagkadis, K. A., & Demetzos, C., (2006). Preparation and characterization of lyophilized liposomes with incorporated quercetin. *J. Liposome Res.* https://doi.org/10.1080/08982100500528594.

44. Batzri, S., & Korn, E. D., (1973). Single bilayer liposomes prepared without sonication. *Biochim. Biophys. Acta - Biomembr., 298*(4), 1015–1019. https://doi.org/10.1016/0005-2736(73)90408-2.

45. Pons, M., Foradada, M., & Estelrich, J., (1993). Liposomes obtained by the ethanol injection method. *Int. J. Pharm., 95*(1–3), 51–56. https://doi.org/10.1016/0378-5173(93)90389-W.

46. Mozafari, M. R., Reed, C. J., Rostron, C., Kocum, C., & Piskin, E., (2002). Construction of stable anionic liposome-plasmid particles using the heating method: A preliminary investigation. *Cell. Mol. Biol. Lett., 7*(3), 923–927.

47. Lasic, D. D., (1988). The mechanism of vesicle formation. *Biochem. J.*, (256), 1–11.

48. Bozzuto, G., & Molinari, A., (2015). Liposomes as nanomedical devices. *Int. J. Nanomedicine, 10*, 975–999. https://doi.org/10.2147/IJN.S68861.

49. Burgess, S. *Liposome Preparation*. https://www.sigmaaldrich.com/technical-documents/articles/biology/liposome-preparation.html (accessed on 17 June 2021).

50. Avanti Polar Lipids. *Lipids for Liposome Formation*. https://avantilipids.com/tech-support/liposome-preparation/lipids-for-liposome-formation (accessed on 17 June 2021).

51. Frazier, R. A., (2011). Food chemistry. In: Campbell-Platt, G., (ed.), *Food Science and Technology*. John Wiley & Sons: New York, NY.

52. Serrano, G., Almudéver, P., Serrano, J. M., Milara, J., Torrens, A., Expósito, I., & Cortijo, J., (2015). Phosphatidylcholine liposomes as carriers to improve topical ascorbic acid treatment of skin disorders. *Clin. Cosmet. Investig. Dermatol., 8*, 591–599. https://doi.org/10.2147/CCID.S90781.

53. Moeller, H., Martin, D., Schrader, K., Hoffmann, W., & Lorenzen, P. C., (2018). Spray-or freeze-drying of casein micelles loaded with vitamin D2: Studies on storage stability and *in vitro* digestibility. *LWT, 97*, 87–93. https://doi.org/10.1016/j.lwt.2018.04.003.

54. Abbasi, A., Emam-Djomeh, Z., Mousavi, M. A. E., & Davoodi, D., (2014). Stability of vitamin D3 encapsulated in nanoparticles of whey protein isolate. *Food Chem., 143*, 379–383. https://doi.org/10.1016/j.foodchem.2013.08.018.

55. Assadpour, E., & Mahdi, J. S., (2019). A Systematic review on nanoencapsulation of food bioactive ingredients and nutraceuticals by various nanocarriers. *Crit. Rev. Food Sci. Nutr., 59*(19), 3129–3151. https://doi.org/10.1080/10408398.2018.1484687.

56. Kumar, P., Huo, P., & Liu, B., (2019). Formulation strategies for folate-targeted liposomes and their biomedical applications. *Pharmaceutics, 11*(8), 381. https://doi.org/10.3390/pharmaceutics11080381.

57. Wagner, A., & Vorauer-Uhl, K., (2011). Liposome technology for industrial purposes. *J. Drug Deliv., 2011*. https://doi.org/10.1155/2011/591325.

58. Cadena, P. G., Pereira, M. A., Cordeiro, R. B. S., Cavalcanti, I. M. F., Barros, N. B., Pimentel, M. D. C. C. B., Lima, F. J. L., et al., (2013). Nanoencapsulation of quercetin and resveratrol into elastic liposomes. *Biochim. Biophys. Acta - Biomembr., 1828*(2), 309–316. https://doi.org/10.1016/j.bbamem.2012.10.022.

59. Caddeo, C., Teskač, K., Sinico, C., & Kristl, J., (2008). Effect of resveratrol incorporated in liposomes on proliferation and UV-B protection of cells. *Int. J. Pharm., 363*(1, 2), 183–191. https://doi.org/10.1016/j.ijpharm.2008.07.024.

60. Amri, A., Chaumeil, J. C., Sfar, S., & Charrueau, C., (2012). Administration of resveratrol: What formulation solutions to bioavailability limitations? *Journal of Controlled Release* (pp. 182–193). Elsevier. https://doi.org/10.1016/j.jconrel.2011.09.083.

61. Saber, F. R., Abdelbary, G. A., Salama, M. M., Saleh, D. O., Fathy, M. M., & Soliman, F. M., (2018). UPLC/QTOF/MS profiling of two Psidium species and the *in-vivo* hepatoprotective activity of their nano-formulated liposomes. *Food Res. Int., 105*, 1029–1038. https://doi.org/10.1016/j.foodres.2017.12.042.

62. Tekade, R. K., (2019). *Basic Fundamentals of Drug Delivery*. Academic Press.

63. Castile, J. D., & Taylor, K. M. G., (1999). Factors affecting the size distribution of liposomes produced by freeze-thaw extrusion. *Int. J. Pharm., 188*(1), 87–95. https://doi.org/10.1016/S0378-5173(99)00207-0.

64. Meure, L. A., Foster, N. R., & Dehghani, F., (2008). Conventional and dense gas techniques for the production of liposomes: A review. *AAPS PharmSciTech, 9*(3), 798. https://doi.org/10.1208/s12249-008-9097-x.

65. Huang, C. H., (1969). Studies on phosphatidylcholine vesicles. Formation and physical characteristics. *Biochemistry, 8*(1), 344–352. https://doi.org/10.1021/bi00829a048.

66. Xu, A. X., West, E. A. L., & Rogers, M. A., (2020). Encapsulation of nutraceuticals. In: Spagnuolo, P. A., (ed.), *Nutraceuticals and Human Health: The Food-to-Supplement Paradigm*. The Royal Society of Chemistry.

67. Perret, S., Golding, M., & Williams, W. P., (1991). A simple method for the preparation of liposomes for pharmaceutical applications: Characterization of the liposomes. *J. Pharm. Pharmacol., 43*(3), 154–161. https://doi.org/10.1111/j.2042-7158.1991. tb06657.x.

68. Chen, H., Wu, J., Sun, M., Guo, C., Yu, A., Cao, F., Zhao, L., et al., (2012). N-trimethyl chitosan chloride-coated liposomes for the oral delivery of curcumin. *J. Liposome Res., 22*(2), 100–109. https://doi.org/10.3109/08982104.2011.621127.

69. Zu, Y., Overby, H., Ren, G., Fan, Z., Zhao, L., & Wang, S., (2018). Resveratrol liposomes and lipid nanocarriers: Comparison of characteristics and inducing browning of white adipocytes. *Colloids Surfaces B Biointerfaces, 164*, 414–423. https://doi.org/10.1016/j.colsurfb.2017.12.044.

70. Patrignani, F., & Lanciotti, R., (2016). Applications of high and ultra high pressure homogenization for food safety. *Front. Microbiol., 7*, 1132. https://doi.org/10.3389/fmicb.2016.01132.

71. Koroleva, M. Y., & Yurtov, E. V., (2012). Nanoemulsions: The properties, methods of preparation and promising applications. *Russ. Chem. Rev., 81*(1), 21–43. https://doi.org/10.1070/RC2012v081n01ABEH004219.

72. Cho, S. C., Choi, W. Y., Oh, S. H., Lee, C. G., Seo, Y. C., Kim, J. S., Song, C. H., et al., (2012). Enhancement of lipid extraction from marine microalga, *Scenedesmus* associated with high-pressure homogenization process. *J. Biomed. Biotechnol., 2012*(2014). https://doi.org/10.1155/2012/359432.

73. Jahn, A., Vreeland, W. N., DeVoe, D. L., Locascio, L. E., & Gaitan, M., (2007). Microfluidic directed formation of liposomes of controlled size. *Langmuir, 23*(11), 6289–6293. https://doi.org/10.1021/la070051a.

74. Ran, R., Sun, Q., Baby, T., Wibowo, D., Middelberg, A. P. J., & Zhao, C. X., (2017). Multiphase microfluidic synthesis of micro- and nanostructures for pharmaceutical applications. *Chem. Eng. Sci., 169*, 78–96. https://doi.org/10.1016/j.ces.2017.01.008.

75. Li, T., Yang, S., Liu, W., Liu, C., Liu, W., Zheng, H., Zhou, W., & Tong, G., (2015). Preparation and characterization of nanoscale complex liposomes containing medium-chain fatty acids and vitamin C. *Int. J. Food Prop., 18*(1), 113–124. https://doi.org/10.1080/10942912.2012.685683.

76. Uluata, S., McClements, D. J., & Decker, E. A., (2015). Physical stability, autoxidation, and photosensitized oxidation of ω-3 oils in nanoemulsions prepared with natural and synthetic surfactants. *J. Agric. Food Chem., 63*(42), 9333–9340. https://doi.org/10.1021/acs.jafc.5b03572.

77. Ding, Y., & Kan, J., (2017). Optimization and characterization of high pressure homogenization produced chemically modified starch nanoparticles. *J. Food Sci. Technol., 54*(13), 4501–4509. https://doi.org/10.1007/s13197-017-2934-8.

78. Jafari, S. M., (2019). *Lipid-based Nanostructures for Food Encapsulation Purposes* (Vol. 2). eBook. ISBN: 9780128156742; Paperback ISBN: 9780128156735.

79. Martínez-Monteagudo, S. I., Yan, B., & Balasubramaniam, V. M., (2017). Engineering process characterization of high-pressure homogenization—from

laboratory to industrial scale. *Food Eng. Rev., 9*(3), 143–169. https://doi.org/10.1007/ s12393-016-9151-5.

80. Gupta, C., Chawla, P., & Arora, S., (2015). Development and evaluation of iron micro encapsules for milk fortification. *CyTA-J. Food, 13*(1), 116–123. https://doi.org/10.10 80/19476337.2014.918179.

81. Mathur, R., & Capasso, P., (1997). *Nonphospholipid Liposomes: Properties and Potential Use in Flavor Encapsulation* (pp. 219–230). No. Mlv. https://doi. org/10.1021/bk-1995-0610.ch017.

82. Shin, G. H., Chung, S. K., Kim, J. T., Joung, H. J., & Park, H. J., (2013). Preparation of chitosan-coated nanoliposomes for improving the mucoadhesive property of curcumin using the ethanol injection method. *J. Agric. Food Chem., 61*(46), 11119–11126. https://doi.org/10.1021/jf4035404.

83. Lasic, D. D., (1995). Mechanisms of liposome formation. *J. Liposome Res., 5*(3), 431–441. https://doi.org/10.3109/08982109509010233.

84. Maherani, B., Arab-Tehrany, E., Mozafari, M. R., Gaiani, C., & Linder, M., (2011). Liposomes: A review of manufacturing techniques and targeting strategies. *Curr. Nanosci., 7*(3), 436–452. https://doi.org/10.2174/157341311795542453.

85. Wagner, A., Platzgummer, M., Kreismayr, G., Quendler, H., Stiegler, G., Ferko, B., Vecera, G., et al., (2006). GMP production of liposomes: A new industrial approach. *J. Liposome Res., 16*(3), 311–319. https://doi.org/10.1080/08982100600851086.

86. Wagner, A., Stiegler, G., Vorauer-Uhl, K., Katinger, H., Quendler, H., Hinz, A., & Weissenhorn, W., (2007). One step membrane incorporation of viral antigens as a vaccine candidate against HIV. In: *Journal of Liposome Research* (Vol. 17, pp. 139–154). Taylor & Francis. https://doi.org/10.1080/08982100701530159.

87. Mozafari, M. R., (2005). *A New Technique for the Preparation of Non-Toxic Liposomes and Nanoliposomes: The Heating Method* (pp. 91–98). Trafford Publishing.

88. Mozafari, M. R., Reed, C. J., & Rostron, C., (2002). Development of non-toxic liposomal formulations for gene and drug delivery to the lung. *Technol. Heal. Care, 10*(3, 4), 342–344.

89. Khosravi-Darani, K., & Mozafari, M. R., (2010). Nanoliposome potentials in nanotherapy: A concise overview. *Int. J. Nanosci. Nanotechnol., 6*(1), 3–13.

90. Mozafari, M., Reed, C., Rostron, C., & Martin, D., (2004). Transfection of human airway epithelial cells using a lipid-based vector prepared by the heating method. *J. Aerosol Medicat., 17*(1), 100.

91. Mozafari, M. R., Reed, C. J., & Rostron, C., (2007). cytotoxicity evaluation of anionic nanoliposomes and nanolipoplexes prepared by the heating method without employing volatile solvents and detergents. *Pharmazie, 62*(3), 205–209. https://doi.org/10.1691/ ph.2007.3.6045.

92. Mortazavi, S. M., Mohammadabadi, M. R., Khosravi-Darani, K., & Mozafari, M. R., (2007). Preparation of liposomal gene therapy vectors by a scalable method without using volatile solvents or detergents. *J. Biotechnol., 129*(4), 604–613. https://doi. org/10.1016/j.jbiotec.2007.02.005.

93. Rasti, B., Jinap, S., Mozafari, M. R., & Yazid, A. M., (2012). Comparative study of the oxidative and physical stability of liposomal and nanoliposomal polyunsaturated fatty acids prepared with conventional and Mozafari methods. *Food Chem., 135*(4), 2761–2770. https://doi.org/10.1016/j.foodchem.2012.07.016.

94. Rasti, B., Jinap, S., Mozafari, M. R., & Abd-Manap, M. Y., (2014). Optimization on preparation condition of polyunsaturated fatty acids nanoliposome prepared by Mozafari method. *J. Liposome Res., 24*(2), 99–105. https://doi.org/10.3109/08982104.2013.839702.

95. Rasti, B., Erfanian, A., & Selamat, J., (2017). Novel nanoliposomal encapsulated omega-3 fatty acids and their applications in food. *Food Chem., 230*, 690–696. https://doi.org/10.1016/j.foodchem.2017.03.089.

96. Chen, X., Zou, L. Q., Niu, J., Liu, W., Peng, S. F., & Liu, C. M., (2015). The stability, sustained release and cellular antioxidant activity of curcumin nanoliposomes. *Molecules, 20*(8), 14293–14311. https://doi.org/10.3390/molecules200814293.

97. Zou, L., Liu, W., Liu, W., Liang, R., Li, T., Liu, C., Cao, Y., Niu, J., & Liu, Z., (2014). Characterization and bioavailability of tea polyphenol nanoliposome prepared by combining an ethanol injection method with dynamic high-pressure micro fluidization. *J. Agric. Food Chem., 62*(4), 934–941. https://doi.org/10.1021/jf402886s.

98. Nacka, F., Cansell, M., Méléard, P., & Combe, N., (2001). Incorporation of α-tocopherol in marine lipid-based liposomes: *In vitro* and *in vivo* studies. *Lipids, 36*(12), 1313–1320. https://doi.org/10.1007/s11745-001-0846-x.

99. Cansell, M., Nacka, F., & Combe, N., (2003). Marine lipid-based liposomes increase *in vivo* FA bioavailability. *Lipids, 38*(5), 551–559. https://doi.org/10.1007/s11745-003-1341-0.

CHAPTER 2

Liposome Preparation and Related Techniques

AMEEDUZZAFAR ZAFAR,[1] NABIL K. ALRUWAILI,[1]
KHALID SAAD ALHARBI,[2] NASSER HADAL ALOTAIBI,[3] and
MOHAMMED JAFAR[4]

[1]*Department of Pharmaceutics, College of Pharmacy, Jouf University,
Aljouf, KSA*

[2]*Department of Pharmacology, College of Pharmacy Jouf University,
Aljouf, KSA*

[3]*Department of Clinical Pharmacy, College of Pharmacy Jouf University,
Aljouf, KSA*

[4]*Department of Pharmaceutics, College of Clinical Pharmacy, Imam
Abdulrahman Bin Faisal University, Dammam, KSA*

ABSTRACT

The main challenge of pharmaceuticals is to deliver the therapeutic agent
into the body in an appropriate form to enhance the therapeutic index,
stability, and lower the side effect. Liposome the novel tool for delivery of
drugs as well as nutraceuticals via topically, orally, and parenterally to the
body. It more versatile technique for the delivery of nutraceuticals. It is a
self-assemble delivery system upon contact with an aqueous phase. It is a
biocompatible, biodegradable, non-toxic, non-immunogenic formulation as
well as easily reached and penetrates to the target site. It is composed of lipids
(soy, egg, semisynthetic, and synthetic lipids) and cholesterol (Chol). In this
review, we discussed about composition, various techniques for preparation,
characterization, stability as well application of liposomes for nutraceutical
delivery into the body for treatment of various acute and chronic diseases.

2.1 INTRODUCTION

Liposomes are a small vesicular system in which drug is incorporated to improve the stability as well as delivery. When lipid (cholesterol (Chol), phospholipids, and surfactant) dispersed into the water system and formed the vesicles called the liposome [1]. Both types of drugs (gydrophilic and lipophilic drugs can be delivered into the body by liposomes. Liposomes are unique, simple, microscopic, attractive, and advanced approaches for delivery of drugs via various rout of administration like oral, parenteral, transdermal, nasal, ocular, pulmonary. Parenteral delivery of liposome has a drawback, and it is laminated from systemic circulation rapidly by opsonization following segregation from blood as well as by reticuloendothelial system (RES) of spleen liver, lung [2]. But this drawback has been solved by coating with polyethylene glycol (PEG, pegylated-liposome). Pegylated liposomes escape RES and attain prolong circulation into the body [3]. Liposomes are discovered by Bangham in 1980 for the delivery of drug in biological system [4]. There are several mechanisms by liposome in topical and inside the body, i.e.: (i) attached to biological membrane and amalgamated with them; (ii) Phospholipids fused with biological membrane and released drug; (iii) the phagocyte capture lysosome act on phospholipid and released drug. Liposomes have gorgeous significance in biological, pharmaceutical, nutraceutical, and diagnostic fields as carrier. It is a most appropriate carrier for delivery of various therapeutic and non-therapeutic agent, i.e., antidiabetic [5], anticancer [6], anti-inflammatory [7], herbal [8], protein-peptide [9], antibiotics [10], antifungal [11], and nutraceutical [12]. There are different types of liposomes on the basis of structure and size, i.e., large unilamellar vesicles (LUV), small unilamellar vehicle (SUV) and multilamellar vesicle (MVV) and multi-vesicular vesicle, in which MVV apply parenteral routs only and remaining other used for flexible routes. Systemic liposomes have many advantages, i.e., biodegradable in nature, low systemic toxicity, protection of bio-sensitive drug, improve pharmacokinetic and pharmacodynamic. Moreover, topical application of liposome has advantages, i.e., significant accumulation of drug at designated site, decreases the undesirable systemic side effect, and decreases the incompatibilities [13]. Various researches investigated for improvement of stability and distribution of ingredients *in vivo* by using of some long-chain polymer like PEG and polyvinyl alcohol. Liposomes have property to clear the reticular endothelial system and opsonization and enhance the permeation effect (EPR). It increases permeability of therapeutic agent to the target site like tumor [14, 15]. Repeated dose of Pegylated liposome reduced the long circulation properties and clearance from the blood (accelerated blood clearance) and more amounts of drug deposited in liver and spleen demonstrated by Dams and its associates [16].

Liposomes are comprising of conventional liposomes and modified (novel) liposomes. The conventional liposome is a primitive and attractive for drug delivery. But this liposome has drawbacks like low residence (circulation) time in blood and accumulated into liver and spleen [17], so it is not suitable as targeted delivery. Now a day modified liposomes commonly applied for tissue or cell targeting because they exhibited prolonged circulation time. Barenholz developed the first FDA-approved Pegylated liposome (Doxil) liposome (Doxil) and showed the long circulation time and high amount of drug deposited into target side without effect on other or normal cell [3]. Liposomal delivery system has many advantages like: (i) biocompatible and suitable for control delivery; (ii) suitable for localized and targeted action; (iii) increase the therapeutic activity and reduced the toxicity of drug; (iv) no antigenicity and improve the pharmacokinetic profile.

2.2 CHEMICAL COMPOSITION OF LIPOSOMES

There are many chemicals required for the development of liposomes, i.e., phospholipids, cholesterol, which are discussed in subsections.

2.2.1 PHOSPHOLIPIDS

Phospholipids are surface-active and amphiphilic molecule, and it contains head (hydrophilic/polar) and tail (hydrophobic/nonpolar). It is used for the manufacturing of different types of pharmaceutical preparation like liposome, micelle, emulsion, suspension.

The rigidity of the layer is an important considering parameter for the preparation of liposomes, and it is provided by phospholipids. A variety group of phospholipids used for the manufacturing of liposome given below:

1. **Phospholipids from Natural** *Source:* It obtained from the vege-
   table and animal sources like soybean, wheat germ, sunflower,
   flaxseed, egg yolk, and milk. Egg lecithin, soybean lecithin,
   phosphatidyl-choline (PC), phosphatidyl-ethanolamine (PE),
   phosphatidyl-inositol (PI) are of natural source phospholipids.
2. **Phospholipids Modified form:** Alkyl phospholipids (APL).
3. **Semisynthetic Phospholipids:** Some parts of molecules derived
   from natural phospholipids.
4. **Enzyme Modified Natural Phospholipids:** Monoacyl-phosphati-
   dylcholine.
5. **Fully Synthetic Phospholipids:** Dimyristoyl phosphatidic acid
   (DMPA), dipalmitoyl phosphatidic acid (DPPA), Distearoyl
   phosphatidic acid (DSPA), Dilauryl phosphatidylcholine (DLPC),
   Dimyristoyl phosphatidylcholine (DMPC), dipalmitoyl phosphatidylcholine (DPPC), Distearoyl phosphatidylcholine (DSPC),
   Dioleolyl phosphatidylcholine (DOPC), Palmitoyl-2-oleoyl
   phosphatidylcholine (POPC), Didecanoyl phosphatidylcholine
   (DDPC), Dimyristoyl phosphatidylglycerol (DMPG), dipalmitoyl
   phosphatidylglycerol (DPPG), Distearoyl phosphatidylglycerol
   (DSPG), Palmitoyl-2-oleoyl phosphatidylglycerol.

Theoretically, stiffer phospholipids like hydrogenated and saturated
phospholipids make stiff liposomal membranes and enhance GIT absorption. In addition, unsaturated phospholipids (soy) build more elastic
membranes and good for intraoral absorption [19].

Celsion Corporation developed the doxorubicin parental liposome
using the DPPC, MSPC, and DSPE-PEG2000, and it under phase-3 of a
clinical trial, it exhibited significant activity on hepatocellular carcinoma
cells.

Sphingomyelin is a membrane component of animal and have cis-
double bonds in amide linked acyl chains. Sphingomyelin is used for the
development of liposomes in place of phospholipids and found a promising effect. Sphingomyelin have the power to entrapped the high amount
of drug by formation of intermolecular and intermolecular hydrogen bond
and fast release of drug as well as high stability as compared to DSPC
liposome [20].

2.2.2 CHOLESTEROL (CHOL)

Chol is a fatty substance, and it has many advantageous and important participants for manufacturing of liposomes. It stabilized and acted as fluidity for liposome. It interpolates between the phospholipids and altered the freedom of carbon molecules to acyl chain length. It improved the package of phospholipids resulting diminished the phase transition temperature (Tc) and improve the stability of liposomal formulation *in vitro* and *in-vivo*. It restricts the conversion of Trans to gauche conformation and decrease the permeability at high temperature. It influences on release of hydrophilic drug, when increases the concentration of Chol, the release of the drug will be fast [21]. On the other side the Chol concentration decrease, leakage of drug from liposome takes place [22]. Mallick and its associated developed the liposome using DOPE and Chol and exhibited high stability and improve the delivery of antimycin-A to mitochondrial cell [23]. Chol concentration also effects on entrapment efficiency of drug, increases the concentration, decrease the entrapment efficiency, it may be due to steric hindrance between Chol and lipophilic drug in space between bilayer of phospholipids [21]. Spectrum pharmaceutical was developed vincristine liposome using Chol and under phase-III clinical trial and exhibited significant lower the toxicity in geriatric patient. Hammoud and its associated developed the essential oil loaded liposome using Chol and found high that increases the solubility, entrapment, and control releases profile over an extended period of time [24]. Chol also increases the positive zeta potential of the liposome because it increases the fluidity of liposome and prevents aggregation of vesicles. Aramakia and its associates developed the cationic liposome using cationic surfactant, lipid, and Chol and found that high shift of the negative-positive potential as compared to cationic surfactant [25].

2.3 TECHNIQUES FOR PREPARATION OF LIPOSOME [20, 26]

Techniques that are used for the development of different type of preparation of liposomes given in Table 2.1. There are various methods present for development of liposome and every method following step:

- Evaporation of organic solvent for drying of lipid;

- Hydrion of dispersion of lipid in to water;
- Purification of liposome;
- Quantification and analysis of liposome.

TABLE 2.1   Liposomal Preparation Method and Type of Liposome Form

SL. No.	Name of Method for Preparation of Liposomes	Type of Liposome Form
Mechanical Method		
1.	Thin-film hydration	MLVs
2.	Membrane extrusion	OLVs, LUVs
3.	French cell extruder	SUVs
4.	high pressure homogenizer	SUVs
5.	Sonication	SUVs
Dispersion Method		
6.	Ether injection	MLVs, OLVs, LUVs
7.	Alcohol injection	LUVs
8.	Reverse phase evaporation	LUVs, OLVs, MLVs
Detergent Removal Method		
9.	Gel exclusion chromatography	SUVs
10.	Dialysis	LUVs, OLVs, MLVs

2.3.1   LIPOSOME PREPARATION METHOD AND LOADING OF DRUG

- Loading of drug through passive method;
- Loading of drug through active method.

2.3.1.1   LOADING OF DRUG THROUGH PASSIVE METHOD

It includes three methods for preparation of liposomes:

1. **Mechanical Dispersion Method:** There are the following sub-methods under mechanical dispersion:
   i. **Sonication:** This method was used for breaking of liposome and formation of large multilamellar vesicle (LMV) to small

unilamellar vehicle (SUV, 15–50 nm). Sonication was done by probe and probe sonicator. The liposomal dispersion was placed into a suitable container (beaker or test tube and soni-cated for 5–10 min.

ii. **Extruder or French Pressure Cell Method:** This method is used for the development of SUL. The aqueous dispersion of excipients (lipid, Chol) was passed through a small orifice of French pressure cell at 20000 psi pressure. Approx. 70% of SUL formed in single pass. This method is rapid, simple, and reproducible method as compared to sonication method and produced 30–80 nm depends upon pressure applied. In this method, problem to stabilize to the required temperature and handle only a small volume of sample (50 ml) [27].

iii. **Freeze-Thaw Extrusion Method:** In this method, liposome developed by film hydration method vortexed with encap-sulated material until all film suspended and frozen in warm water and vortexed. Then formulation subjected for freeze-thaw (2 cycles, three times) and vortexed again, followed by six cycles of freeze-thaw and extrusion. This technique is used to increase the entrapment efficacy of material [28].

iv. **Thin Film Hydration Method (Hand and Non-Hand Shaking Method):** It is the most commonly used technique for manufacturing of liposome and multi-lamellar vesicle formed. It is a very cheap and adoptive method but have drawn back like large size particle form due to aggregation of particle. The excipients like Chol and lipid were dissolving in a suitable evaporated solvent (chloroform, methanol) and transfer into the round bottom flask. Evaporate organic solvent by handshaking or by rotatory evaporator (40–45°C) for making the thin film. After evaporation, the sample was kept in a desiccator to remove moisture. The thin film hydrated with water/buffer, swelling, and hydration take place and liposome form. The therapeutic agent added in aqueous media or organic phase [29, 30].

v.   **Emulsification (Micro):** In this method the micro-fluidizer machine will be used for preparation of liposome. By this method can prepare small vesicle (50–200 nm) with 75% entrapment efficiency.

    vi. **Membrane Extrusion:** In this method, vesicle content are exchange with dispersion medium and passed through polycarbonate membrane. Less pressure (>100 psi) is required as compared to the French pressure cell method. It is used for the processing of MLVs, LUVs. Tortuous (zigzag) and nucleation trach membrane used for manufacturing of liposome.

2.   Solvent Dispersion Method:

    i.   **Ether Injection Method:** In this technique, the lipids dissolve in ether or mixture of ether and ethanol. The lipid solution is injected slowly into an aqueous solution of therapeutic agent or other agent require to encapsulation under reduced pressure or 55–65°C (above the BP of ether). The ether was evaporated or removed under vacuum on contact with water formation of unilamellar liposome take place. The liposome is free from ether. The main disadvantage of this method is the unequal size of liposome form and exposure to high temperature [31].

    ii.   **Ethanol Injection Method:** In this method, the liposomal excipients (phospholipids and Chol) dissolved in ethanol. The lipid solution is injected using a pressure syringe in an aqueous solution of agent under continuous stirring. The liposome automatically formed immediately on contact with water. The sample was kept on stirring for 15 min after complete injecting of lipid solution. This method is very fast, simple, and reproducible. SUL formed without any application of sonication or extruder. In this method less chances of alteration or oxidation [32, 33].

    iii. **Reverse Phase Evaporation Method:** This technique is emulsification type in which the large unilamellar vesicle liposome formed. The lipid phase system dissolved in the organic solvent (diethyl ether, isopropyl ether or mixture of isopropyl ether and chloroform) and added into aqueous phase with sonication, oil in water emulsion form. The organic solvent removes by heating or by under reduced pressure using a rotatory evaporator. This method is used for encapsulation of small and large molecules and can be achieved high encapsulation efficiency (EE) (70%). This method has a drawback like organic solvent exposure and sonication [34]. Liposome prepared by different techniques explains in Table 2.2.

3. Removal of Un-Encapsulated Material (Detergent Removal Method):
   i. Dialysis;
   ii. Detergent removal of mixed micelles;
   iii. Gel permeation chromatography.

TABLE 2.2 Liposomal Preparation Using Various Techniques

Technique Name	Therapeutic Agent	Outcome	References
Thin-film hydration	Besifloxacin	436.8 ± 23.4 nm, and high corneal permeation as compared to eye drop	[35]
Thin-film hydration	Bevasiranib	Significant deposition and cellular uptake in SKBR-3 breast tumor cell line	[36]
Thin-film hydration	HIF-1α siRNA	Significantly reduce the production of HIF-1α-associated protein hypoxia-tolerant melanoma cells	[37]
Thin-film hydration	Dexamethasone	Liposomes are non-toxic, and significant uptake by human Adult Retinal Pigment Epithelial cell line-19 (ARPE-19) cells, and significantly reduce inflammation	[38]
Thin-film hydration	Hydroxycampto-thecin	155.6 ± 2.6 nm and a zeta and significant on the HepG2 cells. Significant increases area under curve and mean residence time as well as a decrease in plasma clearance ($p < 0.05$)	[39]
Thin-film hydration	Lamivudine	Particle size 276.20 ± 13.36 nm, entrapment efficiency 60.20 ± 2.86% with PdI 0.291 ± 0.053. Significantly long-time circulation with accumulation in tissue	[40]
Ethanol Injection	si-RNA	Liposome exhibited insignificant cytotoxicity and improved cellular organization and gene silencing capacity	[41]
Ether injection	Itraconazole	Formulation exhibited significant release in control manner	[42]
Reverse phase evaporation	Ganciclovir	Vesicle size >100 nm, liposome exhibited significant inhibitory effect on cytomegalovirus infected cell	[43]

TABLE 2.2   *(Continued)*

Technique Name	Therapeutic Agent	Outcome	References
Reverse phase evaporation	Cisplatin	Liposomal formulation showed 4.8-fold higher efficacy (11 ± 0.5 and 2.3 ± 0.1 mm³) with significant decreases (3.3-fold) toxicity than plan drug	[44]
Reverse phase evaporation	Guarana powder	Exhibited significant antioxidant activity	[45]
Ethanol injection	Cyclosporine	Particle size 111 ± 1.62 nm, entrapment efficiency of 93 ± 2.12%, and significantly reduce psoriasis in imiquimod induced psoriatic plaque model	[30]

2.4   CHARACTERIZATION OF LIPOSOME

2.4.1   PARTICLE SIZE

The particle size of liposome was measured by the Malvern zeta sizer (DLS scattering method). Diluted sample of liposome placed into cuvette and placed into instrument and measured at 90° scattering. It can also measured by using a transmission electron microscope, atomic force microscope. The particle size is an important parameter for the delivery of drugs through various routes (oral, parenteral, topical, Inhaler). It is monitored by different method, i.e., sonication, homogenization, and extrusion.

2.4.1.1   SONICATION

The LMV can convert to SUL (20–50 nm) by using sonication. Sonication was done by probe and probe sonicator. The liposomal dispersion was placed into a suitable container (beaker or test tube and sonicated for 5–10 min, but liposome prepared by using of (dioctadecyl amido glycylspermine (DOGS), and neutral lipid (dioctadecyl diammonium bromide (DOBAD) cannot be decreased <130 nm [47].

2.4.1.2 HOMOGENIZATION

High-pressure homogenizer applied for breaking of large liposome. This method is used for large-scale production (10 ml to 10 L). The formulation subjected at high pressure with different cycles (2–3 cycles) to obtained desired size [48].

2.4.1.3 EXTRUSION

Extrusion is the technique which is used for decreasing the size of nanoformulation. It is done by using of polycarbonate membrane filter at different size (0.2 μm). The MLVs was passed through the membrane filter by applying pressure. The size of the liposome is controlled by applying pressure or by the size of the membrane filter. When applying pressure increases and pore size of membrane filter decrease the liposomal size and size distribution decreases. It is proven by a study on the passing of MLVs suspension through 0.4 to 0.2 μm resulted in the development of 110–120 nm [49].

2.4.2 ZETA POTENTIAL

Zeta potential is a charge (potential difference) between electrical double layers. This charge developed on the surface of the particle. Zeta potential is directly affected on formulation stability like aggregation. The Zeta potential of $\geq \pm 30$ mV is considered as electrical stable nanoformulation, which means nanoparticle or liposome dispersed uniformly in non-aggregated form. It was measured by placing the sample in an electrical cuvette and applying the scattering laser light on the particle. It is affected by ionic strength, pH, the concentration of the sample [50].

2.4.3 ENCAPSULATION EFFICIENCY (EE)

The entrapment efficacy of the drug in liposomes is mainly measured by two methods, i.e., indirect method and direct method. In the indirect method, the liposomal suspension centrifuges or passed through the dialysis membrane and separated the free drug. The concentration of free drug is determined by the suitable analytical technique like UV, HPLC,

GC, LCMS, etc., in the direct method for determination of entrapment efficiency, liposome separation from suspension, and dissolution with organic solvent (methanol, chloroform). The drug comes out from the liposome into an organic solvent, the organic solvent evaporates, and the concentration of the drug was measured by suitable analytical technique after appropriate dilution. The EE determine by the following equation:

2.4.4   DRUG RELEASE STUDY

Release of drug from liposome can do by using the dialysis membrane bag. Suitable specification of dialysis membrane (molecular cut off) selected and placed the liposomal formulation and sealed both sides. The bag was immersed in releases media (phosphate buffer) with the mimic condition of *in-vivo*. The whole system was placed at 37°C with continuous stirring. At a predetermined time interval, a specific volume of aliquots withdrawn and simultaneously replaced with the same volume with fresh release media. The concentration of the drug was measured by a suitable analytical technique like UV-Vis spectrophotometer, HPLC, HPTLC, LCMS, etc. The release kinetics of drug determines by putting the release data into different models, i.e., zero-order, first-order, Higuchi, Korsmeyer peppas, Hixon-Crowell model. On the basis of regression co-efficient value, select the best kinetic release model.

2.4.5   CHEMICAL CHARACTERIZATION

2.4.5.1   ESTIMATION OF PHOSPHOLIPIDS

Phospholipids content in liposomes can be estimated by two methods:
1. **Bartlett Assay:** It is based on the determination of inorganic phosphate by colorimetric method at 830 nm. It is estimated based on the demolition of phospholipids using perchloric acid (70%) in inorganic phosphate. Then add ammonium molybdate (5%) and covert to phosphor-molybdic acid, and reduced into the blue color complex by 4-amino-2 naphthyl-4-sulfonic acid (reducing agent) during heating. This method can be used when liposomes are prepared in phosphate buffer because it interferes with the

estimation of phospholipids. The heating block does not go above >200°C; otherwise, perchloric acid evaporated and not <180°C.

2. **Steward Assay:** The drawback Bartlett assay can be overcome by steward method and phosphate buffer not interfere in estimation and used only for known lipids. In this method, phospholipids form the complex with ammonium ferrothiocynate. First of all, prepare a standard curve of lipid (0.1 mg/ml) in chloroform with ammonium ferrothiocynate reagent (0.1 M) with vortexing (20 min) and centrifugation at 1000 rpm (10 min). Then the sample is also prepared by the same procedure. The absorbance is measured by a spectrophotometer at 485 nm. The concentration of lipid was measured from the standard plot [51].

2.4.5.2 CHOLESTEROL (CHOL) ESTIMATION

Reverse-phase high-performance liquid chromatography method can be used for estimation of qualitative as well as quantitated analysis of Chol [52]. It can also detect by calorimeter by treating the sample with a reagent having a mixture of ferric perchlorate, ethyl acetate, and sulfuric acid, and color (purple colure complex) was measured at 610 nm [53].

2.4.6 STABILITY STUDY

The stability of liposomes is a critical issue and necessary requires stabilizing the formulation during the preparation and after preparation (storage). The stability (physical and chemical) of formulation directly linked with the biological activity as well as physiochemical characterize of formulation (particle size, polydispersibility index, entrapment efficiency, drug release, etc.). It also affects on the shelf life of the product. The physical stability of liposomes can check visually like color change, rancidity, precipitation, microscopic evaluation (transmission electron microscopy, atomic force microscopy (AFM)), surface charge and particle size by zeta sizer. For storage of formulation must be optimized the storage condition (temperature and humidity). Physical instability the leakage and or fusion of liposome occur during preparation due to imperfection in lattice structure of lipid. Chemical stability of liposome is critical issue if phospholipids is unsaturated in nature because C=C group is prone to oxidation and hydrolysis. Chemical stability can be checked by study of hydrolysis, oxidation on phospholipid by using liquid chromatography, thiobarbituric acid (TBA) test (lipid oxidation) and spectroscopic method. Oxidation of phospholipid can be protecting by exposure from metal ions, peroxide during manufacturing and use of high-quality lipid and other excipients. Using of inner gas during manufacturing, storage at low temperature, and away from light and oxygen can be reduced or prevent from oxidation. The use of antioxidants (αtocopherol, butylhydroxy-toluene) can be protecting from oxidation. Freeze drying of liposomes can be protected from hydrolysis and stable for a long time because it forms glassy matrix which prevents the phase transition and crystalliza-tion, may damage the membrane [54]. The stability of liposomes *in-vivo* is affected by blood products like lipoprotein, albumin [55]. Mady reported that adding of Sphingomyelin in artificial viral envelopes, decrease lysis in serum and increase the stability of liposome [56]. Vermehren and its associates reported that addition of N-acyl-phosphatidyl ethanolamine into liposome increase the stability of liposome against blood component and increase the circulation time of vesicle [57].

2.5 APPLICATION OF LIPOSOME FOR NUTRACEUTICAL DELIVERY

Nutraceutical is a substance which provides the nutrition as well as medi-cine. It has physiological benefits and provides the defense from chronic disease like hyperlipidemia, cancer, inflammation, diabetes, hypertension, etc. It also uses to recover the health, increase the life anticipation, and support the anatomical and physiology of the body [58, 59]. Nutraceutical like isoflavones, carotenoids, flavones, anthocyanidins, flavones for treat-ment of cancer, hypertension, infectious disease, diabetes, etc. Delivery of nutraceutical in the body is the major challenge to researchers as well as for industry of stability in biological system because of low bioavail-ability, stability, solubility, and first passes metabolism effect [60, 61]. Encapsulation is a good choice for protecting the nutraceutical and bioac-tive compounds from deterioration and oxidation. It preserves and sustains the release of nutraceuticals, thereby sustain the color, flavor stability, and shelf life of nutraceutical products and help in body functional benefits. There is less information reported about the application of liposomes in nutraceutical. Mostly thin film hydration technique employed for development of MLVs and future process to formed SUVs of nutraceutical product. But now recently the high-pressure homogenization is used for direct development of SUVs of nutraceutical. Mainly nutraceutical liposome is prepared by using soy or egg lipid. Liposome prevents the encapsulated material to outside environment like light, chemical, oxygen, and humidity. Nutraceutical-loaded liposomal formulation can prevent the wastage of material during processing and enhance the financial side as well as reducing or finish the toxicity problem. Encapsulation with liposomes can be limited the unpleasant smell of nutraceuticals and enhance the pleasant attribute by using of suitable flour. Encapsulated ascorbic acid-loaded liposome would diminish the prevent ascorbic acid by other nutraceutical substances and assure maximum generation of α-tocopherol [62]. Rovoli and its associates developed the vitamin-loaded liposome (SUVs) for oral delivery and protected it from oxidation and other degradation environments. The liposomal system exhibited significant enhanced absorption due to resemble membrane of liposome to biological membrane and protected from degradation as well as external environment like UV-light, air, chemical, etc., [63]. Lukawski and its associates developed vitamin-C loaded liposome for oral delivery for preservation body homeostasis, and it exhibited significant enhancement the oral bioavailability [64]. Davis and its associates developed vitamin-C loaded liposome for enhancement of oral bioavailability because it is less bioavailable orally than parenteral. The study was conducted on human volunteers. It was found that vitamin-C loaded liposome exhibited a significant concentration than plan vitamin C in the same dose (4 g) but less than intravenous and prevented the oxidative stress-mediated by ischemia-reperfusion [65]. A carotenoid is a nutraceutical and is susceptible to oxidation and light. Liposomal formulation can provide prevention against oxidation and other agent which effect on the carotenoids activity. Tan and its associates developed MLVs of Carotenoid (Lutein and β-carotene) by thin-film evaporation method. It exhibited sustained release in the stomach as well as intestinal fluid [66]. Quacertine loaded liposome exhibited significant anticancer and wound healing activity [67–69]. Liposomal delivery of nutraceuticals also used for topical application as wound healing and other treatment. Clares and its associates developed retinyl palmitate for skin protection from sunlight. Skin permeation study was conducted on human skin and exhibited significantly high permeation and flux as compared to other lipid-based formulation (SLN,

NLCs (nanostructured lipids carriers), NEs) without any destruction of corneum stratum confirmed by histopathological examination [69]. Many of the researchers have been reported nutraceutical loaded liposome which is explained in Table 2.3.

TABLE 2.3 Nutraceutical Loaded Liposome

Liposomal Technique	Nutraceutical	Outcome	References
Ether injection	Lutein	Prevent the oxidation of formulation	[64]
Extrusion	Curcumin	Formulation was well distributed in bone marrow, liver, spleen, and demonstrating macrophage targeting	[70]
Thin-film hydration	Copper(II)-quercetin	Quercetin exhibits a significant IC_{50} of >10 μM when tested against cancer cell lines and lung circulation	[66]
Thin-film hydration	Quercetin	High entrapment efficiency significant release of quercetin in wound healing	[67]
Thin-film hydration	Quercetin	The quercetin/PEG2000-DPSE formulation showed significantly more effective than pure quercetin inhibiting the growth of glioma cancer cells.	[71]
Lyophilization	1-Methyl xanthine	Significant high permeation than other lipid-based formulation without effect on stratum corneum	[72]
Ethanol injection, film dispersion-homogenization method	Vitamin-D3	Liposomes significantly improve the stability of vitamin D3. Liposomes exhibited 1.65 times higher retention than and repaired the skin	[73]
Reverse phase evaporation	Simple pegylated resveratrol liposome	Resveratrol loaded liposome significantly improved inhibited cancer cells, possibly owing to the increases the concentration of resveratrol in cancerous cells (human hepatocellular carcinoma (Hep G2 and HeLa cell)	[74]

TABLE 2.3   *(Continued)*

Liposomal Technique	Nutraceutical	Outcome	References
Film hydration with membrane extrusion	Resveratrol liposome	Resveratrol liposome significantly reached to mitochondrial cell	[75]
Thin-film hydration	Fisetin	47-fold increase in relative bioavailability of liposomal fisetin in intra-peritoneal administration compared to its free form	[76]
Thin-film hydration	Genistein	Significantly enhanced drug solubilization shelf-life stability, and sustained release profile, significantly showed high anticancer efficacy in murine and human cancer cell	[77]
Ultrasonication and supercritical carbon dioxide	β-Carotene	Exhibited high entrapment efficiency and good stability	[78]
Thin-film hydration	Berberine	Significantly internalized of liposome and effectively modulated RANKL/p-GSK3β pathway and decreased the osteoclast-mediated bone erosion by post-transcriptional gene silencing via miR-23a	[79]
Thin-film hydration	Berberine	Exhibited high entrapment efficiency, high stability and significantly increased in apoptotic cell of 4T1	[46]
Thin-film hydration	Curcumin	Significantly improve the bioavailability	[18]

2.6   CONCLUSION AND FUTURE PROSPECT

Liposomes are a novel tool as carrier of nutraceuticals and drugs for improving stability as well as a therapeutic index. Researchers are doing work about the last four decades due to their exclusive characteristics like non-immunogenicity, biocompatible to the cellular membrane, biodegradable, and non-toxic) and diminished the systemic toxicity of sensitive drugs, protect the drug from the external environment as well as the biological environment. It is a unique technique for target delivery as well as improving pharmacokinetic properties. It has a wide range of applications like targeted delivery through surface modification of liposomes (ligand-mediated). Due to their versatility in nature, it is unfeasible to cross some biological membranes because of their vesicle size. On the other hand, by application of advanced liposomal technology, will participate an extra significant efficient job in the clinical environment. Enhancing the future researches on the existing platforms and will find the mechanism of transport of nutraceuticals across the biological membrane and interaction with various types of the biological cell as well will investigate the appropriate technique to find out small size of the particle to easily penetrate the biological membrane, high entrapment efficacy, and high drug loading as well as other formulation challenges.

CONFLICT OF INTEREST

The authors declare no conflict of interest.

KEYWORDS

- alkyl phospholipids
- dilauryl phosphatidylcholine
- encapsulation efficiency
- liposome
- polyethylene glycol
- small unilamellar vehicle

REFERENCES

1. Yang, Z., Lu, A., Wong, B. C. K., Chen, X., Bian, Z., Zhao, Z., Huang, W., et al., (2013). Effect of liposomes on the absorption of water-soluble active pharmaceutical ingredients via oral administration. *Curr. Pharm. Des., 19*(37), 6647–6654.
2. Palchetti, S., Colapicchioni, V., Digiacomo, L., Caracciolo, G., Pozzi, D., Capriotti, A. L., et al., (2016). The protein corona of circulating PEGylated liposomes. *Biochim. BiophysActa., 1858,* 189–196.

3. Barenholz, Y., (2012). Doxil®. The first FDA-approved nano-drug: lessons learned. *J. Control Release, 160*, 117–134.

4. Bangham, A. D., (1980). *Liposomes in Biological Systems*. John Wiley and Sons: Chichester.

5. Agrawal, A. K., Harde, H., Thanki, K., & Jain, S., (2014). Improved stability and antidiabetic potential of insulin-containing folic acid functionalized polymer-stabilized multilayered liposomes following oral administration. *Biomacromolecules., 15*(1), 350–360.

6. Shukla, S. K., Kulkarni, N. S., Chan, A., Parvathaneni, V., Farrales, P., Muth, A., & Gupta, V., (2019). Metformin-encapsulated liposome delivery system: An effective treatment approach against breast cancer. *Pharmaceutics, 11*(11), 559.

7. Verma, A., Jain, A., Tiwari, A., Saraf, S., Panda, P. K., Agrawal, G. P., & Jain, S. K., (2019). Folate conjugated double liposomes bearing prednisolone and methotrexate for targeting rheumatoid arthritis. *Pharm. Res., 36.* https://doi.org/10.1007/s11095-019-2653-0.

8. Zhang, S. L., Ma, L., Zhao, J., You, S. P., Ma, X. T., Ye, X. Y., & Liu, T., (2019). The phenylethanol glycoside liposome inhibits PDGF-induced HSC activation via regulation of the FAK/PI3K/Akt signaling pathway. *Molecules, 24*(18), 3282.

9. Giulimondi, F., Digiacomo, L., Pozzi, D., Palchetti, S., Vulpis, E., et al., (2019). Interplay of protein corona and immune cells controls blood residency of liposomes. *Nat. Commun., 10*, 3686.

10. Tran, T. T., Yu, H., Vidaillac, C., Lim, A. Y. H., Abisheganaden, J. A., Chotirmall, S. H., & Hadinoto, K., (2019). *An Evaluation of Inhaled Antibiotic Liposome versus Antibiotic Nanoplex in Controlling Infection in Bronchiectasis, 559*, 382–392.

11. Ambati, S., Ellis, E. C., Lin, J., Lin, X., Zachary, A., Richard, L. B., & Meagher, (2019). Dectin-2-targeted antifungal liposomes exhibit enhanced efficacy. *Therapeutics and Prevention.* doi: 10.1128/mSphere.00715-19.

12. Bonechi, C., Donati, A., Tamasi, G., Pardini, A., Rostom, H., Leone, G., Lamponi, S., et al., (2019). Chemical characterization of liposomes containing nutraceutical compounds: Tyrosol, hydroxytyrosol and oleuropein. *Biophys. Chem., 246*, 25–34.

13. Akbarzadeh, A., Rezaei-Sadabady, R., Davaran, S., et al., (2013). Liposome: Classification, preparation, and applications [journal article]. *Nanoscale Research Letters, 8*(1), 102.

14. Nehoff, H., Parayath, N. N., Domanovitch, L., Taurin, S., & Greish, K., (2014). Nanomedicine for drug targeting: Strategies beyond the enhanced permeability and retention effect. *Int. J. Nanomedicine, 9*, 2539–2555.

15. Sawant, R. R., & Torchilin, V. P., (2012). Challenges in development of targeted liposomal therapeutics. *AAPS J., 14*(2), 303–315.

16. Dams, E. T., Laverman, P., Oyen, W. J., Storm, G., Scherphof, G. L., Van, D. M. J. W., Corstens, F. H., & Boerman, O. C., (2000). Accelerated blood clearance and altered biodistribution of repeated injections of sterically stabilized liposomes. *J. Pharmacol. Exp. Ther., 292*(3), 1071–1079.

17. Poudel, A., Gachumi, G., Wasan, K. M., Bashi, Z. D., El-Aneed, A., & Badea, I., (2019). Development and characterization of liposomal formulations containing phytosterols extracted from canola oil deodorizer distillate along with tocopherols as food additives. *Pharmaceutics, 11*(4), 185.

18. Tai, K., Rappolt, M., He, X., Wei, Y., Zhu, S., Zhang, J., Mao, L., et al., (2019). Effect of β-sitosterol on the curcumin-loaded liposomes: Vesicle characteristics, physicochemical stability, *in vitro* release and bioavailability. *Food. Chem., 293,* 92–102.

19. Silva, A. C., Santos, D., Ferreira, D., & Lopes, C. M., (2012). Lipid-based nanocarriers as an alternative for oral delivery of poorly water-soluble drugs: Peroral and mucosal routes. *Curr. Med. Chem., 19*(26), 4495–4510.

20. Carter, K. A., Luo, D., Razi, A., et al., (2016). Sphingomyelin liposomes containing porphyrin-phospholipid for irinotecan chemo phototherapy. *Theranostics,* 6(13), 2329–2336.

21. Briuglia, M. L., Rotella, C., McFarlane, A., et al., (2015). Influence of cholesterol on liposome stability and on *in vitro* drug release. *Drug. Deliv. Transl. Res.,* 5(3), 231–242.

22. Sadeghi, N., Deckers, R., Ozbakir, B., et al., (2017). Influence of cholesterol inclusion on the doxorubicin release characteristics of lysolipid-based thermosensitive liposomes. *Int. J. Pharma.* doi: 10.1016/j.ijpharm.2017.11.002.

23. Mallick, S., Lee, S., Park, J. I., et al., (2018). Liposomes containing cholesterol and mitochondria-penetrating peptide (MPP) for targeted delivery of antimycin A to A549 cells. *Colloids and Surfaces B: Biointerfaces, 161,* 356–364.

24. Hammoud, Z., Gharib, R., Fourmentin, S., Elaissari, A., & Greige-Gerges, H., (2019). New findings on the incorporation of essential oil components into liposomes composed of lipoid S100 and cholesterol. *Int. J. Pharma., 561,* 161–170.

25. Aramakia, K., Watanabe, Y., Takahashi, J., Tsuji, Y., Ogata, A., & Konno, Y., (2016). Charge boosting effect of cholesterol on cationic liposomes. *Colloids Surf. A Physicochem. Eng. Asp., 506,* 732–738.

26. Riaz, M., (1996). Liposome preparation method. *Pak. J. Pharm. Sci.,* 9(1), 65–77.

27. Hamilton, R. L., Goerke, J. J., Guo, L. S., Williams, M. C., & Havel, R. J., (1980). Unilamellar liposomes made with the French pressure cell: A simple preparative and semiquantitative technique. *J. Lipid. Res., 21*(8), 981–992.

28. Costa, A. P., Xu, X., & Burgess, D. J., (2014). Freeze-anneal-thaw cycling of unilamellar liposomes: Effect on encapsulation efficiency. *Pharm. Res., 31*(1), 97–103.

29. Ameeduzzafar, Alruwaili, N. K., Imam, S. S., et al., (2020). Formulation of chitosan polymeric vesicles of ciprofloxacin for ocular delivery: Box-Behnken optimization, *in vitro* characterization, HET-CAM irritation, and antimicrobial assessment. *AAPS PharmSciTech, 21,* 167. https://doi.org/10.1208/s12249-020-01699-9.

30. Walunj, M., Doppalapudi, S., Bulbake, U., & Khan, W., (2020). Preparation, characterization, and *in vivo* evaluation of cyclosporine cationic liposomes for the treatment of psoriasis. *J. Liposome. Res., 30*(1), 68–79.

31. Mathai, J. C., & Sitaraman, V., (1987). Preparation of large uni-lamellar liposomes by the ether injection method and evaluation of the physical integrity by osmometry. *Biochemical Education, 15*(3), 147–149.

32. Charcosset, C., Juban, A., Valour, J. P., et al., (2015). Preparation of liposomes at large scale using the ethanol injection method: Effect of scale-up and injection devices. *Chemical Engineering Research and Design, 94,* 508–515.

33. Jaafar-Maalej, C., Diab, R., Andrieu, V., et al., (2010). Ethanol injection method for hydrophilic and lipophilic drug-loaded liposome preparation. *Journal of Liposome Research, 20*(3), 228–243.
34. Gharib, R., Greige, H. G., Fourmentin, S., Charcosset, C., & Auezova, L., (2015). Liposomes incorporating cyclodextrin-drug inclusion complexes: Current state of knowledge. *Carbohydrate Polymers, 129,* 175–186.
35. Bhattacharjee, A., Das, P. J., Dey, S., Nayak, A. K., Roy, P. K., et al., (2020). Development and optimization of besifloxacin hydrochloride loaded liposomal gel prepared by thin-film hydration method using 32 full factorial design. *Colloids and Surfaces A: Physicochemical and Engineering Aspects. Int. J. Pharm., 585,* 124071.
36. Golkar, N., Samani, S. M., & Tamaddon, A. M., (2016). Modulated cellular delivery of anti-VEGF siRNA (bevasiranib) by incorporating supramolecular assemblies of hydrophobically modified polyamidoamine dendrimer in stealth liposomes. *Int. J. Pharma., 510*(1), 30–41.
37. Chen, Z., Zhang, T., Wu, B., et al., (2016). Insights into the therapeutic potential of hypoxia-inducible factor-1α small interfering RNA in malignant melanoma delivered via folate-decorated cationic liposomes. *Int. J. Nanomedicine, 11,* 991.
38. Al-Amin, M. D., Bellato, F., Mastrotto, F., Garofalo, M., Malfanti, A., Salmaso, S., & Caliceti, P., (2020). Dexamethasone loaded liposomes by thin-film hydration and microfluidic procedures: Formulation challenges. *Int. J. Mol. Sci., 21*(5), 1611.
39. Zhou, T., Zhang, W., Cheng, D., Tang, X., Feng, J., & Wu, W., (2020). Preparation, characterization, and *in vivo* evaluation of NK4-conjugated hydroxycamptothecin-loaded liposomes. *Int. J. Nanomedicine., 15,* 2277–2286.
40. Godbole, M. D., Sabale, P. M., & Mathur, V. B., (2020). Development of lamivudine liposomes by three-level factorial design approach for optimum entrapment and enhancing tissue targeting. *J. Microencapsul.* doi: 10.1080/02652048.2020.1778806.
41. Fisher, R. K., Mattern-Schain, S. I., Best, M. D., et al., (2017). Improving the efficacy of liposome-mediated vascular gene therapy via lipid surface modifications. *J. Surg. Res., 219,* 136–144.
42. Virendra, T., Md, U., Md, R., Sumeet, D., & Raghvendra, D., (2019). Preparation and evaluation of itraconazole liposome using ether injection solvent evaporation method *IJPLS, 10*(2), 6091–6097.
43. Asasutjarit, R., Managit, C., Phanaksri, T., Treesuppharat, W., & Fuongfuchate, A., (2020). Formulation development and *in vitro* evaluation of transferrin-conjugated liposomes as a carrier of ganciclovir targeting the retina. *Int. J. Pharma., 577,* 119084.
44. Ghaferi, M., Asadollahzadeh, M. J., Akbarzadeh, A., Shahmabadi, H. E., & Alavi, S. E., (2020). Enhanced efficacy of PEGylated liposomal cisplatin: *In vitro* and *in vivo* evaluation. *Int. J. Mol. Sci., 21*(2), 559.
45. Roggia, I., Dalcin, A. J. F., Ourique, A. F., Da Cruz, I. B. M., Ribeiro, E. E., Mitjans, M., Vinardell, M. P., & Gomes, P., (2019). Protective effect of guarana-loaded liposomes on hemolytic activity. *Colloid. Surface. B.* doi: https://doi.org/10.1016/j.colsurfb.2019.110636.
46. Zhang, R., Zhang, Y., Zhang, Y., Wang, X., Gao, X., Liu, Y., Zhang, X., et al., (2020). Ratiometric delivery of doxorubicin and berberine by liposome enables superior therapeutic index than doxil. *Asian J. Pharm. Sci.* https://doi.org/10.1016/j.ajps.2019.04.007.

47. Maulucci, G., De Spirito, M., Arcovito, G., et al., (2005). Particle size distribution in DMPC vesicles solutions undergoing different sonication times. *Biophysical Journal, 88*(5), 3545–3550.
48. Wang, A., Cui, J., Wang, Y., Zhu, H., Li, N., Wang, C., Shen, Y., et al., (2020). Preparation and characterization of a novel controlled-release nano-delivery system loaded with pyraclostrobin via high-pressure homogenization. *Pest. Manag. Sci.* doi: 10.1002/ps.5833.
49. Ong, S. G. M., Chitneni, M., Lee, K. S., et al., (2016). Evaluation of extrusion technique for nanosizing liposomes. *Pharmaceutics, 18*(4), 36.
50. Bhattacharjee, S., (2016). DLS and zeta potential - What they are and what they are not? *Journal of Controlled Release, 235*, 337–351.
51. Stewart, J. C. M., (1980). Colorimetric determination of phospholipids with ammonium ferrothiocynate. *Anal Biochem., 104*, 10.
52. Carla, B., Kastner, R. E., Stone, P., Lowry, D., & Perrie, Y., (2016). Rapid quantification and validation of lipid concentrations within Liposomes. *Pharmaceutics, 8*(3), 29.
53. Lin, C., Guo, Y., Zhao, M., Sun, M., Luo, F., Guo, L., Qiu, B., et al., (2017). Highly sensitive colorimetric immunosensor for influenza virus H5N1 based on enzyme-encapsulated liposome. *Analytica Chimica Acta, 963*, 112–118.
54. Xing, J., Zhang, X., Wang, Z., Zhang, H., Chen, P., Zhou, G., Sun, C., Gu, N., & Ji, M., (2019). Novel lipophilic SN38 prodrug forming stable liposomes for colorectal carcinoma therapy. *Int. J. Nanomedicine, 14*, 5201–5213.
55. Liu, D., Huang, L., Moore, M. A., et al., (1990). Interactions of serum proteins with small unilamellar liposomes composed of dioleoyl phosphatidylethanolamine and oleic acid: High-density lipoprotein, apolipoprotein A, and amphipathic peptides stabilize liposomes. *Biochemistry, 29*(15), 3637–3643.
56. Mady, M., (2005). Serum stability of non-cationic liposomes used for DNA delivery. *Romanian J. Biophys., 14*(1–4), 89–97.
57. Vermehren, C., Hansen, H. S., Clausen-Beck, B., et al., (2003). *In vitro* and *in vivo* aspects of N-acyl phosphatidylethanolamine- containing liposomes. *Int. J. Pharma., 254*(1), 49–53.
58. Khosravi-Boroujeni, H., Sarrafzadegan, N., Mohammadifard, N., Sajjadi, F., Maghroun, M., Asgari, S., Rafieian-Kopaei, M., & Azadbakht, L., (2013). White rice consumption and CVD risk factors among Iranian population *J. Health. Popul. Nutr., 31*(2), 252–261.
59. Chauhan, B., Kumar, G., Kalam, N., & Ansari, S. H., (2013). Current concepts and prospects of herbal nutraceutical: A review. *J. Adv. Pharm. Technol. Res., 4*(1), 4–8.
60. Fernandes, C., Oliveira, C., Benfeito, S., Soares, P., Garrido, J., & Borges, F., (2014). Nanotechnology and antioxidant therapy: An emerging approach for neurodegenerative diseases. *Curr. Med. Chem., 21*, 4311–4327.
61. Hu, K., Huang, X., Gao, Y., Huang, X., Xiao, H., & McClements, D. J., (2015). Core-shell biopolymer nanoparticle delivery systems: Synthesis and characterization of curcumin fortified zein-pectin nanoparticles. *Food Chem., 182*, 275–281. Elsevier Ltd.
62. Farhang, B., Kakuda, Y., & Corredig, M., (2012). Encapsulation of ascorbic acid in liposomes prepared with milk fat globule membrane-derived phospholipids. *Dairy Science and Technology, 92*, 353–366.

63. Rovoli, M., Pappas, I., Lalas, S., Gortzi, O., & Kontopidis, G., (2019). *In vitro* and *in vivo* assessment of vitamin A encapsulation in a liposome-protein delivery system. *J. Liposome Res., 29*, 142–152.

64. Lukawski, M., Dałek, P., Borowik, T., Forys, A., Langner, M., Witkiewicz, W., & Przybylo, M., (2019). New oral liposomal vitamin C formulation: Properties and bioavailability. *J. Liposome. Res.*, 1–8. doi: 10.1080/08982104.2019.1630642.

65. Davis, J. L., Paris, H. L., Beals, J. W., Binns, S. E., Giordano, G. R., Scalzo, R. L., Schweder, M. M., et al., (2016). Liposomal-encapsulated ascorbic acid: Influence on vitamin C bioavailability and capacity to protect against ischemia-reperfusion injury. *Nutr. Metab. Insights., 9*, 25–30.

66. Tan, C., Xue, J., Lou, X., Abbas, S., Guan, Y., Feng, B., Zhang, X., & Xia, S., (2014). Liposomes as delivery systems for carotenoids: Comparative studies of loading ability, storage stability and *in vitro* release. *Food Funct., 5*(6), 1232–1240.

67. Chen, K. T. J., Anantha, M., Leung, A. W. Y., et al., (2020). Characterization of a liposomal copper(II)-quercetin formulation suitable for parenteral use. *Drug Deliv. Transl. Res., 10*, 202–215.

68. Jangde, R., & Singh, D., (2016). Preparation and optimization of quercetin-loaded liposomes for wound healing, using response surface methodology. *Artif. Cells Nanomed. Biotechnol., 44*(2), 635–641.

69. Clares, B., Calpena, A. C., Parra, A., Abrego, G., Alvarado, H., Fangueiro, J. F., & Souto, E. B., (2014). Nanoemulsions (NEs), liposomes (LPs) and solid lipid nanoparticles (SLNs) for retinyl palmitate: Effect on skin permeation. *Int. J. Pharm., 473*(1, 2), 591–598.

70. Sou, K., Inenaga, S., Takeoka, S., & Tsuchida, E., (2008). Loading of curcumin into macrophages using lipid-based nanoparticles. *Int. J. Pharm., 352*(1, 2), 287–293.

71. Gang, W., Jie, W. J., Ping, Z. L., Ming, D. S., Ying, L. J., Lei, W., et al., (2012). Liposomal quercetin: Evaluating drug delivery *in vitro* and biodistribution *in vivo*. *Expert. Opin. Drug. Deliv., 9*(6), 599–613.

72. Yang, S., Liu, C., Liu, W., Yu, H., Zheng, H., Zhou, W., & Hu, Y., (2013). Preparation and characterization of nanoliposomes entrapping medium-chain fatty acids and vitamin C by lyophilization. *Int. J. Mol. Sci., 14*, 19763–19773.

73. Bi, Y., Xia, H., Li, L., Lee, R. J., Xie, J., Liu, Z., Qiu, Z., & Teng, L., (2019). Liposomal vitamin D3 as an anti-aging agent for the skin. *Pharmaceutics, 11*(7), 311.

74. Lu, X. Y., Hu, S., Jin, Y., & Qiu, L. Y., (2012). Application of liposome encapsulation technique to improve anticarcinoma effect of resveratrol. *Drug. Dev. Ind. Pharm., 38*, 314–322.

75. Wang, X., Li, Y., Yao, H., Ju, R., Zhang, Y., et al., (2012). The use of mitochondrial targeting resveratrol liposomes modified with a dequalinium polyethylene glycol-distearoyl phosphatidyl ethanolamine conjugate to induce apoptosis in resistant lung lcancer cells. *Biomaterials, 32*, 5673–5687.

76. Seguin, J., Brulle, L., Boyer, R., Lu, Y. M., Romano, M. R., et al., (2013). Liposomal encapsulation of the natural flavonoid fisetin improves bioavailability and antitumor efficacy. *Int. J. Pharm., 444*, 146–154.

77. Phan, V., Walters, J., Brownlow, B., & Elbayoumi, T., (2013). Enhanced cytotoxicity of optimized liposomal genistein via specific induction of apoptosis in breast, ovarian, and prostate carcinomas. *J. Drug. Target, 21*, 1001–1011.

78. Tanaka, Y., Uemori, C., Kon, T., Honda, M., Machmudah, S., Kanda, H., & Goto, M., (2020). Preparation of liposomes encapsulating β-carotene using supercritical carbon dioxide with ultrasonication. *Journal of Supercritical Fluids, 161,* 104848.
79. Sujitha, S., & Rasool, M., (2019). Berberine coated mannosylated liposomes curtail RANKL stimulated osteoclastogenesis through the modulation of GSK3β pathway via upregulating miR-23a. *Int. Immunopharmacol., 74,* 105703.

CHAPTER 3

Liposomes as Drug Carriers

MOHAMAD TALEUZZAMAN,[1] FARHEEN FATIMA QIZILBASH,[2]
CHANDRA KALA,[3] DIPAK KUMAR GUPTA,[2] and
SADAF JAMAL GILANI[4]

[1]*Department of Pharmaceutical Chemistry, Faculty of Pharmacy, Maulana Azad University, Jodhpur, Rajasthan, India*

[2]*Department of Pharmaceutics, School of Pharmaceutical Education and Research, Jamia Hamdard, New Delhi, India*

[3]*Department of Pharmacology, Faculty of Pharmacy, Maulana Azad University, Jodhpur, Rajasthan, India*

[4]*Department of Basic Health Sciences, Preparatory Year, Princess Nourah Bint Abdulrahman University, Riyadh, Saudi Arabia*

ABSTRACT

Nanoparticles have unique physicochemical properties compared to a macromolecule. Nanotechnology provides a new direction in the scientific world, widely used to develop new formulations with a high therapeutic index. A liposome is one of the best carriers for drugs and nutraceuticals. Regulated drug delivery enhances the bioavailability of the drug, which provides a high therapeutic index. A liposome is a suitable nonmaterial for the food and medicinal industries. Nanoencapsulation of bioactive food compounds is an emerging application of nanotechnology. Masking the undesirable flavor of the bioactive compounds and enhancing the nutritional value of food ingredients, and protect them from spoilage. Several kinds of liposomal nutraceutical products have been employed for curative and preventive action to control the disease with a high therapeutic index.

3.1 INTRODUCTION

The word liposomes is a resultant of two Greek words that are 'Lipos,' which means fat, and 'Soma,' meaning body [1]. As mentioned in previous studies, it has indicated that lipids are amphiphilic compounds in which one portion of it was found to be hydrophilic (water-loving), while the other portion was hydrophobic (water-hating). A disparaging interface of water-hating sections of the fragment with the solvent resulted in the formation of liposomes [2]. Liposomes are small spherical vesicles whose particle size varies between 30 nm to several micrometers and are encapsulated with a bilayer of lipid. Bangham et al. and his subordinates first revealed liposomes in 1961 [3]. Standish and his co-workers defined liposomes as engorged phospholipid organisms. Various studies have been carried out with liposomes since Bangham's discovery, thus making liposomes the safest and most popular nanocarrier systems [4].

Nowadays, liposomes have become a very vital tool and reagent in enormous scientific divisions, comprising biophysics, chemistry, theoretic physics, biochemistry, colloid science, biology, and mathematics. Subsequently, liposomes have gained a lot of importance in the market. They have become promising drug delivery carriers due to their amphiphilic nature and size. They are comprehensively used in various microscopic entities in the pharmaceutical and cosmeceutical industries. Furthermore, farming and food industries have enormously utilized the use of liposomal coating therapy to produce delivery systems that can trap unbalanced entities (example: antioxidants, antimicrobials, bioactive compounds, and flavors), thus safeguard its mechanism. Liposomes can encapsulate both hydrophilic and hydrophobic molecules, thus circumventing the decaying of the trapped combinations that lead to the delivery of trapped entities at the preferred sites [5]. Doxil (Sequus) and DaunoXome (Gilead, Nexstar) were the first-ever outstanding products produced using liposomal technology that are used in anticancer therapy as active medicaments. In the 1990s, both the mentioned drugs displayed satisfactory results with minimal side effects in clinical trials that were later succeeded by the US Food and Drug Administration (FDA) approval [6]. Liposomes are used for improving the therapeutic index of new or pre-existing medicaments. They do so by:

• Reducing the metabolism;
• Enhancing the biological half-life;

- Diminishing the toxicity associated with it;
- Modification of the drug absorption phenomenon.

Liposomal samples increased the therapeutic effectiveness of the medicaments in preclinical representations and also in humans as compared to the conventional samples. The reason for therapeutic effectiveness was attributed to the modifications of the bio-distribution. The drug encapsulated with liposomes is found to have minimal instant degradation along with the least adverse effects. The main reason for this minimal toxic effect is again attributed to the liposomes, as liposomes are made up of biologically inert, non-immunogenic, and biodegradable lipids [2]. Liposomes are generally designed to accomplish the following characteristics (Figure 3.1) [7]:

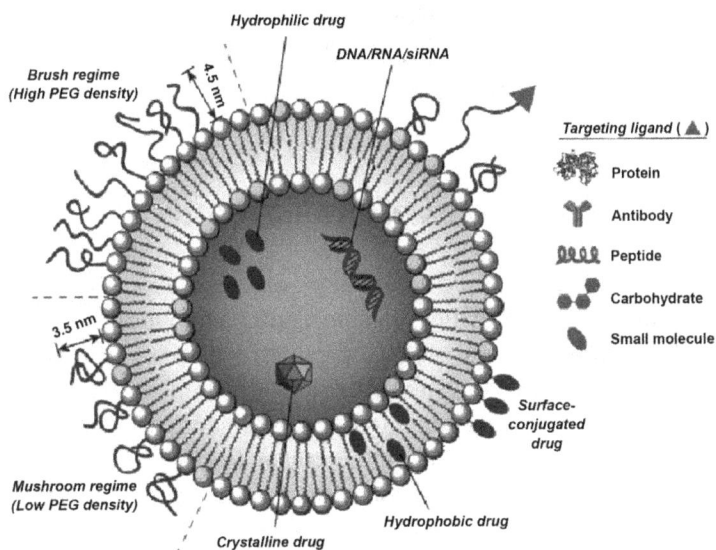

FIGURE 3.1 A schematic representation of liposomal drug delivery.
Source: Reprinted with permission from Ref. [51]. © 2014 Elsevier.

- Loading of drug;
- Controlling of release rate of drug;
- To overcome the instant clearance of liposomes;
- Receptor-Mediated endocytosis of the ligand-targeted liposomes;
- Delivery of nucleic acids and DNA;
- Intracellular delivery of drugs.

Drug delivery based on liposomes has emerged ever since the unpre-tentious growth of them over 3 eras ago. More than a dozen drug delivery systems based on liposomes are currently accepted by the FDA, with several new in the stage of progress and evolvement. The positive opinion of the FDA on liposomes have further aided in the marketing of lipo-somes. Nevertheless, the regulatory guidelines have not kept up with the instant hop of the growth of novel liposomal drug delivery systems. The regulatory demands are high but still are outdated. Therefore, concerning determination and strength on behalf of the industries, academia as well as regulatory bodies is required to uphold, enhance, and promote the overall medical prospective of liposomal drug delivery systems and to maintain transparency in the approval procedure [8].

Furthermore, liposomes possess no antigenic as well as no pyrogenic reactions with diminished antagonistic effects or side effects. Subse-quently, this entire characteristic of liposomes makes liposomes appro-priate, safe, potent, and excellent approach for site-specific delivery of medicaments [2].

3.2 STRUCTURAL COMPONENTS OF LIPOSOME

Liposomes are spherical lipid bilayers of diameter about 50–1000 nm. Liposomes aids as perfect delivery vehicles for compounds that are active biologically. Liposomes are used topically in delivery for anticancer medicaments for diminishing its hazardous effects. Liposomes are also widely used in the field of dermatology. As described in the literature, there is a numerous variety of non-structural and structural constituents of liposomes (Table 3.1). The major components of liposomes are as follows:

1. **Cholesterol (Chol):** A bilayer is not formed by Chol only but is capable to be encompassed into phospholipid membranes in higher concentrations of about 1:1 or 2:1 of the molar ratio of Chol to phosphatidylcholine (PC). Chol includes itself in the membrane with the hydroxyl assembly leaning to the aliphatic chain and aqueous surface. The extraordinary solubility of Chol present in the phospholipid liposome has been credited to hydrophobic and hydrophilic coordination, though there is no vibrant clue for the organization of Chol in the bilayer.

2. **Phospholipids:** These are the main constituent of the biological shell. Two types of phospholipids exist, namely, 'Sphingolipids' and 'Phosphodiglycerides.' Whereas, the utmost mutual phospholipid is the PC molecule. The elements of PC are not solvable in an aqueous medium. The phospholipids are widely utilized constituent of liposome preparation. Some examples of phospholipids are as follows: Phosphatidylserine (PS), Phosphatidylcholine (Lecithin)-PC, phosphatidylinositol (PI), Phosphatidylethanolamine (cephalin)-PE, phosphatidylglycerol (PG).

TABLE 3.1 Advantages and Disadvantages of Liposomes

Advantages	Disadvantages
Non-toxic, biocompatible, flexible.	Low solubility.
Enhances stability via encapsulation.	Half short-life.
Biodegradable, non-immunogenic for non-systemic and systemic administrations.	Seepage and fusion of trapped molecules.
Reduce toxicity of encapsulated agent (Taxol), Site avoidance effect.	Production cost is high.
Liposomes can carry large pieces of DNA, possibly as big as a chromosome.	Phospholipid undergoes oxidation and hydrolysis-like reaction.

Source: Ref. [5].

3.2.1 PREPARATION OF LIPOSOMES

Various techniques are present for the preparation of liposomes, some of which include the use of organic solvents, mechanical techniques, or via elimination of surfactant from the surfactant micelle blends. For the preparation of liposomes, quantity, and type of phospholipid, charges properties of the watery medium are some of the key aspects that conclude the ultimate structure of liposomes [9]. Some of the techniques used for the preparation of liposomes are given in subsections.

3.2.1.1 MULTILAMELLAR VESICLES (MLVS) PREPARATION

In this technique, various steps for the preparation of liposomal production are used as an organic solvent that helps in dissolving the lipid and for drying up the resultant blend. An amalgamation of suitable lipids is taken such as Chol, egg lecithin, and phosphatidylglycerol in the molar ratio of 1.0:0.9:0.1 correspondingly, while a blend of chloroform with methanol is taken in a ratio 2:1. First of all, each one of the components of lipid is made to dissolve separately in the alcoholic medium, succeeded by the successful mixing of lipids with other soluble lipids in the above-mentioned ratios for certifying the constant dispersal of lipids in the blends. Whereas, in an evacuated compartment, the film layer present on the top of lipid is allowed to dry fully to remove the left-out traces of organic solvents. The drying of the film layer of lipid is done for consecutive 4–6 hours [10].

3.2.2.2 UNILAMELLAR VESICLES (ULV) PREPARATION

The unilamellar structure of liposomes permits a constant scattering of the captured particles in an internal single aqueous compartment. The unilamellar vesicles (ULV) can be prepared by numerous techniques, some of which include ethanol injection, detergent technique, extrusion via polycarbonate filters, freeze-thawing, ultrasonication, etc. Bhatia et al., along with subordinates, utilized the blend of various tiny ULV for the preparation of ternary giant unilamellar vesicles (GUV) with uniform property [11].

3.2.2.3 GIANT UNILAMELLAR LIPOSOMES (GUV) PREPARATION

The preparation of GUV can be done by various techniques. The preparation of GUV by these techniques is established on the use of zwitterions or only distilled water. Elevation in the pull between the membranes was produced because of the ions causing a gross charge that in return hinders the segregation of sheets of the membrane throughout various processes such as swelling phenomenon and rehydration procedure. Motta et al., along with his co-workers, explained that numerous techniques are employed for the formation of GUVs, such as electroporation, giant liposomes formation in a speedy way with the aid of physiological buffer, and osmotic shock procedure [12]. Karamdad et al., along with his subordinates, developed a new technique for GUV preparation. They established an innovative microfluidic technique for the generation of compositionally-controlled GUVs where water-in-oil (W/O) droplets were generated in oil flow composed of lipids that were later transferred beyond the bilayer of phospholipid [13].

3.3   CHARACTERIZATION OF LIPOSOMES

Characterization of liposomes is done immediately after the preparation of liposomes. The method of characterization is required to be specific and reproducible. Characterization of liposomes after the preparation and storage of liposomes is necessary for quality control (QC) of the desired product [14].

3.3.1   ESTIMATION OF RESIDUAL ORGANIC PHASE IN PHOSPHOLIPID MIXTURES

During storage, lipids undergo oxidation, therefore to avoid this, chloroform is utilized as a solvent for the protection of egg PC and egg phosphatidylglycerols (EPG) from oxidation. Chloroform not only protects the lipids from oxidation but also promotes an excellent molecular dispersion of numerous lipids during this phenomenon. Though chloroform is an excellent solvent for this procedure, it is important to amputate it from the lipid mixtures in case of human usage. It was indicated that the presence of chloroform might be carcinogenic. Thus, it was well known that the phenomenon of liposome preparation requires the usage of chloroform as a solvent [14].

3.3.2   PERCENT MEDICAMENT ENCAPSULATION

Madden et al., along with their co-workers, explained that the amount of medicament captured in the liposomes could be estimated by the phenomenon called column chromatography. It was indicated that the liposomes are a blend of coated and uncoated potions of medicament. It was explained that the uncoated medicament was called a free medicament. In various studies, it was concluded that free and coated medicaments were separated to determine the concentration of free medicament. Later, the coated medicament was treated with a surfactant for releasing the medicament from liposomes to the aqueous medium [15].

3.3.3 SIZE ANALYSIS

When liposomes are intended for parenteral and oral usage, then the normal size of liposomes acts as a major parameter. Various methods have been employed for the determination of the size and size distribution of liposomes in micrometers, such as size-exclusion chromatography (SEC), static or dynamic light scattering method, and microscopy techniques. Various researches have suggested that a new technique called atomic force microscopy (AFM) is non-invasive, quick, and powerful. AFM analysis was found to be excellent for studying the size, stability, and morphology of liposomes [16].

3.3.4 ZETA POTENTIAL

Zeta potential can be defined as the net charge a particle gains in a particular medium and is a characteristic property of any element present in suspension. For the measurement of zeta potential, a laser is aided for providing a light source within the particles that pass via particles scattering light at an angle of 13°. The data obtained is distributed to a digital signal processor that is then passed to a computer, and thus, the zeta potential of the sample is calculated. Particles with zeta potentials $> +30$ mV or < -30 mV generally marks the stability of the samples [17].

3.3.5 ENCAPSULATION EFFICIENCY (EE)

As discussed above, the preparation of liposomes is a blend of coated and uncoated (free drug) medicament portions [15]. The separation between the coated and uncoated medicament is the initial step of the technique that can be estimated by using a dialysis membrane in which the liposomal sample is immersed in a phosphate buffer solution (PBS) for 120 minutes [18].

3.3.6 LIPID ANALYSIS

This technique is generally aided for the estimation of phospholipid levels in the liposomal sample. Most of lipid analysis technique requires the usage of reagent that constitutes molybdate that further yield to a blue-colored product. One such method of lipid analysis is an ascorbic acid technique in which ammonium molybdate reacts with orthophosphate formed from the digestion of acid to form phosphomolybdic acid. The product obtained is later reduced with the help of ascorbic acid to obtain a blue-colored solution that is later analyzed at 820 nm [19].

3.3.7 IN-VITRO DRUG RELEASE

The *in-vitro* drug release study can be done with the aid of the dialysis tube diffusion method. Some milliliters of the liposomal formulation should be placed in the dialysis bag that should be tied in such a way that air could not pass through it. The dialysis bag should be then placed in the cell containing the suitable aqueous medium and should be maintained at 37°C with continuous magnetic stirring. The cell should be closed to avoid vaporization of the aqueous medium. Samples of the dialysate are then taken out at different time intervals, and at the same time, the same amount of same fresh sample should be added to keep the volume of the cell constant. Withdrawing of samples should be performed in triplicate and should be then assayed for the estimation of medicament via spectro-photometer or high-performance liquid chromatography (HPLC) [16, 20].

3.4 STABILITY OF LIPOSOMES

The stability of liposomes plays a crucial role in steps like storage and delivery. A pharmaceutical dosage form is said to be safe and stable only if it does not react with the active ingredients present in the dosage form thereby, maintaining the physical and chemical reliability of the active ingredients throughout its shelf-life. A formulation is evaluated as safe depending upon well-established study designs including chemical, microbial, and physical factors. Thus, stability analysis of liposomal samples must consist of a unit for the stability of the product during its storage while another unit for the characterization of products. As lipo-somal samples are diverse in size therefore, the total size distribution stability storage. Liposomes have a character trait of growing and fusing into superior vesicles that possess a major problem causing the leakage of medicament from those vesicles. Thus, average size distribution and visual appearance are considered as major factors to determine the physical stability of liposomal preparation [21].

While lipid is the main ingredient used in the liposomal preparation that is derived from synthetic or natural sources of phospholipids that comprise of unsaturated fatty acids that later undergo oxidation. The products obtained after oxidation creates variations in permeability in the bilayer of liposomes. On the other hand, the interaction of phospholipid with the medicament also hinders the chemical stability. Therefore, it is mandatory to follow the protocol of stability over the desired period for the precise estimation of the chemical stability of the medicament. Almost, the majority of the liposomal samples are intended for the parenteral purpose. Thus, there is a need to sterilize the samples to ensure that the product is safe and free from microbial contamination [22].

During manufacture, the outflow, degradation, and liposome's fusion can cause flaws of the lattice in the membrane. Incubation of liposomes at a much higher temperature than the temperature of the phase transition can minimize these defects, thereby making the changes in packing density constant between the edges of the bilayers that are called an annealing process. It was later observed that the problem of fusion and aggregation in the annealing vesicles can even take place for a long period. The reason for the defect of fusion and aggregation was attributed to van der Waals interactions. The solution for this problem was found out to be was the addition of a small number of phospholipids that were negatively charged, such as the addition of 10% of polyethylene glycol (PEG) to the liposomal sample.

Absorptivity of the membranes of the liposomes depends mainly on the coated material and the composition of the lipid membrane. It was observed that the polar or large ions were able to retain more effectively than those with lipid-loving low molecular masses. In a study, it was observed that materials with more saturated and tough membranes led to the formation of the most constant lipid membrane that in turn created the seepage of the coated compound.

Biological stability analysis of liposomal formulation plays a crucial role in several platforms of the delivery of the medicament. Nevertheless, liposomes are found to be a little biologically unhinged when it comes to parenteral preparations. The reason for the biological instability can be attributed to the circulatory clearance via the mononuclear phagocytic system (MPS) that is present in the liver and spleen. The stability of liposomes depends on agents like proteins in it that reacts with them when used. Various methods have been employed to increase the biological stability of liposomes to increase the drug circulation time in blood [23]. The phenomenon complexation, that is of making an atom or compound to form a complex with the other between the liposomes and the polymers was studied to enhance the shelf life of liposomes. In a study, it was indicated that by implanting water-loving polymers to the head groups of phospholipids or fixation of water-miscible polymers with various water-hating groups, there was an enhancement in the circulation time *in-vivo*. It has also shown to hinder the fusion of liposomes. These types of liposomes are known as stealth liposomes or sterically stabilized liposomes. In a study, it was indicated that poly(vinyl pyrrolidone) (PVP) works as an effective steric defender and also displays a high degree of biocompatibility for liposomes. It was indicated that bilayers of the liposomes comprising of lipids being fixed to PEG proved to hinder the cellular as well as the protein interactions with the liposomal samples, thereby enhancing the blood circulation period when administered to the animals [24]. Medicament named Doxil that is used in anticancer therapy is a very popular liposomal doxorubicin that is found to have a shelf life for more than 18 months in a liquid state. The main reason for its stability is attributed to the usage of PEG [2].

3.5 RELEASE MECHANISM OF LIPOSOMES

Liposomes are composed of water-based solution encapsulated in a hydrophobic membrane, and this hydrophobic membrane consists of chemicals that can be effortlessly thawed in the fragments of lipids that make liposomes to transport hydrophobic and hydrophilic entities mutually. Whereas, the rate and extent of position of the medicament are dependent on the composition of lipid and its physicochemical characteristics. Furthermore, a fusion of the lipid bilayers with the bilayers of the cell membrane takes place for the delivery of the content of the liposomes to provide site-specific delivery of the fragments of the medicaments [25].

There are four steps involved in the release mechanism of liposomal delivery that is as follows:

1. **Adsorption:** Liposomes undergo adsorption to the membranes of the cell or fragments that in turn leads to its contact o the membrane of the cell.

2. **Endocytosis:** After undergoing adsorption on the surface of the cell membrane, they undergo engulfment and internalization into the liposomes.

3. **Fusion:** The fusion of with the lipoidal membrane of the cell with the bilayers of lipid with the help of a phenomenon called lateral diffusion as well as blending of lipids leads to the site-specific delivery of the contents of the liposomes into the cytoplasm.

4. **Lipid Exchange:** Because of the resemblance of the lipid membrane of liposomes with membranes of cell's phospholipids, the transmission of the proteins of lipids present in the cell could effortlessly identify liposomes and thereby creating the exchange of the lipids.

Whereas, in the case of cancer cells, the release mechanism of liposomes is still not completely known and is thought to be dependent on three main issues that are as follows [26]:

- Mechanism of action of the loading of the drug;
- Composition of lipid membrane; and
- The microenvironment of the tumor cell.

Another study indicated that in the case of tumor cells, the consumption of tumor cells is more. They consume a large number of fats as an impending basis of nourishment and sustenance to fulfill the need of instant growth. Therefore, they identify liposomes encapsulated with anticancer medicament. Therefore, once they are targeted by liposomes, they instantly get absorbed, thereby causing the release of anticancer medicament from liposomes at the desired site of action. Hence, cancer or tumor cells are executed by the liposome-encapsulated medicament [8].

3.6　PROBLEMS RELATED TO LIPOSOMES DELIVERY

Any foreign substances whenever enter the body come across several defense systems, concentrated at recognition, neutralization, and elimination of invading substances the same happened with liposome. The defense system includes a reticuloendothelial system (RES) opsonization, and immunogenicity [27]. Factor like enhanced permeability and retention (EPR) effects can be used to enhance drug delivery, such effect is the cause for the accumulation of large molecule and nanocarriers within the tumor and forms the basis of passive targeting delivery [28]. The main challenge with passive targeting is a different type of tumor, and it is a various stage that changes the porosity and pore size of tumor vessels, so it not possible to target in all tumors. In solid tumor distribution of nanocarriers, the inhomogeneous way is not possible because the elevated interstitial fluid pressure inhibits it which leads to multiple drug resistance [29]. In humans, first time experimented freshly prepared liposomal drug delivery and after over a year liposome-drug formulation defined its stability and shelf life. In literature, several kinds of preparation methods discussed, only a few of them used in the large-scale production. The maintenance of physical properties of liposomal formulations is very diffi-cult; encapsulated material leakage is one of the main problems. Leakage of content is due to the permeability of the membrane, presence of organic solvent residues, pyrogen control, sterility, size, and size distribution apart from problems in manufacturing process problems are batch to batch reproducibility, stability due to the hydrolytic and oxidative degradation and storage condition [2] During passage gastrointestinal tract, liposomal drug vehicles show extensive leakage of water-soluble drugs this is a problem other than instability. In the case of lipophilic drugs, it almost completely entrapped in the lipid layers chance of loss encountered rarely but challenges in case of lipids molecule have charged it becomes toxic in increased doses; other problems are short shelf life and a problem with encapsulation efficacy [26, 30]. Some formulations can activate immune response following activation of the complement system to activate an acute hypersensitivity syndrome known as complement activation-related pseudoallergy (CARPA). Hypersensitivity reactions in patients have a relative percentage (2–45%) has been reported infusion-related liposomal drug therapy [31].

Liposomal drug delivery in CNS, its overall efficacy is depending upon the properties like pharmacokinetics and bio-distribution, including binding to plasma proteins or degradation of the drug in the blood. Major problems earlier were rapid uptake by the RES and further it removing from circulation but as liposomes had been stabilized with the integra-tion of PEGylated phospholipids into their bilayer, reduced the loss. Fast systemic elimination is not only the limitation but other factors such as metabolic degradation of the phospholipids, vesicle stability, storage, and inability to provide sustained release of drugs. However, some prob-lems have been overcome but it still difficult to ensure if the long term used for the treatment of neurodegenerative diseases [32]. Liposomes formulation has kind of disadvantages like encapsulation efficiency (EE), inability for protection, and inefficiency of skin permeability are the main disadvantages. The major problem generally faces for the development of liposomes in cosmeceutical as well as the topical application is a concern too-decline stability of liposomes over time, problems developed in preparation technique, and limitation of skin permeability [33]. Based on the hydrophobic properties, liposomes can be stuffed with several kinds of drugs. But limitation always with hydrophilic drugs, which mainly suffer its bioavailability at the tumor site because of their very low membrane permeability and thus in the end small amount of drug released at the tumor site. Challenges with highly hydrophobic drugs as they easily bind with the liposomal membrane and exhibit rapid redistribution of the drug to plasma components and leads to lower entrapment stability [34]. The natural antioxidant phenolic compounds Tyrosol, hydroxytyrosol, and oleuropein carried to the targeted site for their maximum effects, but a low bioavailability and accessibility is its limitation [35]. Developmental challenges related to the omega-3 liposomal food fortified diet are substantial, including instability and undesirable flavors/odors. Undesirable odor and rapid deterioration are the limitations of its application [36].

3.7 APPLICATION OF NUTRACEUTICAL LOADED LIPOSOMES TO DIFFERENT DISEASE

The applications of liposomes in nutraceuticals are many have a target to hide sensitive ingredients, enhance the bioavailability of nutrients, increasing the potency of food additives and nasty intern flavor. Nano-carrier is generally employed for entrapment and controlled release delivery system [37]. Carry bioactive molecules to the specific site, reducing side effects, and hiding the active molecule from enzymatic metabolic process. Commercialized liposome products like Liposomal turmeric with fulvic acid, liposomal curcumin (CUR) syrup, and oral CUR liposome syrup have therapeutic effects as an anti-inflammatory. Liposomal glutathione (GSH) capsules and Hemp liposomal syrup have therapeutic effects antioxidants. Apart from this different application also including drug development, food supplements, and food preservatives.

Dietary supplementation of fish oil capsule with omega-3 in countries where fish consumption is very low. Encapsulation of several essential oils (EOs) in liposome nanoparticles overcome the drawback such as instability, degradation in the natural environment, the liposome can protect the fluidity and are stable at 4–5°C at least for 6 months. Zataria multiflora essential oil encapsulated in liposomes employed for antimicrobial therapeutic effects. Bioactive compound, eugenol-loaded liposome protects it from degradation from UV light and maintains the DPPH-scavenging activity of free eugenol [38, 39].

Food or food constituents have properties to prevent and/or cure the tumors and such nutraceuticals have chemopreventive effects. The mechanism that nutraceuticals adopt is generally (a) Inhibit cell proliferation and differentiation (b) Prevent efflux transporters such as breast cancer resistance protein, P-glycoprotein, multidrug resistance protein (MRP) and (c) Decrease the toxicity of chemotherapeutic drugs. Different types of nutraceuticals are used for cancer prevention and treatment of cancer like (a) Dietary Fibers: e.g., soluble fibers and insoluble fibers (b) Fatty acids: e.g., Lecithin, ω-3 fatty acids, conjugated linoleic acids and *Brucea javanica* oil (c) Carotenoids: e.g., Lycopene, Lutein, β-carotene, cryptoxanthin, zeaxanthin, α-carotene (d) Vitamins (e) Minerals (f) Probiotics and Prebiotics (g) Phenolics-quercetin, apigenin, daidzein, naringenin, catechins, myricetin, gallic acid, hesperidin, CUR, resveratrol (RESV), and secoisolariciresinol [40].

Liposomes are an appropriate carrier for the delivery at the site of action because it delays the clearance duration and longer duration intravascular circulation of loaded nutraceuticals, so altering their bio-distribution [41].

Some examples are discussed here, CUR Liposome used for the treatment of pancreatic carcinoma, CUR is a potent inhibitor of NF-κB, a transcription factor implicated for the pathogenesis of several malignancies including pancreatic carcinoma, in treatment of Lymphoma, the liposomal vehicle shown more effective to release bioactive's at the cell, in human prostate cancer shown inhibition of cellular proliferation without affecting their viability, squamous cell carcinoma in a dose-dependent manner suppressed NF-κB, colorectal cancer, cervical cancer. RESV liposome on cancerous cells targets through the mitochondria apoptosis pathway. Combination Treatment of CUR and RESV liposomes improved bioavailability and enhance chemopreventive effects for the treatment of prostate cancer, natural flavonoid fisetin loaded in liposome useful in prevention and treatment of cancer by increasing its bioavailability, effects of crocin liposome on cancerous cells in the human brain prevents oxidative stress and possesses. Cyanidin-3-O-glucoside liposome on cancerous cells. It inhibits proliferation and induces apoptosis in cancerous cells, improves bioavailability. Genistein Liposome binds with estrogen receptors which regulated gene expression. Effective against the cancer cells of breast, prostate, cervical, and ovarian [42].

The efficiency of vitamin C drastically increases when encapsulated compared to free ascorbic acid along with the duration of storage [43]. Vitamin A (Retinol) is sensitive to heat and light, degradation reduced by encapsulation more effective as compare to free retinol. Also, the effectiveness of the phosvitin antioxidant was more in case of encapsulated condition as compare to free [44]. The application of bacteriocin with food products has several problems like proteolytic degradation or interactions with food components. Such problems can be overcome by encapsulation with liposomes. Studied the stability effects of nisin Z. Liposome loaded with nisin Z in manufacturing of cheddar cheese, lowered the bacterial cell numbers of Lactococci, fermentation process of cheddar cheese showed more stability as compared to direct mixing of nisin Z [45].

In diabetes management prepared oral liposomal insulin formulations, it was prepared using insulin (Humulin R) or protamine-containing insulin (Humulin N) with Chol, dipalmitoylphosphatidylcholine (egg) (DPPC)-Chol mixture, and mucoadhesive agent (methylcellulose: MC)-added DPPC-Chol mixture. Highly effective reduced blood glucose level, for better efficacy the important role in the formulation is pH and the presence of protamine sulfate [46]. Rutin an antioxidants prepared Phytosomes-PC—rutin complexes. Rutin encapsulated with PC gives maximum physical and chemical stability with fine particle sizes (<100 nm) and 99%, EE. Undesirable things of rutin have been masked due to encapsulation. Used in fortification of food products with water-insoluble nutraceuticals [47]. Effects of olive oil and its derivatives (tyrosol, hydroxyl tyrosol, and oleuropein) for the treatment of osteoarthritis. Side effects of these compounds are overcome by using liposome preparations, as carriers enhance the pharmacological effects of the drugs/nutraceuticals [35]. Liposome encapsulated based on the thin-film hydration (TFH) method, ULV gave better efficiency of 56% for cobalamin, 76% for _-tocopherol, and 57% for ergocalciferol with better stability [48].

3.8 CONCLUSION AND FUTURE PROSPECT

The uses of nutraceuticals have been researched for the prevention and cure of different diseases. The efficacy of the ingredients in a free condition not up to the mark. Its therapeutic effects enhanced with the application of liposome as a carrier. Encapsulation of the active molecules has been done with focusing on its stability, efficacy, and storage condition. Prolonged the self-life of ingredients. Regulated delivery of drugs at the targeted site reduces toxicity and side effects. Liposomal delivery of system used in several fields like medicine, immunology, diagnostics, cosmetics, and the food industry. The overall therapeutic index improved. An established new concept of distribution, binding, absorption, and reducing metabolism [49]. Several clinical studies have performed found the superiority of liposomal drug formulations over conventional delivery systems. Liposomal encapsulation technique is a very foremost and reliable technique to enhance its efficacy and stability for nutraceutical, especially for cancer treatment [50]. The role of Nutraceutical liposomes has been found curative as well as preventive medicines, but the aim always is the former. Several problems require overcoming before to commercialize. Clinical approval as per the guidelines based on the large-scale production. Positive efforts on the part of regulatory guidance require for the optimum drug delivery need to open discussion between academia and industry is very important. Deciding how to change or modify techniques for the preparation procedure from laboratory scale to mass production. A balance between hydrophilic and hydrophobic encapsulation has required research for the betterment of the formulation of therapeutic values.

KEYWORDS

- atomic force microscopy
- giant unilamellar vesicles
- phosphatidylglycerol
- phosphatidylserine
- quality control
- size-exclusion chromatography
- water-in-oil

REFERENCES

1. Daraee, H., Etemadi, A., Kouhi, M., Alimirzalu, S., & Akbarzadeh, A., (2016). Application of liposomes in medicine and drug delivery. *Artificial Cells, Nanomedicine, and Biotechnology, 44*(1), 381–391.

2. Cağdaş, M., Sezer, A. D., & Bucak, S., (2014). Liposomes as potential drug carrier systems for drug delivery. *Application of Nanotechnology in Drug Delivery, 25.* doi: 10.5772/58459.

3. Deamer, D. W., (2010). From "banghasomes" to liposomes: A memoir of *Alec Bangham*, 1921–2010. *The FASEB Journal, 24*(5), 1308–1310.

4. Bangham, A. D., Standish, M. M., & Watkins, J. C., (1965). Diffusion of univalent ions across the lamellae of swollen phospholipids. *Journal of Molecular Biology, 13*(1), 238–252.

5. Akbarzadeh, A., Rezaei-Sadabady, R., Davaran, S., Joo, S. W., Zarghami, N., Hanifehpour, Y., Samiei, M., Kouhi, M., & Nejati-Koshki, K., (2013). Liposome: Classification, preparation, and applications. *Nanoscale Research Letters, 8*(1), 102.

6. Wagner, A., & Vorauer-Uhl, K., (2011). Liposome technology for industrial purposes. *Journal of Drug Delivery, 2011.*

7. Yadav, M., & Kumar, A., (2017). An Indian outlook on role clarity, organizational citizenship behavior, and gender relationship: Multiple group confirmatory factor analysis (MGCFA) approach. *Jindal Journal of Business Research, 6*(1), 63–75.

8. Zylberberg, C., & Matosevic, S., (2016). Pharmaceutical liposomal drug delivery: A review of new delivery systems and a look at the regulatory landscape. *Drug Delivery, 21, 23*(9), 3319–3329.

9. Li, J., Wang, X., Zhang, T., Wang, C., Huang, Z., Luo, X., et al., (2015). A review on phospholipids and their main applications in drug delivery systems. *Asian J. Pharm. Sci., 10*(2), 81–98.

10. Rieder, A. A., Koller, D., Lohner, K., & Pabst, G., (2015). Optimizing rapid solvent exchange preparation of multilamellar vesicles. *Chem. Phys. Lipids, 186*, 39–44.

11. Bhatia, T., Husen, P., Brewer, J., Bagatolli, L. A., Hansen, P. L., Ipsen, J. H., & Mouritsen, O. G., (2015). Preparing giant unilamellar vesicles (GUVs) of complex lipid mixtures on demand: Mixing small unilamellar vesicles of compositionally heterogeneous mixtures. *Biochimica et Biophysica Acta, 1848*(12), 3175–3180.

12. Motta, I., Gohlke, A., Adrien, V., Li, F., Gardavot, H., Rothman, J. E., & Pincet, F., (2015). Formation of giant unilamellarproteo-liposomes by osmotic shock. *Langmuir, 30, 31*(25), 7091–7099.

13. Karamdad, K., Law, R. V., Seddon, J. M., Brooks, N. J., & Ces, O., (2015). Preparation and mechanical characterization of giant unilamellar vesicles by a microfluidic method. *Lab on a Chip., 15*(2), 557–562.

14. Vemuri, S., & Rhodes, C. T., (1995). Preparation and characterization of liposomes as therapeutic delivery systems: A review. *Pharmaceutica Acta Helvetiae, 70*(2), 95–111.

15. Maddan, T. D., Harrigan, P. R., Tai, L. C. L., Bally, M. B., Mayer, L. D., Redelmeier, T. E., Loughrey, H. C., et al., (1990). The accumulation of drugs within large unilamellar vesicles exhibiting a proton gradient: A survey. *Chem. Phys. Lipids, 53,* 37.

16. Laouini, A., Jaafar-Maalej, C., Limayem-Blouza, I., Sfar, S., Charcosset, C., & Fessi, H., (2012). Preparation, characterization and applications of liposomes: State of the art. *Journal of Colloid Science and Biotechnology, 1*(2), 147–168.

17. Hunter, R., & Midmore, J., (2001). Zeta potential of highly charged thin double-layer systems. *Colloid Interf. Sci., 237,* 147.

18. Padamwar, M. N., & Pokharkar, V. B., (2006). Development of vitamin loaded topical liposomal formulation using factorial design approach: Drug deposition and stability. *International Journal of Pharmaceutics, 320*(1–2), 37–44.

19. Stewart, J. C., (1980). Colorimetric determination of phospholipids with ammonium ferro thiocyanate. *Analytical Biochemistry, 1, 104*(1), 10–14.
20. Reddy, L. H., Vivek, K., Bakshi, N., & Murthy, R. S., (2006). Tamoxifen citrate loaded solid lipid nanoparticles (SLN): Preparation, characterization, *in vitro* drug release, and pharmacokinetic evaluation. *Pharmaceutical Development and Technology, 11*(2), 167–177.
21. Plessis, J., Ramachandran, C., Weiner, N., & Müller, D. G., (1996). The influence of lipid composition and lamellarity of liposomes on the physical stability of liposomes upon storage. *International Journal of Pharmaceutics, 127,* 273–278.
22. Casals, E., Galán, A. M., Escolar, G., Gallardo, M., & Estelrich, J., (2003). Physical stability of liposomes bearing hemostatic activity. *Chemistry and Physics of Lipids, 125*(2), 139–146. https://doi.org/10.1016/s0009-3084(03)00086-0.
23. Simões, S., Moreira, J. N., Fonseca, C., Düzgüneş, N., & De Lima, M. C., (2004). On the formulation of pH-sensitive liposomes with long circulation times. *Advanced Drug Delivery Reviews, 23, 56*(7), 947–965.
24. Allen, T. M., Hansen, C., Martin, F., Redemann, C., & Yau-Young, A., (1991). Liposomes containing synthetic lipid derivatives of poly(ethylene glycol) show prolonged circulation half-lives *in vivo*. *Biochimica et Biophysica Acta, 1066*(1), 29–36. https://doi.org/10.1016/0005-2736(91)90246-5.
25. Torchilin, V. P., (2010). Passive and active drug targeting: Drug delivery to tumors as an example. In: *Drug Delivery* (pp. 3–53). Springer, Berlin, Heidelberg.
26. Yadav, D., Sandeep, K., Pandey, D., & Dutta, R. K., (2017). Liposomes for drug delivery. *Journal of Biotechnology & Biomaterials, 7*(4).
27. Willis, M., & Forssen, E., (1998). Ligand-targeted liposomes. *Adv. Drug Deliv. Rev., 29,* 249–271.
28. Sawant, R. R., & Torchilin, V. P., (2012). Challenges in development of targeted liposomal therapeutics. *AAPSJ, 14,* 303–315.
29. Heldin, C. H., Rubin, K., Pietras, K., & Ostman, A., (2004). High interstitial fluid pressure: An obstacle in cancer therapy. *Nat. Rev. Cancer., 4*(10), 806–813.
30. Mufamadi, M. S., Pillay, V., Choonara, Y. E., Du Toit, L. C., Modi, G., Naidoo, D., & Ndesendo, V. M., (2011). A review on composite liposomal technologies for specialized drug delivery. *Journal of Drug Delivery,* 1–19.
31. Moghimi, S. M., Hamad, I., Andresen, T. L., Jørgensen, K., & Szebeni, J., (2006). Methylation of the phosphateoxygenmoiety of phospholipid methoxy(polyethyleneglycol)conjugate prevents PEGylated liposome- mediated complement activation and anaphylatoxin production. *FASEBJ, 20,* 2591–2593.
32. Lai, F., Fadda, A. M., & Sinico, C., (2013). Liposomes for brain delivery. *Expert Opinion on Drug Delivery, 10*(7), 1003–1022.
33. Van, T. V., Moon, J. Y., & Lee, Y. C., (2019). Liposomes for delivery of antioxidants in cosmeceuticals- Challenges and development strategies. *Journal of Controlled Release, 7.*
34. Barenholz, Y., (1998). Design of liposome-based drug carriers: From basic research to application as approved drugs. In: Lasic, D. D., & Papahadjopoulos, D., (eds.), *Medical Applications of Liposomes* (pp. 545–565). New York: Elsevier.
35. Bonechi, C., Donati, A., Tamasi, G., Pardini, A., Rostom, H., Leone, G., Lamponi, S., Consumi, M., Magnani, A., & Rossi, C., (2019). Chemical characterization

of liposomes containing nutraceutical compounds: Tyrosol, hydroxytyrosol and oleuropein. *Biophysical Chemistry, 246,* 25–34.

36. Iafelice, G., Caboni, M. F., Cubadda, R., Criscio, T. D., Trivisonno, M. C., & Marconi, E., (2008). Development of functional spaghetti enriched with long-chain omega-3 fatty acids. *Cereal Chem., 85,* 146–151.

37. Liu, W., Ye, A., & Singh, H., (2015). Progress in applications of liposomes in food systems. In: *Microencapsulation and Microspheres for Food Applications* (pp. 151–170). Academic Press.

38. Sebaaly, C., Jraij, A., Fessi, H., Charcosset, C., & Greige-Gerges, H., (2015). Preparation and characterization of clove essential oil-loaded liposomes. *Food Chem., 178,* 52–62.

39. Sherry, M., Charcosset, C., Fessi, H., & Greige-Gerges, H., (2013). Essential oils encapsulated in liposomes: A review. *J. Liposome Res., 23,* 268–275.

40. Arora, D., & Jaglan, S., (2016). Nanocarriers based delivery of nutraceuticals for cancer prevention and treatment: A review of recent research developments. *Trends Food Sci Technol., 54,* 114–126.

41. Thangapazham, R. L., Puri, A., Tele, S., Blumenthal, R., & Maheshwari, R. K., (2008). Evaluation of a nanotechnology-based carrier for delivery of curcumin in prostate cancer cells. *Int. J. Oncol., 32,* 1119–1123.

42. Moses, J. A., Dutta, S., & Anandharamakrishnan, C., (2019). Encapsulation of nutraceutical ingredients in liposomes and their potential for cancer treatment. *Nutrition and Cancer,* 1–16.

43. Taylor, T. M., Weiss, J., Davidson, P. M., & Bruce, B. D., (2005). Liposomal nanocapsules in food science and agriculture. *Crit. Reve. Food Sci. Nutri., 45,* 587–605.

44. Lee, S. C., Yuk, H. G., Lee, D. H., Lee, K. E., Ludescher, Y. I., & Ludescher, R. D., (2002). Stabilization of retinol through incorporation into liposomes. *J. Biochem. Mol. Biol., 35,* 358–363.

45. Da Silva, M. P., Daroit, D. J., & Brandelli, A., (2010). Food applications of liposome-encapsulated antimicrobial peptides. *Trends Food Sci. Technol., 21,* 284–292.

46. Tuncer, D. I., Gümüşe, B., Değim, Z., Özçelikay, T., Tay, A., & Güner, S., (2006). Oral administration of liposomal insulin. *J. Nanosci. Nanotechnol., 6*(9, 10), 2945–2949.

47. Babazadeh, A., Ghanbarzadeh, B., & Hamishehkar, H., (2017). Phosphatidylcholine-rutin complex as a potential nanocarrier for food applications. *J. Funct. Foods, 33,* 134–141.

48. Bochicchio, S., Barba, A. A., Grassi, G., & Lamberti, G., (2016). Vitamin delivery: Carriers based on nanoliposomes produced via ultrasonic irradiation. *LWT Food Sci. Technol., 69,* 9–16.

49. Lasic D. D., (1998). Novel applications of liposomes. *Trends in Biotechnology, 16*(7), 307–321.

50. Moghimi, S. M., & Patel, H. M., (1993). Current progress and future prospects of liposomes in dermal drug delivery. *Journal of Microencapsulation, 10*(2), 155–162.

51. Noble, G.T., Stefanick, J.F., Ashley, J. D., Kiziltepe, T., Bilgicer, B. Ligand-targeted liposome design: challenges and fundamental considerations, Trends in Biotechnology, 32(1). 2014, 32-45.

CHAPTER 4

Microencapsulation Liposomal Technologies in Bioactive Functional Foods and Nutraceuticals

SHUQIN XIA,[1] CHEN TAN,[2,3] XINGWEI WANG,[1] and CHUNLI FAN[1]

[1]*School of Food Science and Technology, Collaborative Innovation Center of Food Safety and Quality Control in Jiangsu Province, Jiangnan University, Lihu Road 1800, Wuxi, Jiangsu – 214122, People's Republic of China*

[2]*School of Food and Health, Beijing Technology and Business University, Beijing – 100048, China*

[3]*Department of Food Science, Cornell University, Stocking Hall, Ithaca NY – 14853, USA*

ABSTRACT

Liposome is a very promising microencapsulation technology for the bioactive functional foods and nutraceuticals, which has the advantages of protecting sensitive bioactive ingredients, improving the solubility, enhancing the bioavailability, and endowing the sustained-release property. The chapter includes a discussion on emerging preparation methods to reduce or even completely avoid the use of organic solvents in the liposomal microencapsulation. According to the difference of hydrophobicity and hydrophilicity of bioactive core material, special emphasis is given to recent progress in the application of liposomal microencapsulation. It also highlights the modulation strategy for stability and bioavailability of liposomes.

4.1   INTRODUCTION

Considering the global health problems, it is of great significance to use bioactive functional foods and nutraceuticals to effectively prevent disease and promote health. However, some bioactive ingredients and nutrients are prone to oxidation and degradation due to exposure to environmental factors such as oxygen, heat, light, and extreme pH values. Microencapsulation technology can be employed to protect the chemical structure and function of food material and nutraceuticals and to increase the bioavailability and shelf life of the product. One of the most applied encapsulations and controlled release technologies is liposomes with the particle size ranging from tens of nanometers to microns, which has been widely concerned in the fields of medicines, cosmetics, nutraceuticals, and functional foods [1, 2].

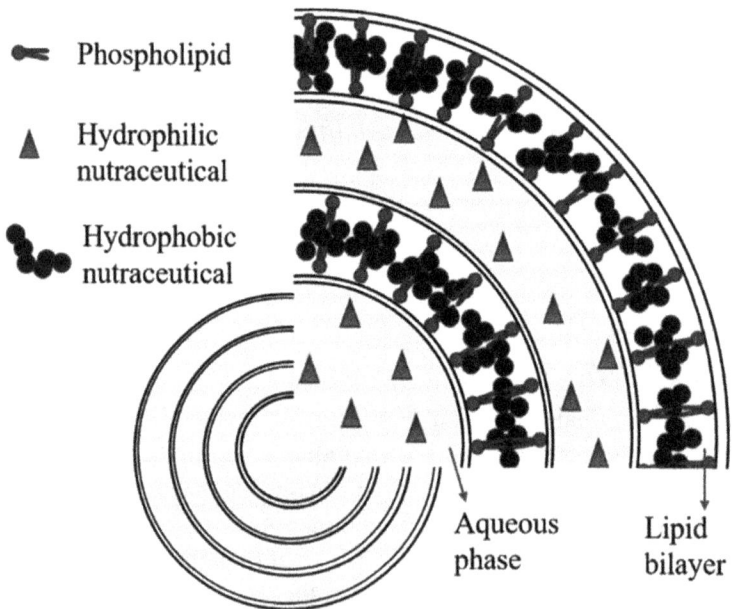

FIGURE 4.1   Cross-section schematic structure of multilamellar liposomes encapsulating hydrophilic and hydrophobic nutraceuticals.

The liposome is a very promising microencapsulation technology for the bioactive functional foods and nutraceuticals, which has the advantages of protecting sensitive bioactive ingredients, improving the solubility, enhancing the bioavailability, and endowing the sustained-release property [3, 4]. In particular, the hydrophobic and hydrophilic material might be encapsulated in the bilayer delivery system at the same time, providing a synergistic effect (Figure 4.1). A lot of researches have shown that liposomes can be used to encapsulate antioxidants [5], vitamins [6], polyphenols [7], antimicrobials [8], fatty acid [9], minerals [10], and enzymes [11], etc. In recent years, various reviews and book chapters on the sustainable preparation and application of liposomes in food have been published [1–4, 12–15].

This chapter reviews various aspects of microencapsulation liposomal technologies in bioactive functional foods and nutraceuticals, including the physicochemical properties, generally employed preparation methods, the application in bioactive ingredients and nutraceuticals, stability modulation strategy and bioavailability assessment.

4.2 PREPARATION METHODS OF LIPOSOMES

In the laboratory, liposomes can be prepared by a variety of methods such as thin-film hydration (TFH), reverse phase evaporation, freeze-thawing, and calcium fusion, etc., and the appropriate preparation method is generally selected according to the properties of the embedded core material. Traditionally, TFH (the Bangham method) is used to encapsulate hydrophobic components, and reverse-phase evaporation vesicles (REV) is employed to encapsulate hydrophilic components. Membrane extrusion, sonication, and homogenization are then used to narrow the size distribution of the final liposomes after TFH and REV. However, the conventional methods for liposome preparation are restricted due to the concern of organic solvent toxicity, which is not suitable for large-scale preparation. In addition, narrow size distribution, high encapsulation efficiency (EE), and good storage stability are important indexes for the selection of preparation parameters. In recent years, several new techniques have been attempted to reduce or even completely avoid the use of organic solvents in the liposomal microencapsulation, including microfluidics, ethanol injection, heating method (HM), rapid expansion of supercritical solutions (RESS), supercritical reverse phase evaporation (SCRPE), and several dense gas processes [4, 13].

Ethanol injection method is suitable for mass production in liposome microencapsulation of active functional ingredients and nutraceuticals. Ethanol injection method can not only avoid the use of toxic solvents (such as chloroform, methanol), expensive detergents (cholate) and other controversial raw materials but also can control the particle size, layer number, and stability of liposomes, which is easy to operate and has good reproducibility. Finally, ethanol can be removed by ultrafiltration, dialysis, reverse osmosis, or vacuum evaporation to meet the requirements of preparing liposomes.

Based on the principle of ethanol injection method, a kind of lipid preparation technology which can be amplified-cross flow injection was invented [16–18]. The results showed that the size of liposomes prepared by this technology was controllable and monodisperse, with certain physical and chemical stability, good reproducibility, and the whole process can ensure aseptic and pyrogen-free. This technique has been successfully used for embedding recombinant human Cu/Zn superoxide dismutase (SOD) [17].

Critical or supercritical fluid has the characteristics of high density, low viscosity, and high diffusion coefficient (similar to the diffusion rate of gas), which can be used to replace ethanol for dissolving the raw materials (phospholipid, Chol, and fat-soluble active components). Dense gas techniques using supercritical carbon dioxide (SC-CO_2) as the phospholipid-dissolving agent can provide a green solution for reduction or avoidance of organic solvent use. High-quality liposomes can also be prepared with supercritical fluid processes [15, 19]. The SC-CO_2 method has potential for scalable production of liposomal nanovesicles with desirable characteristics and free of organic solvents. Vitamins C and E as model hydrophilic and lipophilic compounds have been encapsulated in the integrated liposomes [20].

High pressure homogenization technology such as high-pressure homogenizer or microfluidic processor is also suitable for mass production of liposomes [21–24]. Multilayer liposomes or unhydrated lipids can be added to the homogenizer. After several cycles, the satisfactory size can be obtained. Generally speaking, the average particle size can reach 100 ~ 200 nm after one cycle, but it is related to the lipid and hydration medium, which constitute the bilayer of liposomes. This method has been successfully used in the preparation of liposomes for fish oil [21], black carrot extract [22], and anthocyanin-rich hibiscus extract [25].

After obtaining multilayer liposomes, extrusion method can also be used to obtain small monolayer nanoliposomes [26–28]. This method has many advantages: (1) compared with ultrasound and micro jet, it can avoid the damage of high energy consumption on lipid and make the product stable. (2) According to the need, we can choose different sizes of filter membranes to prepare liposomes of any size. (3) The scale of preparation is easy to scale up, and the maximum production capacity of commercial extruders can reach 50 L. (4) After repeated extrusion, the particle size of liposomes was more uniform, and the reproducibility was good. (5) In the preparation process, the sterilization can be achieved after 220 nm membrane filtration. (6) Liposomes with a high concentration of phospholipids can be prepared.

4.3 LIPOSOME ENCAPSULATION OF HYDROPHOBIC ACTIVE INGREDIENTS AND NUTRIENTS

The hydrophobic active ingredients and nutrients are commonly located in the bilayer phase of liposomes. Furthermore, the EE of hydrophobic components is usually higher than that of hydrophilic components due to the difference in the partition coefficient between the aqueous and lipid environment. One of the main advantages of nanoliposomes used in the functional food and nutraceutical industry is the fortification of water-based food with water-insoluble bioactive ingredients without adversely affecting the sensory attributes of the original product [6, 29]. In addition, improving its bioavailability and efficacy is also an attractive aspect for application. Typical water-insoluble nutrients, including carotenoids, vitamins, polyphenols, active ingredients of essential oil and their liposome composition and preparation methods are summarized in Table 4.1.

4.3.1 CAROTENOIDS

Carotenoids are naturally occurring antioxidants with eight isoprene units where the π-electrons are centrally located in the conjugated double bonds. However, this unique structure also makes carotenoid molecules prone to geometric isomerization and oxidative degradation toward oxygen, light, and heat. Additionally, the practical applications of carotenoids are limited

TABLE 4.1 Liposome Encapsulation of Hydrophobic Active Ingredients and Nutrients

Nutraceuticals	Liposome Composition	Methods	Results	References
Vitamin A	Lecithin, cholesterol	Thin layer hydration	The optimized nanoliposomes had the average size of 67 nm and vitamin A encapsulation efficiency of 15.8%	[30]
Vitamin A	Phosphatidylcholine, cholesterol	Thin-film hydration	Liposomes enhanced the in vivo bioavailability of vitamin A in mice	[31]
Vitamin D	Lecithin, sesame oil, glycerol	Homogenization, ultrasonication	Liposomes loading vitamin D can be used for yogurt fortification.	[6]
Vitamin E	Lecithin, glycerol, gamma oryzanol, PEG-400	Thin-film hydration	Gamma oryzanol led to the highest encapsulation efficiency stability of vitamin E	[32]
Vitamin E	Phosphatidylcholine, cholesterol, chitosan	Ultrasonication, surface coating	Chitosan coating greatly enhanced the stability of vitamin E liposomes during the 8 weeks storage at 4°C	[33]
Vitamin E	Soybean phosphatidylcholine, cholesterol, PEG	Thin-film hydration, surface coating	PEG-coated liposomes can effectively protect vitamin E at 4°C for 15 days storage	[34]
Vitamin E	Phosphatidylcholine, cholesterol, β-lactoglobulin	Thin-film hydration, surface coating	The incorporation of β-lactoglobulin improved the storage stability of liposomes at 4°C	[35]
Vitamin E	Soy phosphatidylcholine, stearic acid, calcium stearate	Thin-film hydration	Liposomes had a protective effect on antioxidant activity of vitamins before and after pasteurization	[36]
Vitamin E	Egg yolk phospholipid, cholesterol, Tween-80, chitosan	Ethanol injection, surface coating	Chitosan-coated nanoliposomes can be employed to protect vitamin E for extending shelf-life and enhancing thermostability.	[37]

TABLE 4.1 *(Continued)*

Nutraceuticals	Liposome Composition	Methods	Results	References
β-carotene	Marine phospholipids	Thin-film hydration	The nanoliposomes showed higher stability at 4°C for 70 days	[38]
β-carotene	Hydrogenated soy phosphatidylcholine	Proliposome hydration	The spray-dried liposome can preserve 90% of the encapsulated β-carotene for 60 days at 4°C	[39]
β-carotene	Hydrogenated soy phosphatidylcholine	Proliposome hydration	Liposomes were capable of protecting β-carotene from degradation for a period of 90 days	[40]
β-carotene	Hydrogenated soy phosphatidylcholine	Supercritical carbon dioxide, ultrasonication	The liposomes had the size of 90–150 nm, and good dispersion stability	[41]
β-carotene	Egg yolk phosphatidylcholine, cholesterol	Ethanol injection	Both the bioavailability and bioactivity of β-carotene were enhanced by liposome encapsulation	[42]
β-carotene	Hydrogenated soy phosphatidylcholine	Proliposome hydration, surface coating	Xanthan and guar gum prevented the aggregation of liposomes and enhanced the retention of β-carotene during storage	[43]
Lutein	Hydrogenated soy phosphatidylcholine	Supercritical carbon dioxide	The lutein liposome was obtained with the encapsulation efficiency of more than 90% after hydrating proliposomes	[44]
Lutein	Egg yolk phosphatidylcholine, Tween-80	Ethanol injection	The liposomes with loaded lutein remained stable during preparation, heating, storage, and surfactant dissolution	[45]
Lutein	Soy lecithin	Supercritical carbon dioxide	A higher pressure (200–300 bar) and depressurization rate (90–200 bar/min) promoted a higher encapsulation efficiency of lutein	[19]

TABLE 4.1 (Continued)

Nutraceuticals	Liposome Composition	Methods	Results	References
Lycopene	Soy lecithin, cholesterol, β-cyclodextrin	Thin-film hydration, inclusion complexation	In vivo study indicated significant cardio-protective activity of encapsulated lycopene	[46]
Astaxanthin	Soybean lecithin	Thin-film hydration	Liposomal encapsulation increased the storage stability and cellular uptake of astaxanthin	[47]
Astaxanthin	Soybean phosphatidylcholine, cholesterol	Thin-film hydration	Liposomes can release encapsulated astaxanthin in a controlled manner	[48]
Astaxanthin	Lecithin, cholesterol, chitosan, lactoferrin	Thin-film hydration, surface coating	The encapsulated astaxanthin exhibited higher in vitro bioaccessibility than the free astaxanthin	[49]
β-carotene, lycopene, lutein, canthaxanthin	Egg yolk phosphatidylcholine, Tween-80	Ethanol injection	The antioxidant activity of carotenoids after liposome encapsulation was correlated with their chemical reactivity, and incorporation efficiencies into liposomal membrane	[50]
β-carotene, lycopene, lutein, canthaxanthin	Egg yolk phosphatidylcholine, Tween-80	Ethanol injection	Lutein exhibited the strongest incorporating abilities into liposomes, followed by β-carotene, lycopene, and canthaxanthin.	[51]
β-carotene, lycopene, lutein, canthaxanthin	Egg yolk phosphatidylcholine, Tween-80	Ethanol injection	The highest bioaccessibility of carotenoids after encapsulated in liposomes was found for lutein, followed by β-carotene, lycopene, and canthaxanthin	[52]

TABLE 4.1 (*Continued*)

Nutraceuticals	Liposome Composition	Methods	Results	References
β-carotene, lycopene, lutein, canthaxanthin	Egg yolk phosphatidylcholine, Tween-80, chitosan	Ethanol injection, surface coating	The adsorption of chitosan chains increased the condensation of liposomal structure by rigidification, providing the liposomal membrane and loaded carotenoids with high stability, dispersibility, and controlled release	[53]
β-carotene, lycopene, lutein, canthaxanthin	Egg yolk phosphatidylcholine, Tween-80	Ethanol injection	Carotenoids modulated the dynamics, structure, and hydrophobicity of liposomal membrane, highly depending on their molecular structures and incorporated concentration	[5]
CoQ_{10}	Phosphatidylcholine, cholesterol	Supercritical carbon dioxide	The entrapment efficiency and loading content of CoQ_{10} reached 82.28% and 8.92%, respectively.	[54]
CoQ_{10}	Soy lecithin	Supercritical carbon dioxide	Encapsulation efficiencies between 87.6% and 99.7% were achieved and stable particles with zeta potential of −70 mV were formed	[55]
CoQ_{10}	Phosphatidylcholine, cholesterol	Supercritical carbon dioxide	The encapsulation efficiency of the CoQ_{10} liposome was up to 90% when the enduring pressure was at 200–300 bar.	[56]
CoQ_{10}	Soy phospholipids, cholesterol	Thin-film hydration	The presence of Tween-80 can release encapsulated CoQ_{10} from liposomal suspension	[57]
CoQ_{10}	Soy phospholipids, cholesterol, Tween-80	Ethanol injection	The liposomes had encapsulation efficiency greater than 95% with a retention ratio higher than 90% after storage at 4°C for 90 days	[58]
CoQ_{10}	Soy phospholipids, cholesterol, Tween-80	Ethanol injection	Incorporation of CoQ_{10} affected the liposomal dynamic and structure, depending on the CoQ_{10} loading concentration	[59]

TABLE 4.1 *(Continued)*

Nutraceuticals	Liposome Composition	Methods	Results	References
CoQ_{10}	Soy phosphatidylcholine, cholesterol, Tween-80, chitosan	Ethanol injection, surface coating	Chitosan coated liposomes showed an excellent hydroxyl radical scavenging activity, sustained release, and *in vitro* penetration in rabbit skin	[60]
CoQ_{10}	Soy phosphatidylcholine, cholesterol, Tween-80, chitosan	Ethanol injection, surface coating	Chitosan coated liposomes exhibited strong antioxidant effect and antimicrobial activity	[61]
CoQ_{10}	Dipalmitoylphosphatidylcholine, cholesterol, chitosan	Ethanol injection, surface coating	Chitosan-coated liposomes greatly increase the bioavailability of CoQ_{10} by enhancing and prolonging the oral exposure of CoQ_{10}.	[62]
Curcumin	Phosphatidylcholine	pH-driven	Curcumin liposomes were stable during storage and showed high bioaccessibility	[63]
Curcumin	Bovine milk phospholipids, krill phospholipids	Thin-film hydration	Curcumin-loaded liposomes prepared from krill phospholipids had superior bioavailability	[64]
Curcumin	Soybean lecithin, β-sitosterol	Thin-film hydration	Addition of 20–30 mol% of β-sitosterol improved the curcumin encapsulation efficiency and bioavailability in liposomes	[65]
Curcumin	Soybean lecithin, cholesterol, chitosan	Thin-film hydration, surface coating	High molecular weight chitosan at relatively high concentrations can improve the stability and sustained release of curcumin liposomes *in vitro*	[66]
Curcumin	Soybean phospholipid, hydrogenated soybean phospholipid, cholesterol	Thin-film hydration	Changing the ratios of various phospholipids can modify the liposome particle size, encapsulation efficiency, and stabilizing effects on curcumin	[67]
Curcumin	Phosphatidylcholine, cholesterol, chitosan	Ethanol injection, surface coating	Chitosan was used to protect the structure stability of bilayer and control the sustained release of encapsulated curcumin	[68]

TABLE 4.1 *(Continued)*

Nutraceuticals	Liposome Composition	Methods	Results	References
Curcumin	Phospholipid S75, Carboxymethyl chitosan	Thin-film hydration, surface coating	The hybrid liposomes significantly increased the storage stability, sustained release, and bioavailability of curcumin	[69]
Curcumin	Phospholipid S100, cholesterol, Tween-80	Thin-film hydration	Nanoliposomes improved the stability of curcumin against alkaline pH and metal ions	[70]
Curcumin, Vitamin D$_3$	Phospholipon 90 H, xanthan gum, guar gum	Proliposome hydration, surface coating	Curcumin and Vitamin D$_3$ were loaded into the same vesicles, with retention efficiencies ranging from 89.6 to 93.3% and from 88.3 to 91.3%, respectively, after 42 days of storage	[71]
Curcumin, resveratrol	Egg yolk phosphatidylcholine, Tween-80	Thin-film hydration	Liposomes loading both curcumin and resveratrol displayed a higher ability during preparation, storage, heating, and surfactant shock than those loaded with individual polyphenol	[7]
Curcumin	Egg yolk phosphatidylcholine, cholesterol, guar gum	Thin-film hydration, surface coating	Guar coating can decrease the membrane fluidity and enhance the lateral packing of lipid, thus protecting curcumin from the damage caused by oxidation or heat	[72]
Curcumin	Phosphatidylcholine, whey protein	Ethanol injection, electrospraying	Encapsulation of curcumin within the protein-coated liposomes significantly increased the bioaccessibility of curcumin (1.7-fold) compared to the free curcumin	[73]
curcumin	Soybean phosphatidylcholine, cholesterol, thiolated chitosan	Thin-film hydration, surface coating	Thiolated chitosan-coated liposomes showed slower *in vitro* release of curcumin than bare liposomes at pH 5.5 and pH 7.4, and the higher retention of curcumin would be remained for the following uptake of cells	[74]

TABLE 4.1　(Continued)

Nutraceuticals	Liposome Composition	Methods	Results	References
Curcumin	Egg yolk lecithin, phosphatidic acid, chitosan	Thin-film hydration, surface coating	Chitosan coating improved the intestinal absorption of curcumin.	[75]
Curcumin	Lecithin, carboxymethyl chitosan, quaternary ammonium chitosan	Thin-film hydration, surface coating	The oral absolute bioavailability of chitosan-coated curcumin liposomes was around 6 folds higher than curcumin liposomes	[76]
Quercetin	Dipalmitoyl lecithin, cholesterol	Thin film hydration	The incorporation of quercetin liposomes into cellulose film can control the release of quercetin for improving food shelf life	[77]
Quercetin	Soy lecithin, chitosan	Thin film hydration	The storage stability and antioxidant activity were improved after nanoliposome encapsulation compared with native quercetin	[78]
Quercetin	Phospholipids whey protein isolate	Thin-film hydration	Protein coating can mask the quercetin bitterness in a dairy drink	[79]
Quercetin, luteolin, kaempferol	Egg yolk phosphatidylcholine, Tween-80	Thin-film hydration	Liposomes displayed a stronger retaining ability to quercetin and luteolin than kaempferol during preparation, storage, heating, and pH shock	[80]

by the water insolubility, poor dispersion in aqueous food formulation, and low oral bioavailability. The liposomal delivery systems have been demonstrated to effectively protect them by providing the physical-chemical barriers, and simultaneously make them water-dispersible. One of the most intensively studied carotenoids is β-carotene. Through conventional TFH or ethanol injection method, β-carotene loaded liposomes can be produced with small particle size, homogeneous dispersion, high EE, and high storage stability [39, 40]. The types of phospholipids affect the encapsulation performance of liposomes to β-carotene. It was found that marine phospholipid liposomes exhibited higher EE and storage stability at 4°C for 70 days than egg phosphatidylcholine (PC) liposomes [38]. Additionally, co-encapsulation of β-carotene and coix seed oil into liposomes has also conferred β-carotene with higher bioavailability, anticancer, and antioxidant activity, compared to liposome containing single bioactive [42]. This is interpreted by the synergistic antioxidant effects of β-carotene and coix seed oil, as well as their orientation within the lipid layer.

In addition to β-carotene, the application of liposomes for carotenoids have also been reported in lycopene, lutein, astaxanthin, and canthaxanthin. For example, different concentrations of lutein have been loaded into the liposomes composed of egg yolk PC and Tween-80 by ethanol injection method [45]. When liposomes were loaded with 1 and 2 wt.% lutein, they possessed high stability during preparation, heating, storage, and surfactant dissolution. Further increasing loading content to 5 and 10 wt.% would decrease the stability of lutein liposomes. This phenomenon is probably due to the fact that at low concentrations, lutein molecules preferentially locate in the opposite polar membrane regions vertically to the membrane plane, which in turn confer the liposomal membrane with a high order of lipid hydrocarbon chain and rigidity. However, at high concentrations, a certain fraction of lutein molecules would adopt a parallel orientation due to its rotational terminal ε-ring. In the case of lycopene, a combination encapsulation of liposomes and β-cyclodextrin has been developed [46]. The optimized β-cyclodextrin-liposomes had a lycopene EE of 78.9% and an average particle size of 255.1 nm. These composite systems can prolong the release of lycopene at pH 6.8 in 12 h and exert significant cardio-protective activity.

Our groups have carried out a series of studies to understand the influence of carotenoid types (i.e., lutein, β-carotene, lycopene, and canthaxanthin) on the liposome EE [51], antioxidant activity [50], and bioaccessibility [52]. Results have shown that lutein displayed the strongest incorporating ability into the liposomes, followed by β-carotene, lycopene, and canthaxanthin. A similar trend was also observed in their antioxidant activities against lipid peroxidation and storage stability [50]. In another study, different carotenoid-loaded liposomes showed different trends in the changes of morphology, size distribution, and lipid digestibility when passing through the simulated gastrointestinal tract [52]. *In vitro* release behaviors have shown that lutein and β-carotene were hardly released in a simulated gastric fluid while displaying a slow and sustained release in a simulated intestinal fluid. However, lycopene, and canthaxanthin underwent fast and considerable release in gastrointestinal fluid. This observation indicates that lutein and β-carotene that either located in the phospholipid surface or well incorporated into liposomes are more easily transformed into micelle phase during *in vitro* digestion, and consequently can improve potential bioaccessibility.

It is believed that there is a mutual relationship between carotenoid modulation in the liposomal membrane and the delivery efficiency of liposomes. Carotenoids are able to modulate the dynamics, structure, and hydrophobicity of the liposomal membrane, highly depending on their molecular structures [5]. Generally, β-carotene, lutein, and canthaxanthin can adopt the vertical orientation with respect to the membrane plane (Figure 4.2). This orientation could rigidify the liposomal membrane and stabilize liposome particles against aggregation and fusion. Due to the high hydrophobicity, lycopene locates deeply inside the hydrophobic core, thus exerting a small influence on liposome properties (Figure 4.2). On the other hand, these modulating effects are also affected by the incorporated concentration of carotenoids. The incorporation of β-carotene and lutein can enhance the lipid lateral packing within the concentration range of 0.25%–1.25%, while beyond this range this effect became weaken. In the case of canthaxanthin, its incorporation could perturb the lateral packing of lipid at high concentrations. The modulation induced by carotenoid incorporation directly affects the stabilization of both liposomes and encapsulated carotenoids.

The protective effect of liposomes on carotenoids can be further improved by modifying the membrane surface via forming the polymeric layers. Chitosan is a linear natural polysaccharide widely applied in the biomedical and biotechnological fields due to its biocompatibility, bioadhesion, and biodegradability [81]. The dense mesh covering of chitosan can improve the structural properties of liposomes and nutraceutical delivery efficiency [82, 83]. Fluorescence polarization analysis and Raman spectroscopy have evidenced that the chitosan coating on the liposomal membrane via electrostatic and hydrophobic interaction can restrict the motion freedom of lipid molecules and enhance their ordering at the polar headgroup region and hydrophobic core of the membrane [84, 85]. As a result, the chitosan-coated carotenoid liposomes exhibited high stability against heating, gastrointestinal stress, and centrifugal sedimentation [53].

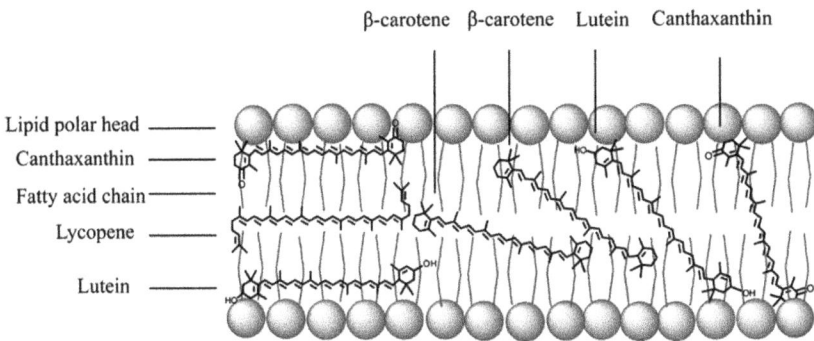

FIGURE 4.2 Schematic representation of the probable patterns of localization and orientation of carotenoid pigments in the egg yolk phosphatidylcholine membrane. *Source:* Reprinted with permission from: Ref. [50]. Copyright (2013), American Chemistry Society.

4.3.2 LIPOSOLUBLE VITAMINS AND VITAMIN-LIKE NUTRIENTS

Liposoluble vitamins include vitamin A, D, E, and K, which are naturally present in food and used in functional foods. These vitamins are sensitive to different factors throughout the process and storage, such as oxygen, temperature, light, pH, moisture content, water activity, degradative enzymes, and metal trace elements [86]. The utilization of liposomes for encapsulation of vitamins is a novel technique to improve their solubility, stability against harmful conditions, and the functional properties [87].

The stability, absorption, and bioavailability of vitamin E can be increased when it is encapsulated in liposomes [88]. With respect to the liposomal delivery systems for vitamin E, their efficient cytotoxic activity *in vitro* and anti-cancer effect *in vivo* on various experimental animal models have also been reviewed [89]. For example, the soy PC liposomes were prepared for vitamin E by the dehydration-rehydration method. Vitamin E was incorporated almost entirely in the liposomal bilayer with EE of 99.73%. The presence of stearic acid and calcium stearate (CaS) in the formulation further helps to stabilize the lipid bilayer by increasing rigidity. As a result, these liposomes had a strong protective effect on the antioxidant activity of vitamin E before and after pasteurization [36]. The high stability of vitamin E is attributed to its deep location in the bilayer hydrophobic core. It is also noted that the incorporation of vitamin E affects the properties of liposomes. In general, vitamin E adopts vertical orientation in the PC membrane and is bound to phospholipid via hydrogen bonds and hydrophobic interactions [90]. This orientation is believed to restrict the molecular motion of both lipids head and hydrophobic core, and simultaneously give protection to unsaturated phospholipids toward oxidation. Thus, the protection from lipid peroxidation owing to the vitamin E antioxidation in turn retains the incorporated vitamin E more effectively.

To further stabilize vitamin E nanoliposomes, different stabilizers have been employed, including gamma oryzanol, lauric acid, and PEG 400 [32]. Results have shown that the addition of hydrophobic stabilizers into nano-liposomes resulted in an increase in the particle size along with a decrease in the polydispersity index (PDI), while the effect of hydrophilic stabilizers on both particle size and their distribution was not significant. Besides, gamma oryzanol displayed the highest EE and stability of nanoliposomes during 40 days of storage at 4°C. In a similar study, the incorporation of β-lactoglobulin improved the EE of vitamin E in liposomes as well as the storage stability at 4°C for 72 h [35]. When the vitamin E liposomes were coated with chitosan, the EE of vitamin E exceeded 99% with the loading content over 27% [33]. The stability of chitosan-coated vitamin E lipo-somes during the 8 weeks storage is over 90% under 4°C. Chitosan can be further cross-linked with sodium tripolyphosphate [37]. By this means, the stability of vitamin E liposomes has been strongly enhanced against temperature, including suppressing particle aggregation and change of membrane fluidity. After heating at 65°C for 30 min and 80°C for 16 s, the retention efficiency of vitamin E in coated liposomes remained around 92% and 97%, respectively, in contrast to the great leakage of vitamin E from uncoated liposomes.

Encapsulation of vitamin A has been performed in lecithin-Chol nano-liposomes [30]. The increase of lecithin concentration improved the EE as well as the stability of nanoliposomes. However, the addition of Chol had a negative effect on the EE of vitamin A. Recently, vitamin A has been efficiently entrapped and delivered in a phospholipid-sterol-protein membrane resembling system [31]. In this study, β-lactoglobulin was incorporated into the phosphatidylcholine-Chol liposomes by a dehydra-tion-rehydration method. The stability and bioavailability of vitamin A were significantly enhanced by the incorporation of protein.

Recently, the group of Jafari has explored a functional yogurt powder fortified with vitamin D liposomes [6]. The lecithin liposomes containing vitamin D of around 120 nm were prepared, followed by spray-drying with various wall matrixes, including milk protein concentrate, modified starch (MS), gum Arabic, and maltodextrin. The best formulation was the maltodextrin matrix (22%) with the spherical shape and even distribution of vitamin D liposomes. Vitamin D_3 has also been encapsulated with curcumin (CUR) together into the multilamellar liposomes [71]. With the help of xanthan and guar gums, the retention efficiencies of vitamin D_3 ranged from 88.3 to 91.3% after 42 days of storage. In another study, the nanoliposomes loaded with three different vitamins (vitamin B_{12}, vitamin E, and vitamin D_2) have been prepared via a TFH method followed by sonication [91]. The highest EE was found for vitamin E (76%), followed by vitamin D_2 (57%) and vitamin B_{12} (56%). The stability assays also revealed that nanoliposomes remained stable after 10 days of incubation at simulated extracellular environment conditions.

Coenzyme Q_{10} (CoQ_{10}), also known as ubiquinone, is a vitamin-like nutrient that functions as an antioxidant and membrane stabilizer. Our group has prepared a food grade nanoliposomes for the delivery of CoQ_{10} by a combination of ethanol injection and sonication methods [58]. The nanoliposomes with around 68 nm have shown excellent EE higher than 95% and retention efficiency higher than 90% after storage at 4°C in the dark for 90 days. The CoQ_{10} nanoliposomes can be also produced in a pilot-scale with ethanol injection and ultra-high-pressure homogenization (HPM), as illustrated in Figure 4.3. Another method for CoQ_{10} liposomes in the food industry is supercritical CO_2 anti-solvent technology that does not require the use of organic solvents, buffers, or surfactants. The particle size, polydispersity index, EE of CoQ_{10} liposomes can be modified by controlling the pressure, temperature, and pressurization rate of CO_2 [54].

At an optimized condition (30 MPa, 40°C, and 6 MPa/min), the EE of CoQ_{10} were achieved between 87.6% and 99.7% [55]. In another study, the particle size of CoQ_{10} liposomes was only 20–40 nm when the pressure was at 200–300 bar [56]. The advantage of supercritical CO_2 is that the CO_2 molecules can form the membranes between the CoQ_{10} liposome particles, preventing the agglomeration of liposome with each other.

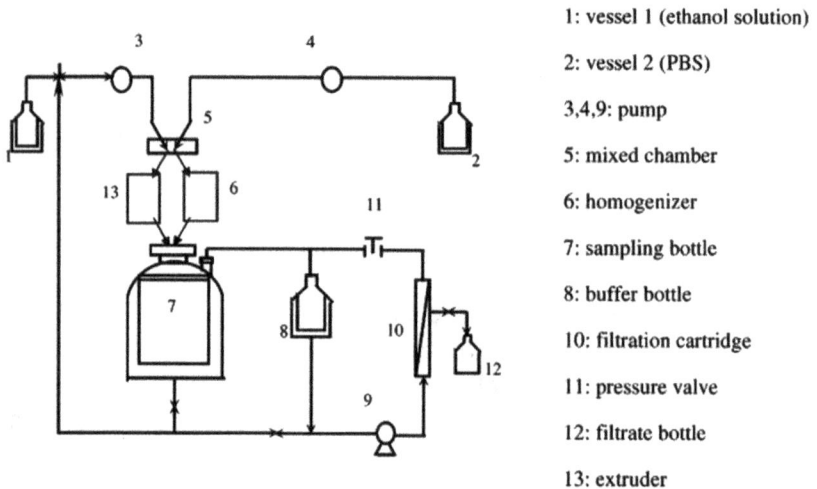

1: vessel 1 (ethanol solution)

2: vessel 2 (PBS)

3,4,9: pump

5: mixed chamber

6: homogenizer

7: sampling bottle

8: buffer bottle

10: filtration cartridge

11: pressure valve

12: filtrate bottle

13: extruder

FIGURE 4.3 Schematic sketch of the pilot plant for the production of liposomes. The buffer solution was pumped from vessel 2 to mixed chamber 5, where the ethanol solution was pumped from vessel 1 and injected into chamber 5 and immediately diluted. The liposome suspensions were processed with a high-pressure homogenizer 6. Ethanol was removed by cross-flow ultrafiltration 7. The samples were then transferred into the sterilized container by extruder 13.
Source: Reprinted with permission from: Ref. [58]. Copyright (2006), American Chemistry Society.

The incorporation of CoQ_{10} can modulate the physical and chemical stability of liposomes. A further study has revealed the interactions and location of CoQ_{10} in the food grade liposomes [59]. Results have shown that CoQ_{10} adjusted the fluidity of lipid bilayer and gave rise to a more ordered arrangement of lipid molecules in liposomes. Additionally, part of the CoQ_{10} molecules might locate parallel to the lipid hydrocarbon chains, and the methyl substituents of quinone rings played the key role. This location endows the CoQ_{10} encapsulated in liposomes with high antioxidant activity against lipid peroxidation induced by Fe(III)/ascorbate.

Chitosan coating further reinforces the stability and intestinal absorption of CoQ_{10} liposomes. It has shown that the chitosan-coated liposomes loading CoQ_{10} exhibited a high EE of 97% and long-term stability at room temperature for 60 days [60]. These liposomes were biocompatible with nonsignificant cytotoxicity and simultaneously exhibited strong antioxidant activity by clearing reactive oxygen species (ROS) from H_2O_2 [61]. Another study further demonstrated that the cellular uptake of CoQ_{10} in Caco-2 cells in chitosan-coated liposomes was about 30-fold greater than the untreated powder formulation. Based on the oral pharmacokinetic studies, the coating of chitosan enhanced the systemic exposure of CoQ_{10} by 3.4 folds as compared to the untreated powder and displayed the extended CoQ_{10} release profile for up to 24 h in rats [62]. On the other hand, the N-trimethyl chitosan chloride (TMC) has been used to stabilize the CoQ_{10} liposomes over a wide pH range [92]. These results confirm that chitosan-coated liposomes are effective oral delivery platforms to improve the stability and oral bioavailability of poorly absorbable CoQ_{10}.

4.3.3 POLYPHENOLS

CUR is a natural hydrophobic polyphenol that has numerous health benefits, such as antioxidant, antimicrobial, anti-inflammatory, and antiproliferative properties. The major challenges for CUR applications are the low solubility in water at acidic or neutral pH, easy degradation, and low oral bioavailability [93, 94]. A number of liposome-based delivery systems have been developed to improve the solubility, stability, and bioavailability of CUR. For example, 57.1% of CUR has been encapsulated in the nanoliposomes consisting of phospholipid, Chol, Tween-80 [70]. These CUR liposomes also exhibited good stability against alkaline pH and metal ions than the free CUR. The composition of phospholipids affects the liposomal encapsulation and stabilization of CUR. Liposomes composed of hydrogenated soybean phospholipid and soybean phospholipids at 1:1 have shown higher CUR EE and better storage stability than those with soybean phospholipids alone [66]. In another study, CUR liposomes prepared from bovine milk phospholipids have demonstrated better stability than liposomes from krill phospholipids under harsh storage conditions (alkaline conditions, oxygen, high temperature, and relative humidity). To the contrast, CUR -loaded liposomes prepared from krill phospholipids had superior bioavailability compared to that prepared from bovine milk phospholipids [64].

Chitosan can stabilize the CUR liposomes against degradation and control the sustained release rate of encapsulated CUR [68]. Due to the mucoadhesive properties of chitosan, chitosan coating enables the enhanced adsorption of CUR liposomes [75]. The chitosan derivative, such as thiolated chitosan has also been used to coat the CUR liposomes and given rise to a slower *in vitro* release at pH 5.5 and 7.4 than bare CUR liposomes [74]. Furthermore, the carboxymethyl and quaternary ammonium chitosan are applied to coat CUR liposomes via a layer-by-layer deposition technique. Resultantly, the chitosan derivatives coated liposomes displayed around 6 folds higher oral bioavailability of CUR than bare CUR liposomes [76]. In addition to chitosan, the cationic guar gum is introduced to modify the CUR liposomes and protect them from the damage caused by oxidation and heat [72]. Fluorescence, Raman spectra, and X-ray diffraction analysis have demonstrated that the cationic guar gum rigidified the liposomal membrane and enhanced the lateral packing of lipids.

Co-encapsulation of CUR and other nutraceuticals could exert synergistic stabilizing effects on liposomes. In a recent study, the incorporation of β-sitosterol of 20–33 mol% has improved the EE and sustained release of CUR in liposomes [65]. This is because sitosterol increased the packing density of lipids and densified the hydrophobic environment inside liposomes. In a similar manner, liposomes loading both CUR and resveratrol (RESV) displayed high stability during preparation, storage, heating, and surfactant shock, compared to those loaded with individual polyphenol [7]. These stabilizing effects are attributed to the fact that CUR and RESV can locate in the hydrophobic core and polar headgroup region, respectively, thus rigidifying the entire lipid bilayers.

Quercetin (QUE) is another popular nutritional supplement due to the numerous biological activities. It is the most abundant flavonoid naturally occurring in a variety of fruits and vegetables, including onion, apple, broccoli, grape, tea, and red wine. However, its low solubility, bitter taste, poor bioavailability, and low chemical stability are the main drawbacks that limit its use in the food industry [95]. Liposomes composed of dipalmitoyl lecithin and Chol have been prepared for the delivery of

QUE with EE of 88.9% and retained them during 21 days of storage [77]. These liposomes were then incorporated into the carboxymethyl cellulose edible films and controlled the release of QUE with antioxidant activity for improving the food shelf life. A recent study has investigated the protective effects of liposomes on a variety of flavonoids including QUE, luteolin, and kaempferol. Their interaction mechanisms were also discussed [80]. Results have shown that liposomes loading QUE exhibited the highest stability and antioxidant capacity, followed by luteolin and kaempferol. The stability and antioxidant assays revealed a mutually protective relationship between incorporated flavonoids and liposomes. That is, these flavonoids can modulate the dynamics and structures of liposomal membranes, highly depending on their molecular structures and incorporation concentration.

QUE liposomes can be also coated with chitosan using a facile electrostatic deposition method. After chitosan coating, the storage stability and antioxidant activity of QUE liposomes have been significantly improved [78]. Meanwhile, chitosan-coated QUE liposomes did not affect the cytotoxicity of QUE on HepG2 cells. Another coating material for QUE liposomes is whey protein (WP) isolate that can enhance the stability of QUE liposomes during storage and at stomach conditions [79]. Whey coating also facilitates to spray-dry and re-dissolve the QUE liposomes in a whey permeate matrix. No bitter taste of QUE and oil mouthfeel is perceived in the WP coated liposomes. Thus, this whey-liposome formulation can be developed as a functional drink based on sweat whey permeate and enriched in QUE.

4.3.4 ACTIVE INGREDIENT OF ESSENTIAL OIL

Essential oil is a kind of important natural spices, usually in the leaf of aromatic plants, flowers, roots, seeds, buds, fruits, such as tissue or secretions as raw material, adopting squeezing, cold mill, solvent extraction, steam distillation, supercritical fluid extraction and other methods, the extract obtained by terpene, aliphatic, alicyclic consisting of a mixture [96, 97]. Essential oils (EOs) have a unique aroma, therefore, in the food industry, it could directly be added to the food as a flavoring agent. Besides, EOs have strong antimicrobial and antioxidant potential, which can be used as natural preservatives to meet consumer needs for safe, healthy, and nutritious foods. However, EOs are often poor solubility in water, highly volatile, and sensitive to environmental conditions such as oxygen, light, and temperature, which seriously affects their functions and applications [98].

In order to improve the application prospect of EOs in food, liposomes prepared from lipids inedible materials such as soybean and egg are the most commonly used delivery system. It has been found that liposomes with lower water-soluble essential oil components have higher embedding efficiency, and hydroxylated essential oil components have better embedding efficiency than non-hydroxylated components in liposomes. In addition, the particle size of the liposome, the load rate of the essential oil, the position of the liposome bilayer, and the incorporation rate of Chol in the liposome are the key parameters affecting the release of the essential oil [99].

The liposome encapsulation could not only improve the stability and bioavailability of EOs and their active components but also affect their ability to inhibit microorganisms to varying degrees. The antibacterial properties of liposome encapsulating essential oil and its main components are shown in Table 4.2.

Although a large number of studies have concluded that liposomes embedded with EOs have a strong antibacterial ability, some studies usually take blank liposomes or liposome-modified materials as the control of antibacterial experiments, and the results can only indicate that liposome encapsulation improves the stability and the long-term antibacterial properties of EOs. On the other hand, the researches using free essential oil as a control to investigate the effect of liposome encapsulation on the antibacterial properties of EOs reported two opposite conclusions. Some studies have found that the bacteriostatic performance of essential oil-loaded liposomes is higher than that of the free EOs [100, 101]. On the contrary, some studies have found that although liposome encapsulation achieved the slow release of essential oil and long-term antibacterial effect, it significantly reduced its rapidly antibacterial effect [8, 102]. The difference of these results may be related to the preparation method and wall material of liposomes, and the essential oil composition [102].

In recent years, studies have found that bacteria such as *Staphylococcus aureus* and *Listeria monocytogenes* could produce pore-forming toxins and form large pores in the bilayer of liposomes, thereby triggering the release of internal essential oil, which not only improves the bacteriostatic efficiency of liposomes but also achieves the targeting and controlled release of essential oil [103–106]. However, although this discovery improves the rapid bacteriostatic performance of liposomes, it is not difficult to see that the application scope is limited by the strains. Moreover, these studies did not use free EOs as a direct control to reflect the difference in antibacterial activity between liposomes and free EOs. However, the experimental data showed that liposome encapsulation still inhibited the antibacterial activity of essential oil, because the incubation time after encapsulation was extended from several hours to 24 hours or even several days.

In order to improve the characteristics and expand the applicability, the surface modification of liposomes with polymers is also a research hotspot in recent years. In addition, due to its intrinsic antimicrobial activity, chitosan coatings could be used to deliver lipophilic bioactive compounds and have a potential synergistic bactericidal effect with incorporated essential oil [107–110]. Therefore, it may be considered to use chitosan and other materials with bacteriostatic properties to decorate the surface of the liposomes, thereby improving the rapid bacteriostatic properties of the essential oil-loaded liposomes, and also avoiding the restriction of bacterial species [111, 112].

4.4 LIPOSOME ENCAPSULATION OF HYDROPHILIC ACTIVE COMPONENTS AND NUTRIENTS

Compared with liposome encapsulation of hydrophobic components, the liposome encapsulation effect of hydrophilic components is poor. This is because the oil-water partition coefficient of the water-soluble components is greatly affected by the pH and ionic strength of the medium, and the encapsulation conditions are difficult to grasp. However, with the development of liposome technology, more and more hydrophilic active ingredients and nutrients are successfully encapsulated and prepared into products. Common water-soluble nutrients, including water-soluble iron fortifiers, water-soluble natural plant extracts, enzymes, peptides, proteins, and water-soluble vitamins, and their liposome composition and preparation methods are summarized in Table 4.3.

TABLE 4.2 The Antibacterial Properties of Liposome Encapsulating Essential Oil and Its Main Components

Essential Oil/Active Ingredient	Classification of Compounds	Wall Materials	Preparation Method	Bacterial Species	Major Findings	References
Clove oil	Phenol	Soy lecithin and cholesterol	Film extrusion	*Staphylococcus aureus and Escherichia coli*	• Liposome encapsulation improved the stability of *Clove oil*; • Compared with *E. coli*, the controlled release of *Clove oil* from liposomes was successfully achieved by PFTs secreted from *S. aureus*.	[103]
Cinnamaldehyde	Aldehyde	Egg yolk lecithin and Tween-80	Ethanol injection	*Staphylococcus aureus*	• Cinnamaldehyde liposomes with high core-wall ratio have better antibacterial activity against *S. aureus* during storage; • Liposomes improve the stability and durability of cinnamaldehyde against *S. aureus*.	[8]
Cinnamon oil	Aldehyde	Soy lecithin and cholesterol	Film ultrasonic dispersion	*Staphylococcus aureus*	• The application of liposomes further improved the stability of cinnamon oil and extended the bacteriostasis time.	[113]

TABLE 4.2 (*Continued*)

Essential Oil/ Active Ingredient	Classification of Compounds	Wall Materials	Preparation Method	Bacterial Species	Major Findings	References
Nutmeg oil	Terpenoids	Soy lecithin and cholesterol	Film dispersion	*Listeria monocytogenes*	• Encapsulation can retain the antibacterial activity of nutmeg oil and improve its stability; • PFTs could trigger the release of nutmeg oil from liposomes and inhibit bacteria growth steadily.	[105]
Tea tree oil	Terpenoids	Soy lecithin and cholesterol	Film dispersion	*S. enteritidis* and *S. Typhimurium*	• The chitosan nanofibers containing tea tree oil liposomes could inhibit *Salmonella* in chicken and had no impact on the sensory quality of chicken meat.	[114]
Thyme essential oil	Phenol	Soy lecithin and cholesterol	Film dispersion	*Escherichia coli*	• The ε-PLY-coated thyme essential oil liposome improved the stability of thyme essential oil; • Solid liposomes containing thyme essential oil were prepared by using β-cyclodextrins as cryoprotectants and possessed favorable antibacterial effects against *E. coli* O157: H7.	[115]

TABLE 4.2 *(Continued)*

Essential Oil/ Active Ingredient	Classification of Compounds	Wall Materials	Preparation Method	Bacterial Species	Major Findings	References
Cardamom essential oil	Phenol	Lecithin	Thin layer hydration	*Staphylococcus aureus* and *Escherichia coli*	• The inhibitory effect of Cardamom essential oil (CEO) loaded nanoliposomes and CEO emulsion were higher on *E. coli* compared to *S. aureus*; • The nanoliposome system was able to protect the bioactive properties of the CEO during storage time.	[116]
Lemongrass oil	Aldehyde	Soy lecithin, cholesterol, and polyvinyl-pyrrolidone	Film dispersion	*Listeria monocytogenes*	• The LO liposomes displayed satisfactory antibiotic activity against *Listeria monocytogenes* in cheese over the storage period at 4°C. • The LO liposomes could be triggered by listeriolysin O and lead to the leakage of LO.	[106]

TABLE 4.2 *(Continued)*

Essential Oil/ Active Ingredient	Classification of Compounds	Wall Materials	Preparation Method	Bacterial Species	Major Findings	References
Salvia oil	Phenol/ Leftover	Soy lecithin, cholesterol, and polyvinyl-pyrrolidone	Thin-film hydration	*Staphylococcus aureus*	• Controlled release of Salvia oil (SO) from nanoliposomes was successfully achieved by α-toxin secreted from *Staphylococcus aureus*; • The anti-biofilm action time of liposome-entrapped Salvia oil has been prolonged significantly.	[104]
Peppermint essential oil	Alcohol/ Ketone/Ester	Miglyol and Precirol	Hot melt homogenization	*Staphylococcus aureus* and *Escherichia coli*, and so on.	• Under *in vitro* conditions, the peppermint essential oil (PEO) loaded into nanostructured lipid carriers showed considerable antibacterial activity against positive bacteria, and also negative bacteria, but did not show a significant difference in antibacterial activity compared with PEO.	[117]

TABLE 4.2 *(Continued)*

Essential Oil/ Active Ingredient	Classification of Compounds	Wall Materials	Preparation Method	Bacterial Species	Major Findings	References
Zataria multiflora essential oil	Phenol	Soy phosphatidyl-choline and cholesterol	Thin-film hydration	*Escherichia coli* serotype O157:H7	• From the results of MIC concentration for essential oil-loaded nanoliposomes had enhanced inhibitory activity than its pure form; • The subinhibitory concentrations of liposomal essential oil had a higher inhibitory effect on toxin titer than free oil.	[100]
Thymol/carvacrol	Phenol	Soy phosphatidyl-choline and cholesterol	Thin-film hydration	*Salmonella* and *Staphylococcus aureus*	• Free carvacrol and thymol might be suitable to control bacterial populations for short time treatments; • Thymol/carvacrol liposome showed a slower release of the encapsulated compound but reduced the rapid antimicrobial activity of the active ingredient.	[102]

TABLE 4.2 *(Continued)*

Essential Oil/ Active Ingredient	Classification of Compounds	Wall Materials	Preparation Method	Bacterial Species	Major Findings	References
Salvia triloba and *Rosmarinus officinalis* essential oils	Phenol/ Terpene/ Ketone	Phospholipon 90G and cholesterol	Thin-film hydration	*Staphylococcus aureus* and *Escherichia coli*, and so on.	• Liposomes were stable over one month period if stored at 4°C and possessed significant antioxidant, anti-inflammatory, and antibacterial activities; • Liposomes can decrease the volatility of EOs, optimize their biological properties and defeat antimicrobial infections	[118]

TABLE 4.3 Liposome Encapsulation of Hydrophilic Active Ingredients and Nutrients

Nutraceuticals	Liposome Composition	Preparation Method	Main Findings	References
Ferrous sulfate	Ascorbic acid, soybean phosphatidylcholine, soybean lecithin	The pro-liposome and the microfluidization technologies	Liposomes prepared by the microfluidization technology and liposomes from the pro-liposome technology exhibited similar levels of oxidative stability, demonstrating the feasibility of microfluidization-based liposomal delivery systems for large-scale food/nutraceutical applications.	[147]
Ferrous sulfate	Tween-80, whey protein isolate, phosphatidylcholine, cholesterol	Reverse-phase evaporation technique	Despite the high encapsulation efficiency of iron in our liposomes, liposomal encapsulation did not inhibit the pro-oxidant activity of iron.	[121]
Ferrous sulfate	Glycerol phospholipid, the reducing agent Vitamin C	Repetition of the cycle of freezing and thawing	Liposome vesicle morphology and size are comparable with classical liposomal structures. Products are stable during specimen preparation and drying. They have a good ability to penetrate into cells, interacting with cytoplasmic organelles, without inducing, at least apparently, any ultrastructural damage.	[148]

TABLE 4.3 (Continued)

Nutraceuticals	Liposome Composition	Preparation Method	Main Findings	References
Ferrous sulfate	Tween-80, soybean lecithin, cholesterol	Reverse-phase evaporation method	Emulsifying activity index and emulsifying stability index of whey protein (WP) in the presence of ferrous sulfate liposomes (FL) increased.	[122]
			The WP-stabilized emulsion droplet size decreased in the presence of FL.	
			The changes of WP emulsifying properties could be attributed to the effects of liposomes on WP secondary structure and tertiary structure.	
Ferrous amino acid	Tween-80, cholesterol, Egg phosphatidylcholine	Reverse-phase evaporation method	The stability of ferrous glycinate in strong acid environment was greatly improved by encapsulation in liposomes.	[149]
			The bioavailability of ferrous glycinate may be increased.	
Ferrous amino acid	Egg phosphatidylcholine, Tween-80, cholesterol	Reverse-phase evaporation method	The size distribution and zeta potential indicated the stability of the nanoliposome suspension.	[127]
			Ferrous glycinate nanoliposomes were tested in vitro for their stability in simulated gastrointestinal juice.	
Salidroside	Lecithin, cholesterol, Tween-80	Thin-film evaporation, freezing-thawing method, sonication method, reverse phase evaporation, melting method	The encapsulating efficiency of liposomes was highest when prepared by freezing-thawing, followed by thin-film evaporation, then reverse-phase evaporation and the lowest with melting and sonication.	[132]

TABLE 4.3 *(Continued)*

Nutraceuticals	Liposome Composition	Preparation Method	Main Findings	References
Salidroside	Egg yolk phosphatidylcholine, cholesterol, Tween-80	The ethanol injection method	The release study of salidroside in vitro from nano-liposomes exhibited a prolonged release profile as studied over a period of 24 h.	[150]
Salidroside	Lecithin, cholesterol, Tween-80	ammonium sulfate gradient method	Salidroside liposome formulation had the potential to act as effective sustained release vaccine delivery systems.	[129]
Anthocyanins	L-α-Phosphatidylcholine, hydrogenated soy, cholesterol, Tween-80	pH-gradient loading method	RAH liposomes-maintained anthocyanin stability under in vitro physiological conditions for 14 days and exhibited enhanced ROS scavenging activity and skin permeability.	[151]
Anthocyanins	Soy lecithin, cholesterol	Supercritical carbon dioxide (SC-CO_2) method	The anthocyanin release from liposomes was slow (<= 35.9%) in the simulated gastric fluid but rapid in the simulated intestinal fluid, induced by the degradation of the vesicles by pancreatin.	[152]
Anthocyanins	Crude phospholipid, soy lecithin, cholesterol	Improved supercritical carbon dioxide method	Anthocyanin encapsulated into liposomes can be protected from adverse external conditions with potential benefits in food and nutraceutical formulations for improved efficacy and health benefits.	[153]
Tea polyphenols	Cholesterol, phosphoric acid, Tween-80	Improved inverse-phase evaporation method	The mean particle size and amount of surfactants could affect the percutaneous absorption characteristic of TP liposomes modified by surfactants.	[154]

TABLE 4.3 (Continued)

Nutraceuticals	Liposome Composition	Preparation Method	Main Findings	References
Flavorsome	Lecithin, cholesterol, glycerol	Heating method (HM)	The results indicated that the lecithin proportion and the stirring time were the major influential variables for both responses.	[142]
Flavorsome	Lecithin, cholesterol, glycerol	Modified heating method	The most influential factor on proteolysis indices was ripening time, while the content of liposomal enzyme and retention time were also significant (p<0.05).	[139]
Flavorsome	Phosphatidylcholine, cholesterol	Mozafari method	The gradually fulfilling proteolysis of casein by liposomal-loaded Flavourzyme in comparison with the free enzyme.	[141]
Natural dipeptide antioxidant	Phospholipids	Thin-film hydration method	Encapsulation of antioxidant peptides by nanoliposomes can address stability issues (such as proteolytic degradation and potential interactions between peptides and food components that may affect their activity) and improve their efficacy.	[144]
Lactoferrin	Phospholipid, cholesterol, Tween-80, vitamin E	Thin-layer dispersion method	Liposomes may prevent the gastric degradation of Lactoferrin (LF) and reduce the rate of hydrolysis of LF in intestinal conditions.	[145]
Vitamin C	Soybean phosphatidylcholine, rapeseed lecithin	—	The new liposomal formulation of vitamin C free of harmful organic solvents was presented and the bioavailability of vitamin C from the formulation was enhanced in the medical experiment.	[155]

TABLE 4.3 *(Continued)*

Nutraceuticals	Liposome Composition	Preparation Method	Main Findings	References
Vitamin C	Sesame phospholipid	Repetition of freeze-thawing	Sesame liposomes were found to encapsulate almost 80% of vitamin C in their interior cavities. During the 8 days storage, the release of vitamin C occurred gradually from the liposome system, which signifies week interactions in the liposome membranes amongst phospholipid molecules and vitamin C.	[156]
Vitamin C	Natural phospholipids, sodium ascorbate	Provided by empirical labs	The efficacy of oral delivery of vitamin C encapsulated in liposomes can decrease oxidative stress and improve physiological function in adult humans.	[157]

Note: **The** symbol "–" in the table indicated that it was not mentioned in the literature.

4.4.1   IRON FORTIFIER

Various iron supplements have been developed to combat iron deficiency. Iron nutritive fortifier is mainly divided into inorganic iron and organic iron. However, there are many difficulties in strengthening iron in food and nutrition products due to the color, flavor, and precipitation problems of many iron fortifiers. Therefore, special measures are needed to stabilize and distribute it evenly in food and nutrition, and liposome encapsulation is an effective and safe means to protect iron nutrient fortifier.

4.4.1.1   FERROUS SULFATE

Ferrous sulfate is an iron nutritive agent with high bioavailability and low price, which can effectively improve the condition of iron deficiency in the human body, thanks to the good solubility of ferrous ions in a neutral and acidic environment. But this iron salt is active and easily oxidized, depending on temperature and air exposure. Oxidation may change other food ingredients, influence the color of food, and lead to the deterioration of product quality. In addition, it has an irritating effect on the stomach and intestines, and the taste of direct consumption is unacceptable [119].

The ferrous sulfate liposomes (FLs) have no adverse effect on the food substrate, which ensures the availability of ferrous ions. The research directions of FLs can be roughly divided into three categories: (1) preparation, encapsulation rate and stability, (2) effects of liposomes on other components in food, (3) bioavailability of ferrous sulfate. In the preparation of FLs, auxiliary membrane materials (such as Tween-80 and chitosan) are often added to enhance the encapsulation rate and stability. Abbasi et al. [120] studied the effects of ferrous sulfate/lipid ratio and Tween-80 on liposome EE. The results showed that the EE increased with decreasing Tween-80 and core/wall weight ratio in which the highest possible microencapsulation efficiency (85%) was achieved at an iron-to-lipid weight ratio and Tween-80 concentration of 0.04% and 5%, respectively. Cengiz et al. [121] investigated the role of liposome phospholipids on the oxidation stability of the iron enhanced emulsion. The PC liposomes could encapsulate ferrous sulfate efficiently and the resulting liposomes were physically stable. However, in the emulsion, these liposomes did not show higher oxidation stability compared with the emulsion with free ferrous sulfate. This might be caused by the oxidation of liposome membrane material itself. Therefore, to improve the oxidation stability of FLs in the emulsion, more saturated PC could be used in the preparation of liposomes, or cationic polymers such as chitosan could be added to make them adsorbed on the liposome membrane layer by layer and reject the metal cations. Ferrous sulfate is often added to milk as an iron fortifier. When liposomes are used as carriers of ferrous sulfate, it is necessary to understand the effect of the interaction between liposomes and proteins on the emulsification and other functional properties of proteins. The effect of FL on the emulsifying property of WP was explored [122]. It was found that FL could improve the emulsification activity index and the emulsification stability index of WP. The change of emulsification performance was related to the effect of liposome on the secondary structure, tertiary structure, Zeta potential, and particle size of the powder. The iron absorption in liposomes was evaluated by Caco-2 cells [123]. *In vitro* digestion and absorption of different liposome formulations were different, and the digestion routine liposome composed of soybean phosphatidylcholine (SPC), hydrogenated phosphatidylcholine (HSPC) or PC and Chol had the highest iron absorption. Notably, all iron liposome formulations protected iron from oxidation and improved iron absorption by intestinal cells compared with ferrous sulfate solutions.

4.4.1.2   AMINO ACID CHELATED IRON

The special form of chelated iron is very effective for the treatment of iron deficiency anemia in adults, youth, and children [124, 125]. Only in the chelating form, that is, only the real amino acid chelates can play its advantages, the higher the chelating rate of the chelate, the higher the biological titer, the better the corresponding effect. However, under the strong acid environment such as gastric juice, amino acid chelated iron would be dissociated into inorganic iron salts, which reduced its absorption and bioavailability [126].

The use of liposomes to encapsulate ferrous glycinate could significantly improve the stability of iron compounds in a strong acid environment and prevent the destruction of ferrous glycinate from the extracapsular environment [127]. The effects of storage time, ultrasonic treatment, and boiling water on the stability of ferrous glycinate nanoliposomes and the stability of ferrous glycinate nanoliposomes in simulated gastric juice *in vitro* were investigated. The iron transport of ferrous glycinate liposomes was evaluated by the Caco-2 cell model [10]. The results showed that the iron transport volume of ferrous glycinate liposomes was significantly higher than that of ferrous glycinate. The iron transport of ferrous glycinate liposomes decreased with the increase of particle size.

4.4.2 WATER-SOLUBLE NATURAL PLANT EXTRACT

4.4.2.1 SALIDROSIDE

Rhodiola Rosea is an Oriental folk herb with extensive functions that has been paid attention to only in recent decades. The main functional components are salidroside and its glycoside. In recent years, many studies have reported that salidroside had an anti-fatigue effect, and had health care effects such as inhibiting blood glucose increase, anti-peroxide effect, enhancing immune response, enhancing memory, anti-depression, and anti-anxiety [128–131]. However, salidroside supplements on the market have many problems, especially low bioavailability. Our group investigated the effects of five different preparation methods on the formation and physicochemical properties of salidroside liposomes [132]. The storage stability showed that the salidroside liposome system prepared by the melting method had better physical and chemical stability. The size of salidroside liposomes increased slowly compared with those without salidroside, which indicated that salidroside played an important role in preventing the accumulation and fusion of liposomes. The preparation conditions of salidroside liposomes were further optimized with a higher encapsulation rate of up to 94.5%, and the immune-enhancing activity as vaccine adjuvant was determined [133].

4.4.2.2 ANTHOCYANINS

Anthocyanin is a kind of strong antioxidant. It is widely found in the tissues of plants such as purple sweet potato, grape, blueberry, strawberry, and mulberry. Anthocyanin is water-soluble, but with poor stability, which is easy to be affected by external conditions [134, 135]. At present, more studies are focused on the microencapsulation of anthocyanins. However, due to its large particle size, it is not conducive to human absorption. How to reduce the particle size remains to be further studied [135, 136]. Liposome encapsulation can solve this problem, and some scholars have carried out some preliminary studies on it. Sealing anthocyanin in the vesicle and isolating it from the outside environment can improve its stability and antioxidant activity, improve its availability in functional products, and promote its physiological function [137, 138].

4.4.3 ENZYMES, PEPTIDES, AND PROTEINS

The encapsulation of proteins, enzymes, and peptides by liposomes is another important research direction of food-grade liposomes. The technology of the liposome embedding enzyme is mainly used to promote the ripening of cheese. Liposome enzymes are added to the cheese before rennet is added, and they disperse evenly throughout the cheese without much prior degradation of the casein [139]. After a clot is formed, the liposomes are surrounded by it, then the liposomes break down and enzymes are released to accelerate proteolysis and cheese ripening, improving the sensory properties of the cheese [140, 141]. Liposomes did not affect the water activity and microorganism of cheese but accelerated cheese ripening to the maximum extent. In addition, liposome technology can be used to encapsulate other food-grade enzyme preparations. Jihadi et al. [142] prepared flavor protease liposomes by heating, and the results showed that the flavor protease liposomes prepared at 45°C and pH 6 were the most stable, with the protease embedding rate and activity of 26.5% and 9.96 LAPU/mL, respectively. People with lactose intolerance may be prevented from absorbing lactose products due to the lack of lactase, especially dairy products, which can be solved by lactase supplementation. If lactase is added directly in dairy production, it will cause the fermentation phenomenon and affect the texture and taste of products. The method of liposome embedding ensures that lactase is only released in the intestinal tract and prevents the enzyme from reacting in the medium [11, 143].

The encapsulation of antioxidant peptides with nanoliposomes can solve their stability problems (e.g., proteolysis, potential interactions between peptides and food components that may affect their activity) and improve their efficacy, making them widely studied as potential biological preservatives in the food industry [144]. The study also suggested that natural dipeptide antioxidants (L-carnosine antioxidants) could solve the related problems of food preservation, such as reducing the oxidation of active substances in the complex reactions of the food system, reducing the deterioration caused by microorganisms on the food surface and oxidative rancidity. Lactoferrin liposomes were prepared by the thin-film dispersion method [145]. It was found that liposome encapsulation could protect lactoglobulin from pepsin hydrolysis and reduce the rate of lactoglobulin hydrolysis in the intestinal tract.

4.4.4   WATER-SOLUBLE VITAMINS

Water-soluble vitamins include the B vitamins and vitamin C (ascorbic acid). Vitamins, as a class of organic compounds with low molecular weight, mostly contain unsaturated double bonds, hydroxyl groups, or some components that are extremely sensitive to other chemical reactions, and are easy to be oxidized or reduced. Liposomes based on soybean PC as vitamin carriers were added to orange juice, and the antioxidant activity of vitamins before and after pasteurization could be maintained [36]. The liposome formula, the combination of vitamins and orange juice did not change its sensory characteristics and showed microbial stability after pasteurization and storage at 4°C for 37 days. Vitamin E (V_E) and vitamin C (V_C) were co-encapsulated in SPC-based liposomes and added to chocolate milk for functional foods and pasteurization [9]. All formulations showed that liposomes remained remarkably stable and protected V_C even after pasteurization.

4.5   STABILITY AND REGULATION STRATEGY OF LIPOSOMES

As a transport system, liposome has many advantages, but its application is limited by its stability. If the core material of liposome leaks rapidly *in vitro*, or leaks from the liposome before reaching the target tissue *in vivo*, its effectiveness as a carrier will be greatly reduced. The stability of liposomes usually includes physical stability, chemical stability, colloidal stability, and biological stability.

4.5.1   PHYSICAL STABILITY

The physical stabilities of liposomes mainly include the change of liposome structure (particle size distribution and shape) and the leakage of core material, and these are closely related to the mechanical properties of liposome bilayer and the thermodynamic properties of the system.

4.5.1.1   THE CHANGE OF LIPOSOME PARTICLE STRUCTURE

Particle size, polydispersity, and size distribution are important parameters to characterize the structural characteristics of liposomes because they could affect the function and storage stability of liposomes. The size of liposomes could be analyzed using a variety of methods, including electron microscopy, fluorescence microscopy, dynamic light scattering, nanoparticle tracking analysis, flow cytometry, and dimensional exclusion chromatography. In addition, the Zeta potential of liposomes could be measured by dynamic light scattering and laser Doppler velocimeter. Higher Zeta potential could prevent the accumulation of liposomes by charge repulsion, thus maintaining the stability of liposomes. Generally, higher potentials (absolute value higher than 30 mV) could enhance the repulsive force between liposomes to prevent aggregation, thereby improving the stability of the liposomes. Therefore, charged components such as phosphatidyl glycerin, phosphatidyl acid, and steramine can be added to the membrane material of liposomes to increase Zeta potential and thus reduce the change of particle size during storage.

4.5.1.2   LEAKAGE OF ENCAPSULATED CORE MATERIAL

The stability of liposomes is directly related to the composition of membrane materials and the stabilities of liposomes formed by different phospholipids are different. When the phospholipid bilayer reaches the phase transition temperature (Tc), the phase separation will occur, which will increase the membrane permeability and lead to the leakage of the encapsulated core material. Because the liposomes formed from single component phospholipids have uniform phase Tc, which are more prone to phase separation and extremely unstable. Therefore, liposomes are often prepared by mixing various phospholipids and adding a proper amount of

Chol [158], phytosterol [159], and other amphiphilic substances [160] to further improve the stability of the bimolecular layer.

The stability of liposomes is closely related to the properties of the encapsulated materials. In general, a substance with excellent fat or water solubility is the best core material for liposome encapsulation. Conversely, substances with poor fat solubility or water solubility are not easily encapsulated in liposomes.

Environmental factors such as pH, temperature, oxygen, and enzymes could cause the accumulation of liposomes, change the surface behavior, and even damage the structural integrity, thus affecting the stability of liposomes. In view of this, thermosensitive liposomes [161–163], pH-sensitive liposomes [164–166], and other stimulation-responsive liposomes [167, 168] have been designed to improve the adaptability to environmental conditions and realize the controlled release of core materials.

4.5.2 CHEMICAL STABILITY

The chemical stability of liposomes could be divided into two aspects: one is the stability of encapsulating materials in liposomes, and another is the stability of membrane components in liposomes. Liposomes could protect the encapsulated material, increase its stability, and even improve its bioavailability. As reported in the studies, liposome encapsulation could improve the resistance of tea polyphenols to environmental conditions [169] and RESV-loaded liposomes showed more obvious antioxidant effects as compared with free RESV [170]. In addition, the encapsulation of vitamin C with liposomes showed better storage stability and improved sustained release ability and skin permeability [171]. Generally, the main components of the liposome membrane are natural phospholipids, which contain unsaturated fatty acid chains in their molecules. Therefore, there are two main ways of chemical degradation, including oxidation of unsaturated fatty acyl chains and hydrolysis of ester bonds between fatty acids and glycerol.

The chemical stability and physical stability of liposomes are closely related. High mechanical strength and compact bilayer of liposomes could reduce the penetration of oxidant or hydrolytic agents and reduce the degree of chemical degradation of liposome membrane material. Conversely, the chemical degradation of the liposome membrane also affects the particle size, size distribution, fusion, and aggregation degree, and mechanical properties of the bilayer. In general, the oxidation and hydrolysis of phospholipid could reduce the fluidity of the liposome membrane and intensify the leakage of core material, which lead to physical instability such as accumulation and rupture of liposome particles.

Compared with unsaturated fatty acids, saturated fatty acids are less susceptible to oxidation, so the use of saturated lipids or hydrogenated fatty acids could reduce the degree of oxidative degradation of the membrane materials [172]. The resistance of liposomes to oxidation could also be improved by low temperature and dark storage under inert gas conditions, adding antioxidant, metal ion complexing agent (EDTA), and photooxidation quenching agent (β-carotene).

4.5.3   COLLOIDAL STABILITY

The liposome is a colloidal dispersion system, and the phospholipid membrane is a dynamic membrane, in which phospholipid molecules can constantly exchange positions, and substances inside and outside the liposome membrane could carry out free transmembrane exchange randomly and selectively. Therefore, the liposome is thermodynamically unstable, and liposome particles are prone to spontaneous accumulation and precipitation.

Colloidal instability in the preparation of liposomes is mainly manifested as particle aggregation and fusion. Aggregation is the formation of large size liposome vesicles, which are still liposomes. Aggregation is the first step of fusion, and is reversible and could be re-dispersed by applying shear force, changing temperature, and other methods. On the contrary, fusion is the formation of new colloidal structures, and the fusion process is irreversible. Ostwald Ripening is the main reason for the colloidal instability of liposomes during storage. This thermodynamic-based process is due to the higher chemical potential of small particles, so it tends to form larger particles during storage to achieve the minimum energy state.

Modifying the surface of the liposome membrane by forming a polymer layer, could improve the structural properties of the liposome, thereby enhancing the colloidal stability and reducing the environmental sensitivity. Moreover, a large number of studies have proved that polysaccharides such as chitosan [53, 66], pectin [174, 175], and carboxymethyl cellulose [176] are usually used as stabilizers and coating materials for liposomes, which could form a protective layer with a certain structure and improve the stability of liposome system.

4.5.4   BIOLOGICAL STABILITY

The ideal delivery system should be stable during ingestion, intravenous injection, the gastrointestinal tract, and before reaching the target tissue. However, conventional liposomes (CL) are quite unstable in the biological environment in practical applications and could be destroyed by digestive fluids and interacting with various enzymes.

Oral liposomes have attracted more and more attention because of their advantages, such as convenient administration and fewer side effects. Moreover, the study of the digestive stability of liposomes in the gastro-intestinal tract is a hot trend in food science. After entering the gastro-intestinal tract, liposomes are easy to be damaged by strong acid and alkali digestives and contents (cholic salt, enzyme, etc.), resulting in leakage of core materials encapsulated in liposomes and reduce of bioavailability. The researches focused on liposomal oral delivery have grown, but substantial breakthroughs are still desired to evolve and commercialize these products for the application. Surface coating or modification of liposomes with cloaking agents such as polyethylene glycol (PEG) [177] and chitosan [178] could overcome most of the obstacles associated with liposome instability, which not only enhance its biological stability but also improve its function efficacy.

4.6   BIOAVAILABILITY OF LIPOSOMES ENCAPSULATING ACTIVE INGREDIENTS

Bioaccessibility refers to the percentage of ingested ingredients that are released from the food substrate and can be used for intestinal uptake but is usually based on *in vitro* assay procedures. Bioavailability, by contrast, is a more comprehensive indicator of the percentage of ingested ingredients measured *in vivo* that can be used for normal physiological functions. Liposomes are often used as carriers of active and functional components. Therefore, in recent years, more and more researchers have focused on the study of bioaccessibility and bioavailability of functional substances encapsulated in liposomes.

4.6.1 EVALUATION METHODS

It is technically difficult to study the actual digestion and utilization process of liposomes in the human body. In addition, this study is not only subject to individual differences, but also restricted by ethical constraints. Therefore, the current researches are basically to characterize the bioaccessibility of liposomes through simulated digestion *in vitro*, while cell and animal models are used as alternative methods for *in vivo* studies to simulate the utilization of liposomes after ingestion and further regulate the bioavailability of liposomes.

4.6.1.1 IN VITRO DIGESTION

Most studies on the *in vitro* digestion behavior of liposomes are based on the static *in vitro* digestion model. Static tools, such as water-bath shaking, air bath rotation, and pH regulator, are used to simulate the digestive environment and peristalsis of the gastrointestinal tract, to predict the release ability and biological accessibility of the active components encapsulated in liposomes. Studies have reported the *in vitro* digestion behaviors of liposomes coated with carotenoids [52, 53], CUR [75], V_E [34], and other active substances. These studies show that the bioaccessibility of liposomes depends strongly on the incorporating ability of encapsulated active ingredients into a lipid bilayer, loading content, and the surface modification of liposomes. Although the static digestion model could simulate the environmental conditions of gastrointestinal digestion *in vitro*, it cannot reproduce the dynamic processes such as gastric evacuation or the continuous changes in pH and secretion velocity during digestion in the human body.

4.6.1.2 CELL MODEL

Caco-2 cells derived from human epithelial colonic rectal adenocarcinoma could mimic human intestinal epithelial cells and are the most commonly used models for studying the absorption of intestinal uptake by drugs and other compounds [69, 70]. There are some common methods and indicators used to monitor the changes of Caco-2 cell monolayer after adding liposomes (Figure 4.4), thus reflecting the absorption of active components by small intestinal epithelial cells.

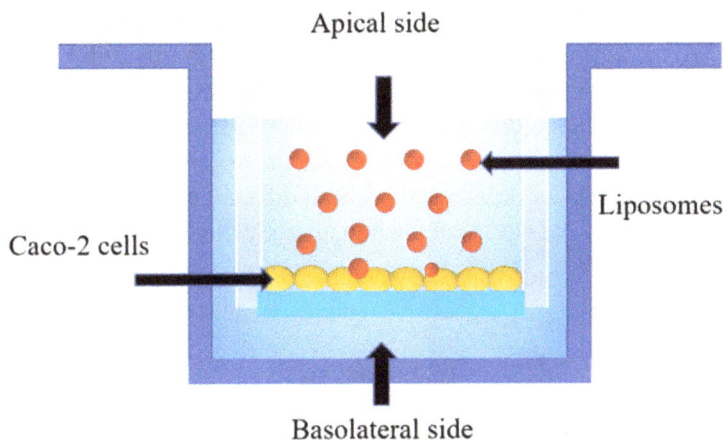

FIGURE 4.4   Schematic diagram of Caco-2 cell monolayer model on Transwell plate.

The integrity of monolayer cells can be determined by measuring their transepithelial resistance (TEER) [184–186]. According to the apparent permeability coefficient (P_{app}) of the active ingredient in Caco-2 cells, the absorbability of the substance in the human body could be predicted [187]. Fluorescence microscopy and flow cytometer could be used to demonstrate and monitor the cellular uptake rate of active substances by Caco-2 cells, and could also be used to analyze the transport mode of encapsulated ingredients in liposomes into cells [69, 187]. Studies have indicated that the liposome encapsulation was shown to significantly enhance the permeability and absorbability of active components through intestinal epithelial cells [184, 185].

In addition, HT29-MTX cells are composed of differentiated goblet cells with mucous secretion characteristics. As a cell model that could form mucous, HT29-MTX cells are also often used in intestinal absorption studies [186]. Moreover, other types of intestinal cells, including M cells and mucosal cells, could also be used as cell models to study the transport and uptake of the liposome [184].

4.6.1.3 ANIMAL MODEL

Animal experiments are the basis of human clinical experiments and the most advantageous *in vitro* model for human digestion study. Using small animals as the research model to conduct *in vivo* simulation experiments is not only economical, rapid, and feasible, but also can realize the artificial control of the diet and drug intake of model animals, and monitor and study the blood samples, liver, and intestinal contents after intake. In recent years, there has been substantial progress in animal model experiments, and many animal models (such as mice [188], rats [183], and rabbits [182]) have been studied. However, due to the differences in dietary habits and the complexity of organisms, it is still difficult to accurately assess the bioavailability of liposomes in the human body directly from the results of animal experiments.

4.6.2 STRATEGY TO IMPROVE BIOAVAILABILITY

The digestion behavior of liposomes depends on the environmental parameters and the characteristics of liposomes. The preparation method of liposomes determines the size, film thickness, and layer number of liposomes, which are related to the EE, storage stability and bioavailability of liposomes, and could further affect the digestion behavior. As the study reported that compared with traditional methods such as thin layer dispersion and ethanol injection, the CUR-loaded liposomes prepared by pH driven method had higher bioavailability [63]. Surface modification compounds could bind with liposomes through electrostatic and hydrogen bond interaction and covering the surface of the liposome (a), inserting the hydrophobic chain in the hydrocarbon chain of the bilayer (b), which were partially embedded and absorbed in lipid double layer (c) or modified phospholipid head groups to form functional liposomes (d) [180, 181], as shown in Figure 4.5. A large number of studies have shown that surface modification could significantly improve the digestive stability and bioavailability of liposomes by forming a contact barrier between phospholipids of liposomes and the external environment [53, 146, 173, 178, 179]. In addition, it has been proved that surface modification of liposomes by PEG could change the fluidity of the phospholipid bilayer and improve the protective capability of the active components [34].

Moreover, the surface of liposomes was changed into positive charge by the modification of cationic chitosan, the permeability of phospholipids bilayer was reduced, and the interaction between digestion fluid and liposomes was also reduced, which improve the absorption and utilization of liposomes in the small intestine [69, 75].

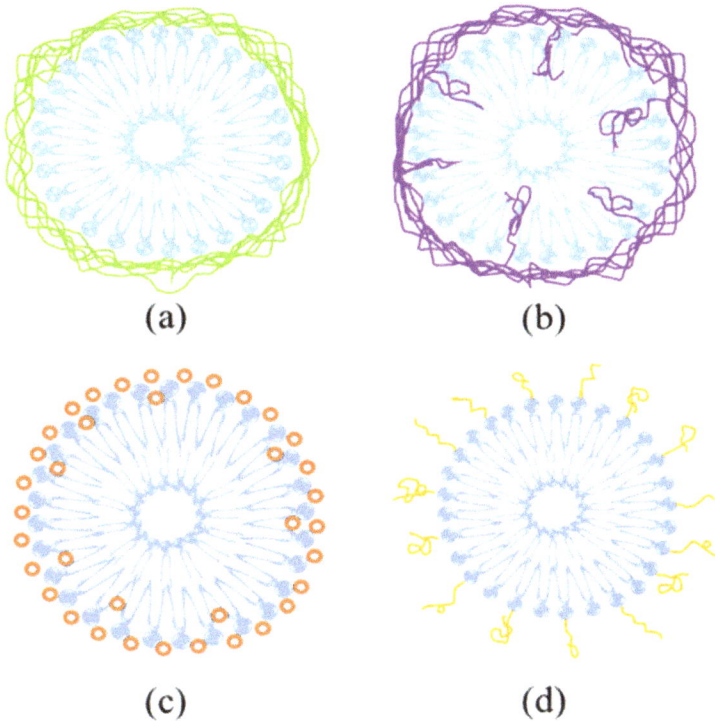

FIGURE 4.5 Schematic diagram of four mechanisms of surface modification of liposomes: (a) coverage; (b) insertion; (c) partially absorption; (d) phospholipid modification.

4.7 CONCLUSION

The great potential of the liposome delivery system in functional foods and nutraceuticals is being rapidly established. In view of global health problems, it is imperative to use novel liposomes for effective disease prevention and health promotion. Although liposomes have made remarkable achievements in laboratory studies for the microencapsulation and efficiency enhancement of bioactive ingredients and nutraceuticals, their application in the food industry is still facing severe challenges. In order to apply liposomes to the food field, future research will focus on the production of liposomes through safe and scalable methods with cheaper commercial phospholipids.

Due to the different absorption routes of various nutrients through the intestine tract, the formulation of liposomes can be rationally designed based on the key factors limiting their absorption. The physicochemical properties (size, potential, layer number, etc.), and release characteristics of liposomes will determine its behavior *in vivo* and its actual usefulness to human health, further research *in vivo* is needed to support the improvement of bioavailability and efficiency. Even more importantly, owing to the complexity of the food system, it is necessary to continue to study the interaction between liposomes and matrix components and to better understand the stability during food processing.

KEYWORDS

- **bioactive ingredients**
- **bioavailability**
- **liposomes**
- **microencapsulation**
- **nutraceuticals**
- **stability**

REFERENCES

1. Liu, W., Ye, A., & Singh, H., (2015). Chapter 8-progress in applications of liposomes in food systems. In: Sagis, L. M. C., (ed.), *Microencapsulation and Microspheres for Food Applications* (pp. 151–170). Academic Press: San Diego.
2. Singh, H., et al., (2012). 11-Liposomes as food ingredients and nutraceutical delivery systems. In: Garti, N., & McClements, D. J., (eds.), *Encapsulation Technologies and Delivery Systems for Food Ingredients and Nutraceuticals* (pp. 287–318). Woodhead Publishing.

3. Islam, S. M. R., et al., (2019). Liposomal delivery of natural product: A promising approach in health research. *Trends in Food Science & Technology, 85,* 177–200.

4. Khorasani, S., Danaei, M., & Mozafari, M. R., (2018). Nanoliposome technology for the food and nutraceutical industries. *Trends in Food Science & Technology, 79,* 106–115.

5. Xia, S., et al., (2015). Modulating effect of lipid bilayer-carotenoid interactions on the property of liposome encapsulation. *Colloids and Surfaces B: Biointerfaces, 128,* 172–180.

6. Jafari, S. M., Vakili, S., & Dehnad, D., (2019). Production of a functional yogurt powder fortified with nanoliposomal vitamin D through spray drying. *Food and Bioprocess Technology, 12,* 1220–1231.

7. Huang, M., et al., (2019). Liposome co-encapsulation as a strategy for the delivery of curcumin and resveratrol. *Food and Function, 10,* 6447–6458.

8. Chen, W., et al., (2019). Modulation effect of core-wall ratio on the stability and antibacterial activity of cinnamaldehyde liposomes. *Chem. Phys. Lipids, 223,* 104790.

9. Marsanasco, M., et al., (2016). Fortification of chocolate milk with omega-3, omega-6, and vitamins E and C by using liposomes. *European Journal of Lipid Science and Technology.*

10. Baomiao, D., et al., (2017). Evaluation of iron transport from ferrous glycinate liposomes using Caco-2 cell model. *African Health Sciences, 17*(3).

11. Song, J. J., (2011). *Preparation of Lactase Liposomes by Thin Film Evaporation-Freeze Drying Method.* Food Industry.

12. Sarabandi, K., et al., (2019). Chapter nine - encapsulation of food ingredients by nanoliposomes. In: Jafari, S. M., (ed.), *Lipid-Based Nanostructures for Food Encapsulation Purposes* (pp. 347–404). Academic Press.

13. Tsai, W. C., & Rizvi, S. S. H., (2016). Liposomal microencapsulation using the conventional methods and novel supercritical fluid processes. *Trends in Food Science & Technology, 55,* 61–71.

14. Mirafzali, Z., Thompson, C. S., & Tallua, K., (2014). Chapter 13 - application of liposomes in the food industry. In: Gaonkar, A. G., et al., (ed.), *Microencapsulation in the Food Industry* (pp. 139–150). Academic Press: San Diego.

15. William, B., et al., (2020). Supercritical fluid methods: An alternative to conventional methods to prepare liposomes. *Chemical Engineering Journal, 383,* 123106.

16. Vladisavljević, G. T., et al., (2014). Production of liposomes using micro engineered membrane and co-flow microfluidic device. *Colloids and Surfaces A: Physicochemical and Engineering Aspects, 458,* 168–177.

17. Vorauer-Uhl, K., Wagner, A., & Katinger, H., (2002). Long term stability of rh-Cu/Zn-superoxide dismutase (SOD)-liposomes prepared by the cross-flow injection technique following international conference on harmonization (ICH)-guidelines. *European Journal of Pharmaceutics and Biopharmaceutics, 54*(1), 83–87.

18. Wagner, A., Vorauer-Uhl, K., & Katinger, H., (2002). Liposomes produced in a pilot-scale: Production, purification and efficiency aspects. *European Journal of Pharmaceutics and Biopharmaceutics, 54*(2), 213–219.

19. Zhao, L., et al., (2017). Encapsulation of lutein in liposomes using supercritical carbon dioxide. *Food Research International, 100,* 168–179.

20. Tsai, W. C., & Rizvi, S. S. H., (2017). Simultaneous microencapsulation of hydrophilic and lipophilic bio-actives in liposomes produced by an ecofriendly supercritical fluid process. *Food Research International, 99*, 256–262.

21. Amnuaikit, T., (2016). Formulation development and preparation of fish oil liposome by using high pressure homogenizer for food supplement product. *Asian Journal of Pharmaceutical Sciences, 11*(1), 126–127.

22. Guldiken, B., et al., (2019). Formation and characterization of spray-dried coated and uncoated liposomes with encapsulated black carrot extract. *Journal of Food Engineering, 246*, 42–50.

23. Guner, S., & Oztop, M. H., (2017). Food grade liposome systems: Effect of solvent, homogenization types and storage conditions on oxidative and physical stability. *Colloids and Surfaces A: Physicochemical and Engineering Aspects, 513*, 468–478.

24. Chung, S. K., et al., (2014). Factors influencing the physicochemical characteristics of cationic polymer-coated liposomes prepared by high-pressure homogenization. *Colloids and Surfaces A: Physicochemical and Engineering Aspects, 454*, 8–15.

25. Gibis, M., Zeeb, B., & Weiss, J., (2014). Formation, characterization, and stability of encapsulated hibiscus extract in multilayered liposomes. *Food Hydrocolloids, 38*, 28–39.

26. Berger, N., et al., (2001). Filter extrusion of liposomes using different devices: Comparison of liposome size, encapsulation efficiency, and process characteristics. *International Journal of Pharmaceutics, 223*(1), 55–68.

27. Shah, V. M., et al., (2019). Liposomes produced by microfluidics and extrusion: A comparison for scale-up purposes. *Nanomedicine: Nanotechnology, Biology and Medicine, 18*, 146–156.

28. Mui, B., Chow, L., & Hope, M. J., (2003). Extrusion technique to generate liposomes of defined size. In: *Methods in Enzymology* (pp. 3–14). Academic Press.

29. Rasti, B., Erfanian, A., & Selamat, J., (2017). Novel nanoliposomal encapsulated omega-3 fatty acids and their applications in food. *Food Chemistry, 230*, 690–696.

30. Pezeshky, A., et al., (2016). Vitamin A palmitate-bearing nanoliposomes: Preparation and characterization. *Food Bioscience, 13*, 49–55.

31. Rovoli, M., et al., (2019). *In vitro* and *in vivo* assessment of vitamin A encapsulation in a liposome-protein delivery system. *Journal of Liposome Research, 29*, 142–152.

32. Amiri, S., et al., (2018). Vitamin E loaded nanoliposomes: Effects of gamma oryzanol, polyethylene glycol and lauric acid on physicochemical properties. *Colloids and Interface Science Communications, 26*, 1–6.

33. Liu, N., & Park, H. J., (2009). Chitosan-coated nanoliposome as vitamin e carrier. *Journal of Microencapsulation, 26*, 235–242.

34. Zhao, L. P., et al., (2011). PEG-coated lyophilized proliposomes: Preparation, characterizations and *in vitro* release evaluation of vitamin E. *European Food Research and Technology, 232*(4), 647–654.

35. Rovoli, M., et al., (2014). *β-Lactoglobulin improves liposome's* encapsulation properties for vitamin e delivery. *Journal of Liposome Research, 24*, 74–81.

36. Marsanasco, M., et al., (2011). Liposomes as vehicles for vitamins E and C: An alternative to fortify orange juice and offer vitamin C protection after heat treatment. *Food Research International, 44*(9), 3039–3046.

37. Xia, S., et al., (2014). Chitosan/tripolyphosphate-nanoliposomes core-shell nanocomplexes as vitamin E carriers: Shelf-life and thermal properties. *International Journal of Food Science & Technology, 49*(5), 1367–1374.

38. Hassane, H. A., et al., (2020). Comparison of β-carotene loaded marine and egg phospholipids nanoliposomes. *Journal of Food Engineering, 283.*

39. Moraes, M., et al., (2013). Liposomes encapsulating beta-carotene produced by the proliposomes method: Characterization and shelf life of powders and phospholipid vesicles. *International Journal of Food Science and Technology, 48*, 274–282.

40. Carvalho, J. M. P., et al., (2015). Physico-chemical stability and structural characterization of thickened multilamellar beta-carotene-loaded liposome dispersions produced using a proliposome method. *Colloid and Polymer Science, 293*, 2171–2179.

41. Tanaka, Y., et al., (2020). Preparation of liposomes encapsulating β-carotene using supercritical carbon dioxide with ultrasonication. *Journal of Supercritical Fluids, 161.*

42. Bai, C., et al., (2019). Development of oral delivery systems with enhanced antioxidant and anticancer activity: Coix seed oil and β-carotene coloaded liposomes. *Journal of Agricultural and Food Chemistry, 67*, 406–414.

43. Toniazzo, T., et al., (2014). *β-carotene-loaded liposome dispersions stabilized with xanthan and guar gums:* Physico-chemical stability and feasibility of application in yogurt. *LWT - Food Science and Technology, 59*, 1265–1273.

44. Xia, F., et al., (2012). Preparation of lutein proliposomes by supercritical anti-solvent technique. *Food Hydrocolloids, 26*, 456–463.

45. Tan, C., et al., (2013). Liposomes as vehicles for lutein: Preparation, stability, liposomal membrane dynamics, and structure. *Journal of Agricultural and Food Chemistry, 61*(34), 8175–8184.

46. Jhan, S., & Pethe, A. M., (2020). Double-loaded liposomes encapsulating lycopene β-cyclodextrin complexes: Preparation, optimization, and evaluation. *Journal of Liposome Research, 30*, 80–92.

47. Sangsuriyawong, A., et al., (2019). Properties and bioavailability assessment of shrimp astaxanthin loaded liposomes. *Food Science and Biotechnology, 28*, 529–537.

48. Pan, L., Wang, H., & Gu, K., (2018). Nanoliposomes as vehicles for astaxanthin: Characterization, *in vitro* release evaluation and structure. *Molecules, 23.*

49. Qiang, M., et al., (2020). Effect of membrane surface modification using chitosan hydrochloride and lactoferrin on the properties of astaxanthin-loaded liposomes. *Molecules, 25.*

50. Tan, C., et al., (2014). Liposome as a delivery system for carotenoids: Comparative antioxidant activity of carotenoids as measured by ferric reducing antioxidant power, DPPH assay, and lipid peroxidation. *Journal of Agricultural and Food Chemistry, 62*(28), 6726–6735.

51. Tan, C., et al., (2014). Liposomes as delivery systems for carotenoids: Comparative studies of loading ability, storage stability and *in vitro* release. *Food & Function, 5*(6), 1232–1240.

52. Tan, C., et al., (2014). Modulation of the carotenoid bio-accessibility through liposomal encapsulation. *Colloids and Surfaces B: Biointerfaces, 123*, 692–700.

53. Tan, C., et al., (2016). Biopolymer-coated liposomes by electrostatic adsorption of chitosan (chitosomes) as novel delivery systems for carotenoids. *Food Hydrocolloids, 52*, 774–784.

54. Xia, F., et al., (2012). Preparation of coenzyme Q10 liposomes using supercritical anti-solvent technique. *Journal of Microencapsulation, 29*, 21–29.

55. Villanueva-Bermejo, D., & Temelli, F., (2020). Optimization of coenzyme Q10 encapsulation in liposomes using supercritical carbon dioxide. *Journal of CO₂ Utilization, 38*, 68–76.

56. Xu, S., Zhao, B., & He, D., (2015). Synthesis of highly dispersed nanoscaled CoQ₁₀ liposome by supercritical fluid. *Materials Letters, 142*, 283–286.

57. Xia, S., & Xu, S., (2006). Improved assay of coenzyme Q10 from liposomes by Tween-80 solubilization and UV spectrophotometry. *Journal of the Science of Food and Agriculture, 86*, 2119–2127.

58. Xia, S., Xu, S., & Zhang, X., (2006). Optimization in the Preparation of Coenzyme Q10 Nanoliposomes. *Journal of Agricultural and Food Chemistry, 54*(17), 6358–6366.

59. Xia, S., et al., (2007). Effect of coenzyme Q10 incorporation on the characteristics of nanoliposomes. *Journal of Physical Chemistry B, 111*, 2200–2207.

60. Zhao, G. D., et al., (2015). Development and characterization of a novel chitosan-coated antioxidant liposome containing both coenzyme Q10 and alpha-lipoic acid. *Journal of Microencapsulation, 32*, 157–165.

61. Zhao, G., Hu, C., & Xue, Y., (2018*). In vitro* evaluation of chitosan-coated liposome containing both coenzyme Q10 and alpha-lipoic acid: Cytotoxicity, antioxidant activity, and antimicrobial activity. *Journal of Cosmetic Dermatology, 17*, 258–262.

62. Shao, Y., Yang, L., & Han, K. H., (2015). TPGS-chitosome as an effective oral delivery system for improving the bioavailability of Coenzyme Q10. *European Journal of Pharmaceutics and Biopharmaceutics, 89*, 339–346.

63. Cheng, C., et al., (2017). Improved bioavailability of curcumin in liposomes prepared using a pH-driven, organic solvent-free, easily scalable process. *RSC Advances, 7*(42), 25978–25986.

64. Wu, Y., et al., (2020). Curcumin-loaded liposomes prepared from bovine milk and krill phospholipids: Effects of chemical composition on storage stability, *in-vitro* digestibility, and anti-hyperglycemic properties. *Food Research International, 136*.

65. Tai, K., et al., (2019). Effect of β-sitosterol on the curcumin-loaded liposomes: Vesicle characteristics, physicochemical stability, *in vitro* release and bioavailability. *Food Chemistry, 293*, 92–102.

66. Tai, K., et al., (2020). The stabilization and release performances of curcumin-loaded liposomes coated by high and low molecular weight chitosan. *Food Hydrocolloids, 99*, 105355.

67. Tai, K., et al., (2020). Stability and release performance of curcumin-loaded liposomes with varying content of hydrogenated phospholipids. *Food Chemistry, 326*.

68. Liu, Y., et al., (2015). Temperature-dependent structure stability and *in vitro* release of chitosan-coated curcumin liposome. *Food Research International*.

69. Peng, S., et al., (2017). Hybrid liposomes composed of amphiphilic chitosan and phospholipid: Preparation, stability, and bioavailability as a carrier for curcumin. *Carbohydrate Polymers, 156*, 322–332.

70. Chen, X., et al., (2015). The stability, sustained release, and cellular antioxidant activity of curcumin nanoliposomes. *Molecules, 20*(8), 14293–14311.

71. Chaves, M. A., et al., (2018). Structural characterization of multilamellar liposomes Coen encapsulating curcumin and vitamin D3. *Colloids and Surfaces A: Physicochemical and Engineering Aspects, 549,* 112–121.

72. Pu, C., et al., (2019). Stability enhancement efficiency of surface decoration on curcumin-loaded liposomes: Comparison of guar gum and its cationic counterpart. *Food Hydrocolloids, 87,* 29–37.

73. Gómez-Mascaraque, L. G., et al., (2017). Microencapsulation structures based on protein-coated liposomes obtained through electrospraying for the stabilization and improved bioaccessibility of curcumin. *Food Chemistry, 233,* 343–350.

74. Li, R., et al., (2017). Liposomes coated with thiolated chitosan as drug carriers of curcumin. *Materials Science and Engineering C, 80,* 156–164.

75. Cuomo, F., et al., (2018). *In-vitro* digestion of curcumin loaded chitosan-coated liposomes. *Colloids and Surfaces B: Biointerfaces, 168,* 29–34.

76. Tian, M. P., et al., (2018). Inducing sustained release and improving oral bioavailability of curcumin via chitosan derivatives-coated liposomes. *International Journal of Biological Macromolecules, 120,* 702–710.

77. Silva-Weiss, A., et al., (2018). Design of dipalmitoyl lecithin liposomes loaded with quercetin and rutin and their release kinetics from carboxymethyl cellulose edible films. *Journal of Food Engineering, 224,* 165–173.

78. Hao, J., et al., (2017). Encapsulation of the flavonoid quercetin with chitosan-coated nano-liposomes. *LWT-Food Science and Technology, 85,* 37–44.

79. Frenzel, M., et al., (2015). Physicochemical properties of WPI coated liposomes serving as stable transporters in a real food matrix. *LWT-Food Science and Technology, 63*(1), 527–534.

80. Huang, M., et al., (2017). Encapsulation of flavonoids in liposomal delivery systems: The case of quercetin, kaempferol and luteolin. *Food & Function, 8*(9), 3198–3208.

81. Hamed, I., Özogul, F., & Regenstein, J. M., (2016). Industrial applications of crustacean by-products (chitin, chitosan, and chitooligosaccharides): A review. *Trends in Food Science & Technology, 48,* 40–50.

82. Chun, J. Y., et al., (2013). Formation and stability of multiple-layered liposomes by layer-by-layer electrostatic deposition of biopolymers. *Food Hydrocolloids, 30,* 249–257.

83. Madrigal-Carballo, S., et al., (2010). Biopolymer coating of soybean lecithin liposomes via layer-by-layer self-assembly as novel delivery system for ellagic acid. *Journal of Functional Foods, 2,* 99–106.

84. Tan, C., et al., (2013). Dual effects of chitosan decoration on the liposomal membrane physicochemical properties as affected by chitosan concentration and molecular conformation. *Journal of Agricultural and Food Chemistry, 61*(28), 6901–6910.

85. Tan, C., et al. (2015). Insights into chitosan multiple functional properties: The role of chitosan conformation in the behavior of liposomal membrane. *Food & Function, 6*(12), 3702–3711.

86. Ball, G. F. M., (2005). Vitamins in foods: Analysis, bioavailability, and stability. *Vitamins in Foods: Analysis, Bioavailability, and Stability,* 1–787.

87. Katouzian, I., & Jafari, S. M., (2016). Nano-encapsulation as a promising approach for targeted delivery and controlled release of vitamins. In: *Trends in Food Science and Technology* (pp. 34–48).

88. Dhakal, S. P., & He, J., (2020). Microencapsulation of vitamins in food applications to prevent losses in processing and storage: A review. In: *Food Research International*.

89. Koudelka, S., et al., (2015). Liposomal delivery systems for anti-cancer analogues of vitamin E. In: *Journal of Controlled Release* (pp. 59–69).

90. Salgado, J., Villalain, J., & Gomez-Fernandez, J. C., (1993). *α-tocopherol interacts with natural micelle-forming single-chain phospholipids stabilizing the bilayer phase. Archives of Biochemistry and Biophysics, 306*, 368–376.

91. Bochicchio, S., et al., (2016). Vitamin delivery: Carriers based on nanoliposomes produced via ultrasonic irradiation. *LWT - Food Science and Technology, 69*, 9–16.

92. Zhang, J., & Wang, S., (2009). Topical use of coenzyme Q10-loaded liposomes coated with trimethyl chitosan: Tolerance, precorneal retention and anti-cataract effect. *International Journal of Pharmaceutics, 372*, 66–75.

93. Araiza-Calahorra, A., Akhtar, M., & Sarkar, A., (2018). Recent advances in emulsion-based delivery approaches for curcumin: From encapsulation to bio-accessibility. In: *Trends in Food Science and Technology* (pp. 155–169).

94. Kharat, M., et al., (2017). Physical and chemical stability of curcumin in aqueous solutions and emulsions: Impact of PH, temperature, and molecular environment. *Journal of Agricultural and Food Chemistry, 65*, 1525–1532.

95. Wang, W., et al., (2016). The biological activities, chemical stability, metabolism and delivery systems of quercetin: A review. In: *Trends in Food Science and Technology* (pp. 21–38).

96. Sherry, M., et al., (2013). Essential oils encapsulated in liposomes: A review. *J. Liposome Res., 23*(4), 268–275.

97. Sedaghat, D. A., et al., (2020). Recent advances in food colloidal delivery systems for essential oils and their main components. *Trends in Food Science & Technology, 99*, 474–486.

98. Prakash, B., et al., (2018). Nanoencapsulation: An efficient technology to boost the antimicrobial potential of plant essential oils in food system. *Food Control, 89*, 1–11.

99. Hammoud, Z., et al., (2019). New findings on the incorporation of essential oil components into liposomes composed of lipoid S100 and cholesterol. *Int. J. Pharm., 561*, 161–170.

100. Khatibi, S. A., et al., (2018). Effect of nanoliposomes containing *Zataria multiflora* Boiss. essential oil on gene expression of Shiga toxin 2 in *Escherichia coli* O157:H7. *J. Appl. Microbiol., 124*(2), 389–397.

101. Liolios, C. C., et al., (2009). Liposomal incorporation of carvacrol and thymol isolated from the essential oil of *Origanum dictamnus* L. and *in vitro* antimicrobial activity. *Food Chemistry, 112*(1), 77–83.

102. Engel, J. B., et al., (2017). Antimicrobial activity of free and liposome-encapsulated thymol and carvacrol against *Salmonella* and *Staphylococcus aureus* adhered to stainless steel. *Int. J. Food Microbiol., 252*, 18–23.

103. Cui, H., Zhao, C., & Lin, L., (2015). The specific antibacterial activity of liposome-encapsulated clove oil and its application in tofu. *Food Control, 56*, 128–134.

104. Cui, H., Zhou, H., & Lin, L., (2016). The specific antibacterial effect of the Salvia oil nanoliposomes against *Staphylococcus aureus* biofilms on milk container. *Food Control, 61*, 92–98.

105. Lin, L., et al., (2016). Liposome containing nutmeg oil as the targeted preservative against Listeria monocytogenes in dumplings. *RSC Advances, 6*(2), 978–986.
106. Cui, H. Y., Wu, J., & Lin, L., (2016). Inhibitory effect of liposome-entrapped lemongrass oil on the growth of *Listeria monocytogenes* in cheese. *J. Dairy Sci., 99*(8), 6097–6104.
107. Wang, L., et al., (2011). Synergistic antimicrobial activities of natural essential oils with chitosan films. *J. Agric. Food Chem., 59*(23), 12411–12419.
108. Pabast, M., et al., (2018). Effects of chitosan coatings incorporating with free or nano-encapsulated *Satureja* plant essential oil on quality characteristics of lamb meat. *Food Control, 91*, 185–192.
109. Feyzioglu, G. C., & Tornuk, F., (2016). Development of chitosan nanoparticles loaded with summer savory (*Satureja hortensis* L.) essential oil for antimicrobial and antioxidant delivery applications. *LWT, 70*, 104–110.
110. Azevedo, A. N., et al., (2014). Response surface methodology for optimization of edible chitosan coating formulations incorporating essential oil against several foodborne pathogenic bacteria. *Food Control, 43*, 1–9.
111. Cui, H. Y., et al., (2016). Anti-listeria effects of chitosan-coated nisin-silica liposome on cheddar cheese. *J. Dairy Sci., 99*(11), 8598–8606.
112. Lin, L., et al., (2019). Characterization of chrysanthemum essential oil triple-layer liposomes and its application against Campylobacter jejuni on chicken. *LWT, 107*, 16–24.
113. Cui, H., et al., (2016). Liposome containing cinnamon oil with antibacterial activity against methicillin-resistant *Staphylococcus aureus* biofilm. *Biofouling, 32*(2), 215–225.
114. Cui, H., et al., (2018). Fabrication of chitosan nanofibers containing tea tree oil liposomes against *Salmonella spp.* in chicken. *LWT, 96*, 671–678.
115. Lin, L., et al., (2018). Improving the stability of thyme essential oil solid liposome by using beta-cyclodextrin as a cryoprotectant. *Carbohydr. Polym., 188*, 243–251.
116. Keivani, N. F., et al., (2019). Investigation of physicochemical properties of essential oil loaded nanoliposome for enrichment purposes. *LWT, 105*, 282–289.
117. Ghodrati, M., Farahpour, M. R., & Hamishehkar, H., (2019). Encapsulation of peppermint essential oil in nanostructured lipid carriers: *In-vitro* antibacterial activity and accelerative effect on infected wound healing. *Colloids and Surfaces A: Physicochemical and Engineering Aspects, 564*, 161–169.
118. Risaliti, L., et al., (2019). Liposomes loaded with *Salvia triloba* and *Rosmarinus officinalis* essential oils: *In vitro* assessment of antioxidant, anti-inflammatory, and antibacterial activities. *Journal of Drug Delivery Science and Technology, 51*, 493–498.
119. Andrews, M., et al., (2014). effect of calcium, tannic acid, phytic acid and pectin over iron uptake in an *in vitro* Caco-2 cell model. *Biological Trace Element Research, 158*(1), 122–127.
120. Abbasi, S., & Azari, S., (2011). Efficiency of novel iron microencapsulation techniques: Fortification of milk. *International Journal of Food Science & Technology, 46*(9), 1927–1933.
121. Cengiz, A., et al., (2018). Oxidative stability of emulsions fortified with iron: The role of liposomal phospholipids. *Journal of the Science of Food and Agriculture.*

122. Yi, X., et al., (2018). Liposomal vesicles-protein interaction: Influences of iron liposomes on emulsifying properties of whey protein. *Food Hydrocolloids*.
123. Hermida, L. G., et al., (2011). Preparation and characterization of iron-containing liposomes: Their effect on soluble iron uptake by Caco-2 cells. *J. Liposome Res., 21*(3), 203–212.
124. Layrisse, M., et al., (2000). Iron bioavailability in humans from breakfasts enriched with iron bis-glycine chelate, phytates and polyphenols. *J. Nutr., 130*(9), 2195–2199.
125. Osman, A. K., & Al-Othaimeen, A., (2002). Experience with ferrous bis-glycine chelate as an iron fortificant in milk. *International Journal for Vitamin & Nutrition Research, 72*(4), 257–263.
126. Shubham, K., et al., (2020). Iron deficiency anemia: A comprehensive review on iron absorption, bioavailability and emerging food fortification approaches. *Trends in Food Science & Technology, 99*, 58–75.
127. Ding, B., et al., (2011). Preparation, characterization and the stability of ferrous glycinate nanoliposomes. *Journal of Food Engineering, 102*(2), 202–208.
128. Palmeri, A., et al., (2016). Salidroside, a bioactive compound of Rhodiola rosea, ameliorates memory and emotional behavior in adult mice. *Journal of Alzheimer's Disease JAD, 52*(1), 65–75.
129. Zhao, X., et al., (2013). Salidroside liposome formulation enhances the activity of dendritic cells and immune responses. *International Immunopharmacology, 17*(4), 1134–1140.
130. Ma, C., et al., (2015). An UPLC-MS-based metabolomics investigation on the anti-fatigue effect of salidroside in mice. *Journal of Pharmaceutical & Biomedical Analysis, 105*, 84–90.
131. Zhang, X. R., et al., (2016). Salidroside-regulated lipid metabolism with down-regulation of miR-370 in type 2 diabetic mice. *European Journal of Pharmacology, 779*, 46–52.
132. Fan, M., et al., (2007). Effect of different preparation methods on physicochemical properties of salidroside liposomes. *Journal of Agricultural & Food Chemistry, 55*(8), 3089–3095.
133. Feng, Y., et al., (2015). Optimization on Preparation conditions of salidroside liposome and its immunological activity on PCV-2 in mice. *Evid. Based Complement Alternat. Med., 2015*, 178128.
134. Reque, P. M., et al., (2014). Cold storage of blueberry (*Vaccinium* spp.) fruits and juice: Anthocyanin stability and antioxidant activity. *Journal of Food Composition & Analysis, 33*(1), 111–116.
135. Chung, C., et al., (2017). Stability improvement of natural food colors: Impact of amino acid and peptide addition on anthocyanin stability in model beverages. *Food Chemistry, 218*, 277–284.
136. Flores, F. P., et al., (2014). Total phenolics content and antioxidant capacities of microencapsulated blueberry anthocyanins during *in vitro* digestion. *Food Chemistry, 153*, 272–278.
137. Guldiken, B., et al., (2018). Physical and chemical stability of anthocyanin-rich black carrot extract-loaded liposomes during storage. *Food Research International, 108*, 491–497.

138. Bian, C., et al., (2015). Study on preparation and antioxidant activity *in vitro* of xanthophyll-anthocyanins liposome. *Cereals & Oils.*

139. Jahadi, M., et al., (2016). Modeling of proteolysis in Iranian brined cheese using proteinase-loaded nanoliposome. *International Journal of Dairy Technology, 69*(1), 57–62.

140. Mohammadi, M., et al., (2013). Application of liposome nanocarrier in cheese production and ripening. *Memoirs of Tokyo Metropolitan Institute of Technology, 10*(1), 109–115.

141. Vafabakhsh, Z., et al., (2013). Stability and catalytic kinetics of protease loaded liposomes. *Biochemical Engineering Journal, 72,* 11–17.

142. Jahadi, M., et al., (2015). The encapsulation of flavor zyme in nanoliposome by heating method. *Journal of Food Science & Technology, 52*(4), 2063–2072.

143. Kim, C. K., (1999). Development of dried liposomes containing β-galactosidase for the digestion of lactose in milk. *International Journal of Pharmaceutics, 183*(2), 185–193.

144. Maherani, B., et al., (2012). Influence of lipid composition on physicochemical properties of nanoliposomes encapsulating natural dipeptide antioxidant L-carnosine. *Food Chemistry, 134*(2), 632–640.

145. Liu, W., et al., (2013). Stability during *in vitro* digestion of lactoferrin-loaded liposomes prepared from milk fat globule membrane-derived phospholipids. *Journal of Dairy Science, 96*(4), 2061–2070.

146. Liu, W., et al., (2017). Kinetic stability and membrane structure of liposomes during *in vitro* infant intestinal digestion: Effect of cholesterol and lactoferrin. *Food Chem., 230,* 6–13.

147. Kosaraju, S. L., Tran, C., & Lawrence, A., (2006). Liposomal delivery systems for encapsulation of ferrous sulfate: Preparation and characterization. *Journal of Liposome Research, 16*(4), 347–358.

148. Battistelli, M., Salucci, S., & Falcieri, E., (2018). Morphological evaluation of liposomal iron carriers. *Microscopy Research and Technique, 81*(11), 1295–1300.

149. Ding, B. M., et al., (2009). Preparation and pH stability of ferrous glycinate liposomes. *Journal of Agricultural and Food Chemistry, 57*(7), 2938–2944.

150. Fan, M. H., et al., (2008). Preparation of salidroside nano-liposomes by ethanol injection method and *in vitro* release study. *European Food Research and Technology, 227*(1), 167–174.

151. Lee, C., & Na, K., (2020). Anthocyanin-loaded liposomes prepared by the pH-gradient loading method to enhance the anthocyanin stability, antioxidation effect and skin permeability. *Macromolecular Research, 28*(3), 289–297.

152. Zhao, L., Temelli, F., & Chen, L., (2017). Encapsulation of anthocyanin in liposomes using supercritical carbon dioxide: Effects of anthocyanin and sterol concentrations. *Journal of Functional Foods, 34,* 159–167.

153. Zhao, L., & Temelli, F., (2017). Preparation of anthocyanin- liposomes using an improved supercritical carbon dioxide method. *Innovative Food Science & Emerging Technologies, 39,* 119–128.

154. Ma, Q. H., et al., (2007). Effect of surfactants on preparation and skin penetration of tea polyphenols liposomes. In: *2007 IEEE/ICME International Conference on Complex Medical Engineering* (Vol. 1–4, pp. 209–212).

155. Lukawski, M., et al., (2019). New oral liposomal vitamin C formulation: Properties and bioavailability. *Journal of Liposome Research.*
156. Hudiyanti, D., et al., (2019). Encapsulation of vitamin C in sesame liposomes: Computational and experimental studies. *Open Chemistry, 17*(1), 537–543.
157. Davis, J. L., et al., (2016). Liposomal-encapsulated ascorbic acid: Influence on vitamin C bioavailability and capacity to protect against ischemia-reperfusion injury. *Nutrition and Metabolic Insights, 9,* 25–30.
158. Kaddah, S., et al., (2018). Cholesterol modulates the liposome membrane fluidity and permeability for a hydrophilic molecule. *Food Chem Toxicol., 113,* 40–48.
159. Zhao, L., et al., (2015). Preparation of liposomes using supercritical carbon dioxide technology: Effects of phospholipids and sterols. *Food Research International, 77,* 63–72.
160. Tai, K., et al., (2018). The effect of sterol derivatives on properties of soybean and egg yolk lecithin liposomes: Stability, structure and membrane characteristics. *Food Res. Int., 109,* 24–34.
161. Xi, L., et al., (2020). Novel thermosensitive polymer-modified liposomes as nano-carrier of hydrophobic antitumor drugs. *J. Pharm. Sci.*
162. Wang, Z. Y., et al., (2016). Preparation, characterization, and efficacy of thermosensitive liposomes containing paclitaxel. *Drug Deliv., 23*(4), 1222–1231.
163. Eleftheriou, K., et al., (2020). A combination drug delivery system employing thermosensitive liposomes for enhanced cell penetration and improved *in vitro* efficacy. *Int. J. Pharm., 574,* 118912.
164. Fan, Y., et al., (2017). Study of the pH-sensitive mechanism of tumor-targeting liposomes. *Colloids Surf. B Biointerfaces, 151,* 19–25.
165. Paliwal, S. R., Paliwal, R., & Vyas, S. P., (2015). A review of mechanistic insight and application of pH-sensitive liposomes in drug delivery. *Drug Deliv., 22*(3), 231–242.
166. Zhou, W., et al., (2012). Characteristics, phase behavior and control release for copolymer-liposome with both pH and temperature sensitivities. *Colloids and Surfaces A: Physicochemical and Engineering Aspects, 395,* 225–232.
167. Qiu, D., & An, X., (2013). Controllable release from magneto liposomes by magnetic stimulation and thermal stimulation. *Colloids Surf. B Biointerfaces, 104,* 326–329.
168. Shah, S. A., et al., (2016). Doxorubicin-loaded photosensitive magnetic liposomes for multimodal cancer therapy. *Colloids Surf. B Biointerfaces, 148,* 157–164.
169. Zou, L. Q., et al., (2014). Characterization and bioavailability of tea polyphenol nanoliposome prepared by combining an ethanol injection method with dynamic high-pressure micro fluidization. *J. Agric. Food Chem., 62*(4), 934–941.
170. Vanaja, K., et al., (2013). Liposomes as carriers of the lipid-soluble antioxidant resveratrol: Evaluation of amelioration of oxidative stress by additional antioxidant vitamin. *Life Sci., 93*(24), 917–923.
171. Zhou, W., et al., (2014). Storage stability and skin permeation of vitamin C liposomes improved by pectin coating. *Colloids Surf. B Biointerfaces, 117,* 330–337.
172. Sebaaly, C., et al., (2016). Effect of composition, hydrogenation of phospholipids and lyophilization on the characteristics of eugenol-loaded liposomes prepared by ethanol injection method. *Food Bioscience, 15,* 1–10.

173. Zou, L., et al., (2015). A novel delivery system dextran sulfate coated amphiphilic chitosan derivatives-based nanoliposome: Capacity to improve *in vitro* digestion stability of (–)-epigallocatechin gallate. *Food Research International, 69*, 114–120.
174. Shao, P., et al., (2018). Environmental stress stability of pectin-stabilized resveratrol liposomes with different degrees of esterification. *Int. J. Biol. Macromol., 119*, 53–59.
175. Lopes, N. A., Pinilla, C. M. B., & Brandelli, A., (2017). Pectin and polygalacturonic acid-coated liposomes as novel delivery system for nisin: Preparation, characterization and release behavior. *Food Hydrocolloids, 70*, 1–7.
176. Wang, X., et al., (2019). Sodium carboxymethyl cellulose modulates the stability of cinnamaldehyde-loaded liposomes at high ionic strength. *Food Hydrocolloids, 93*, 10–18.
177. Mo, L., et al., (2018). PEGylated hyaluronic acid-coated liposome for enhanced *in vivo* efficacy of sorafenib via active tumor cell targeting and prolonged systemic exposure. *Nanomedicine, 14*(2), 557–567.
178. Zhou, F., et al., (2018). Chitosan-coated liposomes as delivery systems for improving the stability and oral bioavailability of acteoside. *Food Hydrocolloids, 83*, 17–24.
179. Mohanraj, V. J., Barnes, T. J., & Prestidge, C. A., (2010). Silica nanoparticle coated liposomes: A new type of hybrid nanocapsule for proteins. *Int. J. Pharm., 392*(1, 2), 285–293.
180. Tan, C., et al., (2015). Biopolymer-lipid bilayer interaction modulates the physical properties of liposomes: Mechanism and structure. *J. Agric. Food Chem., 63*(32), 7277–7185.
181. Liu, W., et al., (2019). Advances and challenges in liposome digestion: Surface interaction, biological fate, and GIT modeling. *Adv. Colloid Interface Sci., 263*, 52–67.
182. Zhang, H., et al., (2012). Therapeutic effects of liposome-enveloped *Ligusticum chuanxiong* essential oil on hypertrophic scars in the rabbit ear model. *PLoS One, 7*(2), e31157.
183. Chang, M., et al., (2018). Comparative analysis of EPA/DHA-PL forage and liposomes in orotic acid-induced nonalcoholic fatty liver rats and their related mechanisms. *J. Agric. Food Chem., 66*(6), 1408–1418.
184. Du, L., et al., (2016). Transport and uptake effects of marine complex lipid liposomes in small intestinal epithelial cell models. *Food Funct., 7*(4), 1904–1914.
185. Xia, S., et al., (2009). Nanoliposomes mediate coenzyme Q10 transport and accumulation across human intestinal Caco-2 cell monolayer. *J. Agric. Food Chem., 57*(17), 7989–7996.
186. Li, Y., et al., (2017). Mucus interactions with liposomes encapsulating bio-actives: Interfacial tensiometry and cellular uptake on Caco-2 and cocultures of Caco-2/HT29-MTX. *Food Res. Int., 92*, 128–137.
187. Zhang, T., et al., (2019). Transepithelial transport route and liposome encapsulation of milk-derived ACE-inhibitory peptide Arg-Leu-Ser-Phe-Asn-Pro. *J. Agric. Food Chem., 67*(19), 5544–5551.
188. Yang, G., et al., (2014). *In vitro* and *in vivo* antitumor effects of folate-targeted ursolic acid stealth liposome. *J. Agri. Food Chem., 62*(10), 2207–2215.

CHAPTER 5

Liposomal Carrier Systems

SYED SARIM IMAM,[1] SULTAN ALSHEHRI,[1] MOHAMMED JAFAR,[2] MOHAMMED ASADULLAH JAHANGIR,[3] MOHAMAD TALEUZZAMAN,[4] and AMEEDUZZAFAR ZAFAR[5]

[1]Department of Pharmaceutics, College of Pharmacy, King Saud University, Riyadh – 11451, Saudi Arabia

[2]Department of Pharmaceutics, College of Clinical Pharmacy, Imam Abdulrahman Bin Faisal University, Dammam, Saudi Arabia

[3]Department of Pharmaceutics, Nibha Institute of Pharmaceutical Sciences, Rajgir, Nalanda – 803116, Bihar, India

[4]Faculty of Pharmacy, Maulana Azad University, Jodhpur – 342802, Rajasthan, India

[5]Department of Pharmaceutics, College of Pharmacy, Jouf University, Sakaka, Aljouf, Saudi Arabia

ABSTRACT

Liposomes are nanometric-sized lipid vesicles made up of one or more lipid bilayers. It consists of single or different amphiphilic lipids. It can entrap a variety of molecules such as drugs, proteins, as well as nutraceuticals. It has been widely used to enhanced the therapeutic efficacy of a variety of molecules for different topical and systemic diseases. It can be prepared by a variety of methods to load the drugs. Due to the unique property of liposomes make them a potential therapeutic carrier to enhance the nutraceutical biological efficacy by increasing the solubility, bioavailability, and cellular uptake. The therapeutic efficacy can be enhanced by delivering the drug at the site of action by maintain the therapeutic concentration at the target area.

5.1 INTRODUCTION

Liposomes word derived from the Greek words, lipo means fat, and soma means structure. It is nano-sized spherical structure consist of a phospholipid bilayer with a liquid core. The size of the liposomes varies from 50 nm to 1 μm in diameter. The hydrophobic tails of the phospholipid groups face each other, and the hydrophilic head face towards the inner core of the liposomes. The hydrophilic substance encapsulates within the liposome core and hydrophobic substance partitioned within the bilayer [63]. It can encapsulate both the hydrophilic and hydrophobic molecules. It is widely accepted in the cosmetic delivery systems. Liposomes improve the therapeutic efficacy, pharmacokinetic profile, drug stability, and also enhance the drug circulation. It also helps to reduce the toxicity of the encapsulated drug. The liposomes have shown great potential for the encapsulation and controlled release of nutritionally active compounds. There are very limited liposomal delivery systems for nutraceuticals that have been reported. There are multiple reasons for the limited availability is the use of organic solvents, stability problems and production of large volume sample. The number of production methods like high pressure homogenization, microfluidization are available to prepare the large volume samples. The several nutraceuticals (vitamins, enzymes, herbal extract) loaded liposomal formulations have been reported to enhance the solubilization and absorption [2].

Liposomes are known small unilamellar (SUVs) or large unilamellar (LUVs) when they contain only a single bilayer membrane. The diameter of SUVs and LUVs range between 0.02 and 0.05 μm and 0.05–0.1 μm, respectively. For liposomes, the theoretical minimum diameter possible is 0.02–0.025 μm, that is, restricted by the surface curvature (responsible for phospholipid head group crowding) [3]. The single bilayer has quite permeable nature, especially to water-soluble molecules. Although the applications of liposomes are vast yet the utilization is limited to the food industry compared to cosmetic and pharmaceutical industries. This is attributed to difficulties in finding safe, solvent-free and lost cost ingredient; also, there is limited low-cost processing methods required for production of large volume of liposomes with consistent characteristics.

Although techniques such as pro-liposomes, microfludization, and high-pressure homogenization (HPM) solve many problems related to low-cost processing methods. For suitable low-cost ingredient, researches are going on for the use of commercially cheaper lecithin fraction. In recent years, liposomes technology has been used for the formulation of several nutraceutical products [2]; with improved nutrient/bioactive solubilization and enhanced absorption. However, there are hurdles such as complexity regarding the stability of these systems during processing and their less understood behavior during gastrointestinal transit. The liposomes resistance to pH and intestinal bile salts seems to be attributed to the composition of the membrane [4]. Research is required to warrant the claims of enhanced bioavailability and efficacy of the liposomes after reaching the intestinal environment. Most of the available research has focused on egg and soy as obvious source of phospholipids, in recent years, however, milk-derived phospholipids as ingredient for liposomes has been of potential interest. Interest in the use of such phospholipids in food applications may also increase due to their increased commercial availability. These milk phospholipids are considered advantageous due to their unique composition and their reported health benefits [5].

This chapter gives insight into the formation, structure, and physico-chemical properties of liposomes, the main characterization methods, and their potential applications from food technology perspective. The applications of liposomes in food systems and their potential uses as carriers for bioactive compounds in foods will be highlighted.

5.2 METHOD OF PREPARATION

The method of preparation of liposomes is accomplished in four stages which are:

- Drying out lipid from organic solvents;
- Liquid is dispersed in aqueous media;
- Resultant liposome is purified; and
- The final product is analyzed.

Method of liposome preparation includes both, i.e., preparing the liposome and loading of the drug into it. Both of these can be achieved by either of the following techniques:

- Active loading;
- Passive loading.

Passive loading techniques involves three different types of methods which are:

- Solvent dispersion method;
- Mechanical dispersion method; and
- Removal of non-encapsulated material or detergent removal method [6, 7].

5.2.1 SOLVENT DISPERSION METHOD

5.2.1.1 ETHANOL INJECTION

A lipid-based ethanolic solution is injected rapidly into buffer which is in excess. Multilamellar vesicles (MLVs) gets formed instantaneously. However, this technique comes with a disadvantage of providing heterogeneous products which ranges between 30 nm and 100 nm, the formed liposomes are dilute and removal ethanol completely is a hectic process as it forms an azeotropic mixture with water and causes the biologically active molecule to get inactivated even under low ethanol content [8].

5.2.1.2 SOLVENT VAPORIZATION OR ETHER INJECTION

Lipid is dissolved in either ether-methanol mixture of diethyl ether and then injected into the aqueous solution of the encapsulated material at a temperature between 55°C and 65°C under reduced pressure. Removal of ether under vacuum forms liposomes. This technique comes with the disadvantage of providing heterogeneous products which range between 70 nm and 200 nm, and it also exposes the encapsulated material to the organic solvent at elevated temperature [9].

5.2.1.3 REVERSED-PHASE EVAPORATION METHOD

It provided a novel platform for liposomal technology as with this approach liposomes were prepared with high aqueous to lipid ratio which allowed to entrap high content of hydrophilic materials. Reverse phase evaporation technique forms inverted micelles. The shapes of the inverted micelles are done by sonicating the buffered aqueous phase having hydro-philic molecules to be encapsulated into the liposomes. The amphiphilic molecules are solubilized into the organic phase. The organic solvent is slowly eliminated which converts the inverted micelles into gel form and viscous state. The gel phase collapses at a critical point and the inverted micelles gets distributed. Phospholipids which are in excess contribute to the formation of lipid bilayer around the residual micelles which eventu-ally forms liposomes. Liposomes prepared by reversed-phase evaporation method provide the advantage of being made from different lipid formula-tions and provides more than four times higher aqueous volume to lipid ratio than liposomes formed by hand-shaken method or in multilamellar liposomes [7, 10].

Firstly, the W/O emulsion is prepared by sonicating the two-phase system which contains phospholipids in organic solvents like that of diethyl ether or isopropyl ether or a mixture of chloroform and isopropyl ether with aqueous buffer. The organic solvents are detached by lowering the pressure, resulting in the formation of viscous gel. The liposomes get its shape during the detachment of the residual solvent by continued rotary evaporation conducted at low pressure. High encapsulation efficiency (EE) of more than 65% can be achieved with a medium like 0.01 M NaCl having low ionic strength. This method is capable of encapsulating small to large macromolecules. However, this technique comes with the disadvantage of exposing the encapsulated material to the organic solvent and the low sonication period [62]. Handa et al. demonstrated modified reverse-phase evaporation and concluded high EE of about 80% with the developed liposomes [11].

5.2.2   MECHANICAL DISPERSION METHOD

The mechanical dispersion method involves some techniques which are discussed in the subsections.

5.2.2.1   FRENCH PRESS CELL EXTRUSION

French pressure cell involves the extrusion of MLV through a narrow orifice. During this process, proteins are not significantly pretentious as they are involved in sonication [12]. The vesicles formed by the French press are capable of recalling the solutes which are entrapped in larger quantities than produced by sonication or detergent removal technique [13, 14]. This method handles the unstable materials gently and provides numerous advantages over sonication techniques [15, 16]. Liposomes prepared by French press cell extrusion are usually larger than sonicated single unilamellar vesicles (ULV). However, this technique comes with the drawback of attaining high temperature, which is a bit difficult to achieve, and also it involves comparatively small working volumes of about 50 ml only [6].

5.2.2.2   SONICATION

Sonication technique is one of the most widely employed methods for preparing single ULV. MLV are prepared using probe or bath type soni-cator in a passive atmosphere. This technique comes with the disadvan-tage of degradation of encapsulated materials and phospholipids, low EE, prone to metal pollution from probe tip, elimination of larger molecules, heterogeneous product containing mixture of multilamellar and single ULV [6]. In probe sonication technique, the tip of the sonicator is directly engaged into lipid dispersion. Local hotness occurs at the tip; thus, it requires engrossing into ice/water bath. In probe sonicators, titanium may get slough off and often pollute the solution. In bath sonication, the temperature control of lipid dispersion is usually easy. The material which is to be sonicated can be protected in a sterile vessel or under an inert atmosphere [17].

5.2.2.3   FREEZE-THAWED LIPOSOMES

Single unilamellar vesicle is rapidly frozen and thawed gradually. The sonication dispersion is short-lived and gets aggregated to form large unilamellar vesicles (LUV). These large vesicles are formed by the fusion of single ULV throughout the thawing and freezing process. This tech-nique may get inhibited by increasing the ionic strength and by increasing the concentration of phospholipids. With this technique the EE of only 20–30% can be obtained [18].

5.2.3   DETERGENT REMOVAL METHOD OR REMOVAL OF NON-ENCAPSULATED MATERIAL

5.2.3.1   DIALYSIS

At critical micellar concentration, detergents can solubilize the lipids. As the detergent gets detached, the micelles become increasingly better-off in phospholipid which further combines to form LUV. The detergent is further removed by dialysis technique [19]. LipoPrep, a commercial product based on dialysis system by Diachema AG, Switzerland, is capable of eliminating detergents. Equilibrium dialysis is performed in dialysis bags engaged in large detergent-free buffers [20].

5.2.3.2   DETERGENT REMOVAL OF MIXED MICELLES

Absorption of detergent is done by shaking micellar solution with beaded organic polystyrene absorbers like Bio-beads SM2, XAD-2 beads, etc. It provides the advantage that they can eliminate detergents with very low critical micellar concentration.

5.2.3.3   GEL PERMEATION CHROMATOGRAPHY

Size simple chromatography is used to deplete detergent. Sephacryl S200-S1000, Sephadex G-50, Sephadex G-1 00, Sepharose 2B–6B can be used for gel filtration. The liposomes do not get penetrated into the pores of the beads which are packed in the column. At a slow flow rate, the separation of liposomes from detergents is better than at a high flow rate. Polysaccharide beads swells and absorbs amphiphilic lipids, thus makes the pre-treatment necessary. Pre-treatment is done by pre-saturating the gel filtration column with lipids [14].

5.2.3.4   DILUTION

The micellar solution of detergent and phospholipid buffer is diluted which increases the micellar size and polydispersity. Dilution beyond mixed micellar phase, vesicles are formed from polydispersed micelles [14].

5.2.4   DRUG LOADING IN LIPOSOMES

Loading of drug can be achieved either by active or passive technique. Lipophilic drugs like Taxol or amphotericin B are allowed to directly combine with liposomes during the formation of vesicles. The retention and amount to be taken up is influenced by drug-lipid interaction. Depending upon the solubility of the drug in the liposomal membrane, higher trapping efficiency can be achieved. Passive encapsulation of hydrophilic drugs is dependent on the ability of liposomes to entrap drugs dissolved in aqueous buffers. Factors like drug solubility and delimited trapped volume influence the trapping effectiveness [14].

5.3   APPLICATION OF NUTRACEUTICAL LOADED LIPOSOMES

To date, the information published in the scientific literature on the use of liposomes in foods is comparatively very low; therefore, it is quite hard at this stage to present a detailed comparison of the likely success of proposed applications. One of the reasons for this less information on liposomes in foodstuffs could be the lack of their commercial value (it could be due to unavailability of suitable methods and excipients for their commercial-scale manufacturing) and the other reasons attributed could be the commercial sensitivity within the nutraceutical and food manufacturing companies, wherein several novel developments are usually kept in-house. However, brief discussions on prospective applications of liposomes in the food manufacturing companies is presented as following, and it includes preservation of vital constituents and improving the efficacy of some ingredients through restricting unwanted exploitation to the inner part of the liposomes. After going through the substantial amount of literature on liposomal manufacturing, it is interesting to record that several investigations have utilized organic solvents to solubilize the lipid stuff, and thin-film rehydration technique to produce MLV or subsequent processing to get SUV. However, the current studies report on utilization of HPM or microfluidization in the preparation of food grade SUV. Unlike synthetic phospholipids used in the pharmaceutical grade liposomal preparations, in the manufacturing of food-grade liposomes, phospholipids of natural material such as egg, soy, or milk are commonly used. The capability of liposomes to protect the enclosed material from the harsh external environmental conditions is the basis of its several potential uses. In several different food formulations, a huge amount of an active food constituents (proteins, vitamins, etc.), are incorporated in the formulation to compensate the losses that usually occur while manufacturing or storage. Substantial decrease in usual loss by the use of liposomes may overcome the exigency for this, ameliorating the economics of the formulation and alleviating potential toxicity issues. Other important applications of liposomes include, two or more incompatible substances that can be used together in the same liposome. Moreover, a bitter taste of substances with nutritional or pharmacological values could be overcome, thus improving the flavor and taste of the final product.

To provide protection by entrapment to the vital antioxidant constituents is commonly cited and reported in several scientific literatures as an important application of liposomes in foodstuffs. The progress towards the utilization of unsaturated fat in place of saturated fat in the diet has led to the high chances of proneness of many fat-containing foods to oxidation, mainly in emulsion-based nutritional products such as mayonnaise, margarine, and spreads. α-tocopherol and ascorbic acid can act synergistically as natural antioxidant combination. As reported by Stahl et al. [21], the α-tocopherol reacts with peroxy radicals in the continuous phase of the food to form α-tocopheroxyl radicals, which are less effective than peroxy radicals in oxidation chain reaction initiation. Ascorbic acid can reduce α-tocopheroxyl radical into α-tocopherol. This new arrangement prolongs and improves the antioxidant activity of the α-tocopherol. Nevertheless, because of its hydrophobic nature α-tocopherol cannot interact with hydrophilic ascorbic acid. There are several derivatives of ascorbic acid, which posses' hydrophobic characteristics, and it is possible to use them in nutritional preparations, but the problem is that their effective dispersion requires high temperatures, increasing the chances of oxidation problems in the food formulations. In one of the reports, it is demonstrated that the liposome entrapped α-tocopherol has been shown to be more effective at preventing oxidation in oil-in-water emulsions than when the free form was dissolved in the oil [22, 23]. Usually, oxidation takes place at the water-oil interface, if the liposome acquires this place; the α-tocopherol in the membrane could drastically minimize the peroxy radicals before the oxidation cycle is initiated by these radicals. The presence of ascorbic acid in the aqueous portion of the liposome could aid in the regeneration of the α-tocopherol. Moreover, in another study, it is shown that liposomal entrapment of the ascorbic acid would decreases the degradation of the ascorbic acid by other food ingredients and also facilitate for the regeneration of high levels of α-tocopherol [24]. Improving the stability and efficacy of flavoring constituents by entrapping them in liposomes is another major field of investigation on liposome uses in food formulations. Presently, this use of liposomes is limited by cost, though the comparatively high price of liposomes is offset somewhat through minimizing the amount of active constituent needed. Prices should be reduced as the technology is highly developed and more widespread and enhances in efficiency. Liposomes are amicable to their surface as well as bilayer composition modification, so by manipulating their bilayer composition (in order to adjust their phase transition temperature (Tc)) it is possible to keep flavors and such other vital constituents protected against degradation during their shelf life, but to be released from the liposomal reservoir in the oral cavity in response to increased temperature of the mouth. Liposome technology could also be utilized in foods also to encapsulate vitamins. The encapsulation of these vital functional and nutritional molecules, not only improve their stability during manufacturing and storage, but it also improves the biological performance of these constituents. Vitamin A (retinol) entrapped in liposomes can be protected from heat or light induced degradation [25]. Vitamin E was also incorporated together with tea polyphenol (water-soluble), and encapsulation efficiencies of 94% and 50% are reported for the hydrophobic and hydrophilic compounds, respectively [26]. The concurrent encapsulation of vitamin E with vitamin C has also been reported [27]. The authors report the potential application of these vitamin carrier systems in juices and other fortified beverages, as the liposomes protected both vitamin E and C from heat degradation. Ascorbic acid has been shown to be more stable at higher concentrations. By entrapping the vitamin in liposomes, it could be held at a much more concentrated level and therefore could have a longer shelf life [3]. Using a model system, Kirby et al. [23] demonstrated that liposome-entrapped ascorbic acid was more stable to storage compared to the free form; as more than 50% was left after 50 days at 4°C, whereas the free ascorbic acid was fully degraded after 20 days. Stability in the presence of degradative substances, such as copper, ascorbic acid oxidase and lysine, can also be improved with encapsulation in liposomes [23]. Recently similar results have been reported for ascorbic acid encapsulated in SUV liposomes prepared with milk phospholipids by microfluidization [28]. In this case, it was demonstrated peroxide values and liposome sizes did not change after storage of the liposomes at pH 7 for 8 weeks at 4°C.

As previously mentioned, the incorporation of hydrophobic compounds in liposomes may be of great potential to deliver bioactive compounds in foods [29]. Polyphenol entrapped in liposomes have been shown to effectively be taken up by cancer cells and inhibit their proliferation [30, 31]. CoQ_{10} was also successfully incorporated in egg phospholipid liposomes containing Tween-80 and cholesterol (Chol) [32]. Liposomes containing CUR showed a significant improvement in the bioefficacy of this lipophilic polyphenol [33]. It was hypothesized that the enhanced antihepatotoxic activity of CUR in liposomes may be due to a synergistic effect of phospholipids and the polyphenol molecules. The incorporation of phytosterols in SUV prepared by high-pressure homogenization using soy phospholipids and with no solvent addition have demonstrated high EE compare to plain liposomes [34].

Similarly, phytosterol-based MLV liposomes have been prepared and used as a substitute of Chol, depicted high efficiencies of encapsulation compared to liposomes with no sterols [35]. Liposomes have also been proposed as potential carriers of minerals in foods. For instance, liposomes can be utilized to minimize unwanted aggregation by controlling the mineral release during heat treatment in dairy products. Liposomes system containing $CaCl_2$ with Tc of 37°C exemplified the triggered release of ions [36]. No changes in the viscosity of the loaded liposomes dispersed in liquid sodium alginate (SA) solution was noted for several days at room temperature, contrary to that, the liquid gels rapidly when heated to 37°C. Apart from their medical and nutraceuticals application, these systems draw interest in controlling gelation in processed foods. Liposomes are also reported to have antimicrobial applications in foods [64]. For instance, antimicrobial loaded liposomes addition in cheese milk decreased bacterial load [37, 38], antimicrobial bacteriocins encapsulation has been reported to assist preservation during storage. Nisin encapsulated liposomes with pure phospholipids has been reported more potently inhibit bacterial growth compared to free nisin [39, 40]. However, there are certain limitations for the use of antimicrobial peptides in liposomes, viz. disruption of membrane due to ability of these molecules to interact with the bilayer and care required to minimize disruption during formulation of the lipid composition of the membrane bilayer [41]. Interaction with the other components in foods or procession condition leads to the inactivation of enzymes. The advantage of liposome entrapment is that it enables the enzyme to retain its activity by isolating the enzyme from the surrounding that would otherwise impede the performance or even cause denaturation of the enzyme. An enzyme can be entrapped at an optimum pH for stability or function, afterwards, external pH can be changed to the desirable value for the food product. Additionally, liposomes can also be utilized for the controlled release. This facilitates the addition of enzyme to a food system much before than when its activity is needed without exerting any negative effects due to early addition of the free enzyme. For instance, addition of liposomes encapsulated with β-galactosidase facilitating the digestion of the lactose in dairy products. Release of the enzyme takes place in the presence of the bile salt in the upper intestine [42]. There are a number of areas where liposomes-based approach has been reported in dairy application technology. For instance, production of cheese with improved ripening times and flavor profile by entrapment of proteolytic enzymes in liposomes [43, 44].

Free enzyme addition to the milk may lead to premature proteolysis and hence, low yield and poor structure of the curd. Proteases entrapped MLVs within the core have been demonstrated to inhibit rapid and unwanted proteolysis of β-casein, imparting firmer appearance to the final cheese [45]. Additionally, the cost of the product is also increased as a significant amount of enzyme is lost in the whey stream and high initial enzyme concentration is required limiting downstream whey procession options. Enzymes entrapped liposomes reduce enzyme requirement to 100 folds, produce cheese with good texture and flavor in half the normal time [46]. It was observed that liposome entrapped with lipase for the production of Cheddar cheese indicated enhanced production of free fatty acids and a more desirable flavor profile [44]. During the last few, more emphasis is given on reduced use of chemical preservatives in the foods, and where possible, substances perceived to be of natural origin are used as substitute of synthetic substances. Liposomes is beneficial in this point also, as during the cheese ripening, liposomes, and microorganism may accumulate in the same microenvironment in the cheese matrix [24] raising the possibility of targeted antimicrobial delivery in the dairy matrices. Such targeting leads to a substantial reduction of the overall concentration of required antimicrobial agents, permitting the utilization of natural substances. For instance, preservation of washed curd cheese such as Edam, Emmental, and Gouda is done by this application. Spore forming bacteria can easily spoil these cheeses. Such spoilage can be limited by addition of nitrate to the milk during processing but the use of nitrates in the food increases the health concern. In such a situation, nitrates can be replaced by use of lysozyme, an enzyme derived from egg white. However, lysozyme binds to the casein present in the milk leading to reduction of its potency and rendering it ineffective at high spore count. The interaction between casein and lysozyme can be prevented by liposome entrapment, targeting the regions in the cheese matrix having more possibility of bacterial accumulation, making lysozyme active against bacteria. Water entrapment is another potential application of liposomes [47]. In the case of products with low water activity, the shelf life can be increased if water can be contained in liposomes. It is reported that such products have substantially better aroma, flavor, bite/chew properties, impression of freshness on reheating when compared with standard products. However, scientific work is required to support this area, and bilayer membranes offer only minimal resistance to water diffusion. Applications of liposomes in foods can be extended by polysaccharides coated liposomes. Recently it has been reported that preparation of polysaccharide coated liposomes and loaded with vitamin A and E loaded and gelling the inner core is possible [48]. Surface of vitamin C and E loaded liposomes can be modified using Chitosan [48, 49]. The significance of such a system is high stability with 90% vitamin E retention after prolonged storage of 8 weeks [49]. In another study, Anthracyanin loaded liposomes were developed by supercritical carbon dioxide (SC-CO$_2$) method [50]. the prepared liposomes showed the mean size of 160 ± 2 nm, PDI 0.26 ± 0.01 and entrapment efficiency of 52.2 ± 2.1%. At high pressure the smaller size uniform particles prepared. From the study it can be concluded that this method of preparation may useful to prepare scalable liposomes loaded with bioactives. Liposomes containing theobromine, caffeine, catechin, epicatechin, and a cocoa extract were prepared by microfluidization and sonication method [65]. The prepared liposomes showed enhanced encapsulation with high physicochemical stability. The liposomes showed the particle size range between 73.9 nm and 84.3 nm at pH 5. The biological activity evaluated in exposing the sample to simulated *in vitro* digestion model and result revealed that flavan-3-ols loaded liposomes enhanced the biological activity, which may be important for the development of nutraceutical-enriched functional foods.

5.4 CONCLUSION

The nutraceutical-based liposomes are being extensively used in health care systems. Liposomes mainly consist of phospholipids, Chol, and nutraceuticals is generally attached to these prior to liposomal formation. The use of nutraceutical loaded liposomes have been reported against a wide range of topical and systemic diseases (Table 5.1). The liposomes delivery systems have a narrow therapeutic window, so an exhaustive characterization required in the early stage of formulation development. The understanding of their *in vivo* behavior during preclinical and clinical trials is a vital source of information for the success of pharmaceutical development so that failures in late phases of research and development are avoided.

TABLE 5.1 Application of Nutraceutical Loaded Liposomes in Different Diseases

Sl. No.	Drug	Activity	Inference	References
1.	Curcumin	Anti-hyperglycemic	Curcumin-loaded liposomes prepared from bovine milk phospholipids have inferior bioavailability compared to that prepared from krill phospholipids.	[51]
2.	Resveratrol	Anti-cancer	Improved cytotoxicity against human colon cancer cells	[52]
3.	Curry plant essential oil	Antimicrobial	Liposome encapsulated curry plant essential oil exhibited efficient antimicrobial activity against *B. cereus*	[53]
4.	Retinol	Stabilization	Reduces the degradation of vitamin A and enhanced the retinol stability	[54]
5.	Polyunsaturated fatty acids from fish oil	Fortification of yogurt	Adding nano-encapsulated fish oil into yogurt gave closer characteristics to the control sample in terms of sensory characteristics than yogurt fortified with free fish oil.	[55]
6.	Curcumin	Nanovesicle behavior	Improved stability and bioavailability	[56]

TABLE 5.1 *(Continued)*

Sl. No.	Drug	Activity	Inference	References
7.	Beta carotene	Physical characteristic and bioavailability improvement	Increases the chemical stability and bioavailability	[57]
8.	Peptides	Antimicrobial	After the encapsulation of the peptides into polymer-coated nano-liposomes, the antibacterial activity increased approximately 2000-fold against that of *L. monocytogenes*	[58]
9.	Flavourzyme	Enzyme activity	Protect the casein from early proteolysis and inhibit the pre maturation of curd	[59]
10.	Curcumin	Anti-cancer	Improved *in-vitro* cytotoxicity against specific cancer cells	[60]
			Enhanced antioxidant activity	

KEYWORDS

- **homogenization**
- **hydrophobic molecules**
- **large unilamellar**
- **liposomes**
- **microfluidization**
- **small unilamellar**
- **solubilization**

REFERENCES

1. Bonechi, C., Donati, A., Tamasi, G., Pardini, A., Rostom, H., Leonea, G., Lamponia, S., Consumia, M., Magnania, A., Rossi, C., (2019). Chemical characterization of liposomes containing nutraceutical compounds: Tyrosol, hydroxytyrosol and oleuropein. *Biophysical Chemistry, 246*, 25–34.

2. Keller, B., (2001). Liposomes in nutrition. *Trends in Food Science and Technology, 12,* 25–31.

3. Reineccius, G., (1995). Liposomes for controlled release in the food industry. In: Risch, S., & Reineccius, G., (eds.), *Encapsulation and Controlled Release of Food Ingredients* (pp. 113–131). American Chemical Society, Washington DC.

4. Hermida, L., Sabés-xamaní, M., & Barnadas-Rodríguez, R., (2009). Combined strategies for liposome characterization during *in vitro* digestion. *J. of Liposome Research, 19,* 207–219.

5. Spitsberg, V. L., (2005). Invited review: Bovine milk fat globule membrane as a potential nutraceutical. *Journal of Dairy Science, 88,* 2289–2294.

6. Riaz, M., (1996). Liposome preparation method. *Pak. J. Pharm. Sci., 9*(1), 65–77.

7. Himanshu, A., Sitasharan, P., & Singhai, A. K., (2011). Liposomes as drug carriers. *IJPLS, 2*(7), 945–951.

8. Batzri, S., & Korn, E. D., (1973). Single bilayer liposomes prepared without sonication. *Biochim. Biophy. Acta, 298*(4), 1015–1019.

9. Deamer, D., & Bangham, A. D., (1976). Large volume liposomes by an ether vaporization method. *Biochim. Biophys. Acta, 443*(3), 629–634.

10. Alamri, A., Huang, Y., & Zhang, H., (2018). Liposomal nanoformulation of piperlongumine for improved aqueous solubility and enhanced anti-tumor activity *in-vitro. Int. J. Pharm. Pharmacol., 2,* 133.

11. Handa, T., Naito, S., Hiramatsu, M., & Tsuboi, M., (2006). Thermal SiO and H^{13}CO$^+$line observations of the dense molecular cloud G0.11–0.11 in the galactic center region. *Astrophys J., 636,* 261–266.

12. Mayer, L. D., Bally, M. B., Hope, M. J., & Cullis, P. R., (1986). Techniques for encapsulating bioactive agents into liposomes. *Chem. Phys. Lipids, 40,* 333–345.

13. Song, H., Geng, H. Q., Ruan, J., Wang, K., Bao, C. C., Wang, J., Peng, X., et al., (2011). Development of polysorbate 80/phospholipid mixed micellar formation for docetaxel and assessment of its *in vivo* distribution in animal models. *Nanoscale Res. Lett., 6,* 354.

14. Akbarzadeh, A., Rezaei-Sadabady, R., Davaran, S., Joo, S. W., Zarghami, N., Hanifehpour, Y., Samiei, M., et al., (2013). Liposome: Classification, preparation, and applications. *Nanoscale Research Letters, 8*(1), 102.

15. Midha, K., Rani, P., Arora, G., et al., (2017). Solid dispersion: A recent update. *Int. J. Pharm. Pharmacol., 1,* 104.

16. Hamilton, R. L., & Guo, L. S. S., (1984). Liposomes preparation methods. *J. Clin. Biochem. Nut., 7,* 175.

17. Kataria, S., Sandhu, P., Bilandi, A., Akanksha, M., Kapoor, B., Seth, G. L., & Bihani, S. D., (2011). Stealth liposomes: A review. *IJRAP, 2*(5), 1534–1538.

18. Liu, L., & Yonetaini, T., (1994). Preparation and characterization of liposome-encapsulated hemoglobin by a freeze-thaw method. *J. Microencapsulation, 11*(4), 409–421.

19. Daemen, T., Hofstede, G., Ten, K. M. T., Bakker-Woudenberg, I. A. J. M., & Scherphof, G. L., (1995). Liposomal doxorubicin-induced toxicity: Depletion and impairment of phagocytic activity of liver macrophages. *Int. Cancer, 61,* 761–721.

20. Shaheen, S. M., Shakil, A. F. R., Hossen, M. N., Ahmed, M., Amran, M. S., & Ul-Islam, M. A., (2006). Liposome as a carrier for advanced drug delivery. *Pak. J. Biol. Sci., 9*(6), 1181–1191.

21. Stahl, W., Junghans, A., De Boer, Driomina, E. S., Briviba, K., & Sies, H., (1998). Carotenoid mixtures protect multilamellar liposomes against oxidative damage: Synergistic effects of lycopene and lutein. *FEBS Letters, 427*, 305–308.

22. Arnaud, J. P., (1995). Liposomes in the agro/food industry. *Agro-Food Industry Hi-Tech, 8*, 30–36.

23. Kirby, C. J., Whittle, C. J., Rigby, N., Coxon, D. T., & Law, B. A., (1991). Stabilization of ascorbic acid by microencapsulation in liposomes. *International Journal of Food Science and Technology, 26*, 437–449.

24. Kirby, C., (1991). Microencapsulation and controlled delivery of food ingredients. *Food Science and Technology Today, 5*, 74–78.

25. Lee, S., Lee, K., Kim, J. J., & Lim, S. H., (2005). The effect of cholesterol in the liposome bilayer on the stabilization of incorporated retinol. *Journal of Liposome Research, 15*, 157–166.

26. Ma, Q. H., (2009). Preparation and characterization of tea polyphenols and vitamin E loaded nanoscale complex liposome. *Journal of Nanoscience and Nanotechnology, 9*, 1379–1383.

27. Marsanasco, M., (2011). Liposomes as vehicles for vitamins E and C: An alternative to fortify orange juice and offer vitamin C protection after heat treatment. *Food Research International, 44*, 3039–3046.

28. Farhang, B., Kakuda, Y., & Corredig, M., (2012). Encapsulation of ascorbic acid in liposomes prepared with milk fat globule membrane-derived phospholipids. *Dairy Science and Technology*. Available on-line. doi 10.1001/ s13594-012-0072-7.

29. Gibis, M., Vogt, E., & Weiss, J., (2012). Encapsulation of polyphenolic grape seed extract in polymer-coated liposomes. *Food and Function*. doi: 10.1039/C1FO10181A.

30. Fang, J., Hung, C., Hwang, T., & Huang, Y., (2005). Physicochemical characteristics and *in vivo* deposition of liposome-encapsulated tea catechins by topical and intratumor administrations. *J. of Drug Target., 13*, 19–27.

31. Narayanan, N. K., Nargi, D., Randolph, C., & Narayanan, B. A., (2009). Liposome encapsulation of curcumin and resveratrol in combination reduces prostate cancer incidence in PTEN Knockout mice. *International Journal of Cancer, 125*, 1–8.

32. Xia, S., Xu, S., & Zhang, X., (2006). Optimization in the preparation of coenzyme Q10 nanoliposomes. *Journal of Agricultural and Food Chem., 54*, 6358–6366.

33. Takahashi, H., Kilamoto, D., Imura, T., Oku, H., Takara, K., & Wada, K., (2008). Characterization and bioavailability of liposomes containing a ukon extract. *Biosciences and Biotechnology Biochemistry, 72*, 1199–1205.

34. Alexander, M., Acero-lopez, A., Fang, Y., & Corredig, M., (2012). Incorporation of phytosterols in nanoliposomes: Encapsulation efficiency and stability. *LWT Food Science and Technology*. doi: 10.1016/j.lwt.2012.01.041.

35. Chan, Y. H., Chen, B. H., Chiu, C. P., & Lu, Y. F., (2004). The influence of phytosterols on the encapsulation efficiency of cholesterol liposomes. *International Journal of Food Science and Technology, 39*, 985–995.

36. Westhaus, E., & Messersmith, P., (2001). Triggered release of calcium from lipid vesicles: A bioinspired strategy for rapid gelation of polysaccharide and protein hydrogels. *Biomaterials, 22,* 453–462.
37. Benech, R. O., Kheadr, E. E., Lacroix, C., & Fliss, I., (2003). Impact of nisin producing culture and liposome-encapsulated nisin on ripening of lactobacillus added-cheddar cheese. *Journal of Dairy Science, 86,* 1895–1909.
38. Taylor, T., Davidson, P., Bruce, B., & Weiss, J., (2005). Liposomal nanocapsules in food science and agriculture. *Critical Reviews in Food Science and Nutrition, 45,* 587–605.
39. Were, L. M., Bruce, B., Davidson, P. M., & Weiss, J., (2004). Encapsulation of nisin and lysozyme in liposomes enhances efficacy against *Listeria monocytogenes. Journal of Food Protection, 67,* 922–927.
40. Da Silva, M. P., Daroit, D. J., Pesce, D. S. N., & Brandelli, A., (2010a). Effect of nanovesicle-encapsulated nisin on growth of *Listeria monocytogenes* in milk. *Food Microbiology, 27,* 175–178.
41. Da Silva, M. P., Daroit, D. J., & Brandelli, A., (2010b). Food applications of liposome-encapsulated antimicrobial peptides. *Trends in Food Science & Tech., 21,* 284–292.
42. Rodríguez-Nogales, J. M., & López, A. D., (2006). A novel approach to develop β-galactosidase entrapped in liposomes in order to prevent an immediate hydrolysis of lactose in milk. *International Dairy Journal, 16,* 354–360.
43. Zeisig, R., & Cammerer, B., (2001). Liposomes in the food industry. In: Vilstrup, P., (ed.), *Microencapsulation of Food Ingredients* (pp. 101–119). Leatherhead Publishing, London.
44. Kheadr, E. E., Vuillemard, J. C., & El-deeb, S. A., (2003). Impact of liposome-encapsulated enzyme cocktails on cheddar cheese ripening. *Food Research International, 36,* 241–252.
45. Law, B. A., & King, J. S., (1985). Use of liposomes for proteinase addition to cheddar cheese. *Journal of Dairy Research., 52,* 183–188.
46. Kirby, C. J., Brooker, B. E., & Law, B. A., (1987). Accelerated ripening of cheese using liposome-encapsulated enzyme. *International Journal of Food Science and Technology, 22,* 355–375.
47. Krotz, R., (1995). Phospholipids and liposomes in bakery products. *Food Ingredients and Food Analysis International, 17,* 14–16.
48. Liu, N., & Park, H. J., (2010). Factors effect on the loading efficiency of vitamin C loaded chitosan-coated nanoliposomes. *Colloids and Surfaces B: Biointerfaces, 76,* 16–19.
49. Nan, L., & Hyun-Jin, P., (2009). Chitosan-coated nanoliposome as vitamin E carrier. *Journal of Microencapsulation, 26,* 235–242.
50. Lisha, Z., & Feral, T., (2017). Preparation of anthocyanin-loaded liposomes using an improved supercritical carbon dioxide method. *Innovative Food Science & Emerging Technologies, 39,* 119–128.
51. Yujie, W., Bolin, M., Shuang, S., Chin-Ping, T., Oi-Ming, L., Cai, S., & Ling-Zhi, C., (2020). Curcumin-loaded liposomes prepared from bovine milk and krill phospholipids: Effects of chemical composition on storage stability, *in-vitro* digestibility and anti-hyperglycemic properties. *Food Research International, 136,* 109301.

52. Soo, E., Thakur, S., Qu, Z., Jambhrunkar, S., Parekh, H. S., & Popat, A., (2016). Enhancing delivery and cytotoxicity of resveratrol through a dual nanoencapsulation approach. *Journal Colloids and Interface Science, 462,* 368–374.

53. Cui, H., Li, W., & Lin, L., (2017). Antibacterial activity of liposome containing curry plant essential oil against *Bacillus cereusin* rice. *J. Food Saf., 37,* 3–7.

54. Shukla, S., Haldorai, Y., & Hwang, S. K., (2017). Current demands for food-approved liposome nanoparticles in food and safety sector. *Front Microbiol., 8,* 1–14.

55. Ghorbanzade, T., Jafari, S. M., Akhavan, S., & Hadavi, R., (2017). Nanoencapsulation of fish oil in nano-liposomes and its application in fortification of yogurt. *Food Chem., 216,* 146–152.

56. Hasan, M., Ben, M. G., & Michaux, F., (2016). Chitosan-coated liposomes encapsulating curcumin: Study of lipid-polysaccharide interactions and nanovesicle behavior. *RSC Adv., 6,* 45290–45304.

57. Hudiyanti, D., Fawrin, H., & Siahaan, P., (2018). Simulant encapsulation of vitamin C and beta-carotene in sesame (*Sesamum indicum* l.) liposomes. In: *IOP Conference Series: Materials Science and Engineering* (Vol. 349, No. 1, p. 0129014).

58. Cantor, S., Vargas, L., Rojas, A., Yarce, C. J., Salamanca, C. H., & Onate-Garzon, J., (2019). Evaluation of the antimicrobial activity of cationic peptides loaded in surface-modified nanoliposomes against foodborne bacteria. *Int. J. Mol. Sci., 20*(3), 1–15.

59. Jahadi, M., Khosravi-Darani, K., & Ehsani, M. R., (2015). The encapsulation of flavorzyme in nanoliposome by heating method. *J. Food Sci. Technol., 52,* 2063–2072.

60. Ibrahim, S., Tagami, T., Kishi, T., & Ozeki, T., (2018). Curcumin marinosomes as promising nano-drug delivery system for lung cancer. *Int. J. Pharm, 540*(1, 2), 40–49.

61. Claudia, B., Alessandro, D., Gabriella, T., Alessio, P., Hanzadah, R., Gemma, L., Stefania, L., et al., (2019). Chemical characterization of liposomes containing nutraceutical compounds: Tyrosol, hydroxytyrosol and oleuropein. *Biophysical Chemistry, 246,* 25–34.

62. Szoka, F. Jr., & Papahadjopoulos, D., (1978). Procedure for preparation of liposomes with large internal aqueous space and high capture by reverse-phase evaporation. *Proc. Natl. Acad. Sci. USA, 75*(9), 4194–4198.

63. Toro-Uribe, S., Elena, I., Eric, A. D., McClements, D. J., Ruojie, Z., López-Giraldo, L. J., & Miguel, H., (2018). Design, fabrication, characterization, and *in vitro* digestion of alkaloid-, catechin-, and cocoa extract-loaded liposomes. *J. Agric. Food Chem., 66*(45), 12051–12065.

64. Singh, H., Thompson, A., Liu, W., & Corredig, M., (2012). Liposomes as food ingredients and nutraceutical delivery systems. *Encapsulation Technologies and Delivery Systems for Food Ingredients and Nutraceuticals* (pp. 287–318). Woodhead publishing series in food science, technology and nutrition.

65. Toro-Uribe, S., Ibañez, E., Decker, E. A., Villamizar-Jaimes, A. R., & López-Giraldo, L. J., (2020). Food-safe process for high recovery of flavonoids from cocoa beans: Antioxidant and HPLC-DAD-ESI-MS/MS analysis. *Antioxidants (Basel), 9*(5), 364.

CHAPTER 6

Formulation and Challenges in Liposomal Technology in Functional Food and Nutraceuticals

MOHAMMED ASADULLAH JAHANGIR,[1] ABDUL MUHEEM,[2]
M. AKIFUL HAQUE,[3] CHETTUPALLI ANANDA,[4]
MOHAMAD TALEUZZAMAN,[5] and CHANDRA KALA[5]

[1]*Department of Pharmaceutics, Nibha Institute of Pharmaceutical Sciences, Rajgir, Nalanda – 803116, Bihar, India*

[2]*Department of Pharmaceutics, School of Pharmaceutical Education and Research, Jamia Hamdard, New Delhi – 110062, India*

[3]*Department of Pharmaceutical Analysis, Anurag University, Venkatapur, Ghatkesar, Hyderabad – 500088, Telangana, India*

[4]*Department of Pharmaceutics, Center for Nanomedicine, School of Pharmacy, Anurag University, Venkatapur, Ghatkesar, Hyderabad – 500088, Telangana, India*

[5]*Faculty of Pharmacy, Maulana Azad University, Jodhpur – 342802, Rajasthan, India*

ABSTRACT

Since its discovery, liposomes have been vastly studied and applied in the field of pharmaceutical, food, and nutraceutical industry. Being a delivery vesicle, liposomes are capable of effectively transporting drugs, food, or nutraceuticals into the body by facilitating absorption directly in the mouth while protecting the encapsulated substance from degradation by the acidic environment of the stomach. Their ability to incorporate both hydrophobic and hydrophilic drugs as well as their biocompatibility, reduced risk of toxic side effects made them the most extensively studied drug delivery system. The application of liposomes in the food industry was first published in the early 1980's. The initial application of liposome included the controlled release of enzymes in the making of cheese. Further, both lipid and aqueous soluble food ingredients were delivered by liposomal technology. However, designing food and nutraceuticals into liposomal delivery system comes with its own challenges. This chapter details about the techniques in developing liposomal based food and nutraceuticals, formulations, and characterization of liposomal products in functional food and nutraceuticals, challenges in formulating functional food and nutraceuticals into liposomal systems, global market of liposomal system in functional food and nutraceutical products and recent patents related to liposomal technology in functional food and nutraceuticals.

6.1 INTRODUCTION

Discovered in 1961, liposomes were introduced to the world in the year 1964 [1]. Since its discovery, it has been one of the most vastly studied and applied technology in the pharmaceutical, food, and nutraceutical industry. Being a delivery vesicle, liposomes are capable of effectively transporting drugs, food, or nutraceuticals into the body by facilitating absorption directly in the mouth and also protects the substance from degradation by the acidic environment of the stomach [2, 3]. Their ability to incorporate both hydrophobic and hydrophilic drugs as well as their biocompatibility, reduced risk of toxic side effects made them the most extensively studied drug delivery system [4]. Liposomes belongs to the family of lipid-based systems, of which the emulsions are the most common forms based on oil and water having time-limited stability. Emulsions uses surfactants which assists in developing oil in water or water in oil systems.

They are single-layered or multi-layered vesicles which completely incorporates the aqueous phase inside a phosphor-lipid membrane. Upon dispersion into the aqueous medium, the phospholipids immediately form vesicles. Liposomes can encapsulate both lipophilic and hydrophilic active agents, however substance which are neither soluble into a lipid nor in aqueous phase cannot be encapsulated inside a liposome. Most of the food ingredients can be encapsulated inside the liposomes, but the flavoring compounds may get dissolved in both the phases. Liposomes can be of varied sizes, which range from a few nanometers to microns. Liposomes can be designed to sufficiently small size to efficiently pass through the smallest of blood capillaries. Apart from the enormous possibilities of liposomes in drug delivery, they have vast opportunities in the food and nutraceutical industries. The application of liposomes in the food industry was first published in the early 1980's. The initial application of liposome included the controlled release of enzymes in the making of cheese. Further, both lipid and aqueous soluble food ingredients were delivered by liposomal technology.

Over the years, nutraceuticals and food industry have extensively exploited different techniques of microencapsulation. Techniques like melt injection, spray cooling, fluid coating spray drying, emulsification, complexation, coacervation, liposome formations have been applied in pharmaceutical, food, and nutraceutical industry. Specialized coating techniques allows controlled release of encapsulated substance. Specific technique like liposome development can also assist in increasing solubility and bioavailability of nutritional and food agents. In the food and nutraceutical industry, liposomes are often used as vehicles for dissolving aqueous soluble, lipid or multiple lipid-soluble materials. These substances are used in a symbiotic way. For example, curcumin (CUR) is a lipid-soluble nutritional substance having high metabolic rate within serum and low solubility which can be overcome by incorporating into liposomes which is coupled into lipid-soluble substance enzyme-like quercetin (QUE) which are capable of blocking the metabolites and eventually increases the bioavailability. Thus, creating a symbiotic association between the liposome and the active agents and provides a valuable tool for the nutraceutical and food industry. Liposome also acts as a protective agent and inhibits hydrolysis, unwanted interactions, and degradation. Nisin within the liposome acts as an anti-microbial agent and controls the growth of bacteria during the storage of cheese. In some studies, it has been reported that encapsulating exogenous enzymes helps in improving the retention time of enzymes in the curd. This technique has also been exploited in the browning and Maillard reaction. The liquid composition of liposomes is not widely applied. The reason is the instability of phospholipids at low pH and high temperature. The liquid liposomes developed in the pharmaceutical industries needs refrigeration or need special processing like freeze-drying technique or development into powdered form [5].

It is not unjustifiable to state that the liposomal technology has limited development in the food industry in comparison to the pharmaceutical industry. It may be attributed to the lack of finding safe, low-cost, solvent-free ingredients as well as processing methods. With the development of advanced high-pressure homogenizers, pro-liposome, and micro fluidization techniques have assisted in overcoming the above-mentioned formulation issues. In the recent times, many nutraceutical and food products like enzymes, herbal extracts, and vitamins are getting formulated in the form of liposomes. Such formulations improve the absorption and nutrient bioactive solubilization. But such systems are complex in terms of maintaining the stability throughout the process [6].

6.2 TECHNIQUES IN DEVELOPING LIPOSOMAL BASED FUNCTIONAL FOOD AND NUTRACEUTICALS

Thin-film dispersion stands among the most unique technique for formulating liposomes, which is based on Bangham's method [7] and involves three steps: Firstly, the lipid is dissolved in organic solvent, which is then evaporated to form a thin film, the process is followed by the addition of the aqueous media into lipid film to form liposomal suspension. However, the method comes with the drawback of producing multilamellar large vesicles having low encapsulation efficiency (EE). In the recent times, several other techniques like ultrasonication, freeze-thawing, extrusion homogenization to further process the liposomes obtained from thin-film technique. These techniques help in improving the stability and provides higher EE and also produces more homogeneous liposomal products. However, application of such techniques involves non-food grade detergents and solvents to effectively dissolve the lipid phase, it also lacks the capability of large-scale continuous production and involves high cost. The residual organic solvents are removed by involving techniques like vacuum, nitrogen flushing, filtration, etc. Heating and microfluidization methods do not require hazardous chemicals and solvents in comparison to dialysis techniques and thus are more appropriate for developing food grade liposomes. Since 1984 liposomes are being prepared using the microfluidization technique. Under this technique, the fluid is forced through the orifice under high pressure, which divides the lipid dispersion into two streams and then colliding the two high fluids under high velocity in an interaction chamber. High frequency vibration, high velocity impact, instant pressure drops, and intense shear forces can get combined and reach up to 200 MPa [7]. By controlling the microfluidization cycles and pressure, the particle size of liposomes can be managed. Thus, making the particle size distribution narrower and the entrapment efficiency gets increased. Microfluidization technique can also be exploited as a way of sterilizing the product pertaining to the high temperature and high pressure involved in the process. This technique is also used in pharmaceutical industry for developing emulsion and liposomal formulation are now being extensively used in the food industry. However, this technique comes with the disadvantage of over-treatment, which may lead to damage of the liposomal structure and eventually leading to the leakage of the encapsulated material. Another method for the development of liposomes is heating or Mozafari technique which prepares liposomes in one step without involving any potentially toxic solvents under very low shear force [7]. It can be used for large scale production of liposomes and requires the ingredients of the carrier system to be hydrated for at least 1 hour with continuous heating and stirring at low speed, this step is followed by the encapsulation of the ingredient in the presence of polyol at temperature between 40°C and 120°C [7]. In an attempt to prepare liposomes by this technique, Mozafari et al. reported particles of anionic liposome-plasmid with an entrapment efficiency of 70.3% [8]. In another study by Colas et al. different phospholipids were used to develop nisin-Z-loaded nanoliposomes. The researchers reported that the prepared liposomes were stable for around 1 year and 2 months at a temperature of 4°C and for 1 year at around 25°C and were found to be efficiently target the bacterial cells. however, this technique comes with the drawback of being unsuitable for the preparation of thermal sensitive liposomes with bioactive compounds, as it may alter their bioactivity under high temperature during processing [8].

Nature of the medium, concentration of the material to be encapsulated, polydispersity, size, shelf-life, and other Physico-chemical characteristics of the ingredients are some of the factors which needs to be considered during the preparation of liposomes. Based on the lamellarity liposomes may be said to be: multilamellar vesicles (MLVs), oligolamellar vesicles (OLV), unilamellar vesicles (ULV), small unilamellar vesicles (SUV), large unilamellar vesicles (LUV) and multivesicular vesicles (MVV).

Nanoliposomes from liposomes are prepared by evaporating methanol/ chloroform from the solution of amphiphilic contents like carbohydrate sterol, phospholipids, protein derivatives, etc. On adding hydrophilic contents to the thin film layer which is followed by the addition of the water-loving compound with applying sufficient amount energy in the form of agitation or shaking, thermal energy, sonication forms bilayer sheets with the inclusion of water-hating compounds which can be separated from the bulk to give nanoliposomes [9–11]. Nanoliposomes can also be formed by entrapping the compound during vesicle formation where the hydrophobic compounds are encapsulated in the lipid while hydrophilic compounds are located in the aqueous part in the bilayer system [11, 12]. It can also be formed by remote or active loading of the bioactive material into the vesicles. By providing additional driving force by ammonium sulfate and calcium acetate, which are weak base and acid, respectively, the bioactive materials are inserted into liposomes. This method is utilized to encapsulate high drug to lipid ratio by the application of pressure or force. This method is used to enhance the controlled release activity of the drug [13, 14]. Nanoliposomes comes with the disadvantage of low EE, lack of control parameter during continuous production at industrial scale, extreme pH condition during processing, usage of non-food grade solvents, application of high force and pressure [11]. Microfluidization technique overcomes all the discrepancies listed above during the production of liposomes or nanoliposomes.

6.3 FORMULATIONS AND CHARACTERIZATION OF LIPOSOMAL PRODUCTS IN FUNCTIONAL FOOD AND NUTRACEUTICALS

6.3.1 LIPOSOMES AS A CARRIER FOR ENZYMES

Liposomes are extensively studied and applied for encapsulating enzymes in food and nutraceutical industry. It has been widely applied for accelerating the ripening of cheese [15]. Normally the ripening of cheese requires 4 weeks to 3 years for soft to hard cheese. Accelerating this process provides economic and technical advantages. It also reduces microbial risk. However, the product cost may increase [16]. They also come with the risk of early proteolysis, which may result in unfavorable texture and flavor. There are reports of poor enzyme distribution upon direct addition of enzyme to the curd. Thus, encapsulating enzymes protects them from the environmental effects and also releases it in a controlled manner. As SUVs have poor EE, so mostly MLV-type liposomes are exploited in the dairy products [17]. In a study by Law and King accelerated proteolysis was reported with delayed-release by MLV-entrapped proteolytic enzyme [17]. Other technique like dehydration-rehydration vesicles with entrapped neutrase was found to accelerate the ripening of cheddar cheese [18]. Other techniques like trypsin-containing microfluidized liposomes and encapsulated fungal proteinase were found to have superior results. Chymosin, cyprosins, Proteinase from *Bacillus subtilis* entrapped inside liposome was found to accelerate ripening of Manchego cheese [18]. The entrapped enzyme is slowly released from the liposomal enzyme formulation allowing catalytic degradation and modifications in the matrix of the cheese during ripening [19]. Cheese fat content and pH stands among some of the factors which influence the rate of enzyme release from the developed liposome [20]. Researchers have also investigated entrapped lipase-based liposome in the manufacturing of cheese. Even a low quantity of entrapped lipase-based liposome increases the production of free fatty acids and flavor in Cheddar cheese. However, a higher concentration of the developed liposome may develop soapy flavor [21]. Entrapped β-galactosidase in liposomes have also been applied in the dairy products especially developed for lactose intolerants, also assist in preventing the change of flavor by lactose hydrolysis [22].

The developed liposomes are evaluated for their entrapment efficiency, which is calculated in terms of entrapped protein [15]. Factors like enzyme type, method of preparation affects the entrapment efficiency of the liposomes carrying enzymes. In the study by Kirby, it was reported that the entrapment efficiency for liposomes developed by DR vesicles was found to be higher than the same obtained by the sonicated vesicles, thin-film hydration (TFH), reverse phase evaporation technique for Neutrase entrapped liposomes [23]. Matsuzaki et al. in their study reported no loss of β-galactosidase specific activity with liposomes developed by reverse-phase evaporation technique. However, significant loss (10–20%) of enzyme activity was reported with liposomes developed by DR technique [24]. Different enzymes encapsulate by the same method will also lead to differences in the entrapment efficiency of the encapsulating technique. Other factors like solute concentration can also influence the entrapment efficiency of the same liposomal technique for different enzymes.

Several reports are published which are based on the stability of liposomal preparation containing enzymes. Matsuzaki et al. reported that the developed β-galactosidase entrapped liposomes were found to retain their stability at lower pH (3.0) with enzymatic activity being retained for one month if stored at 5°C under nitrogen. The researchers also reported that the capacity to resist acidic environment was found to be dependent on the molar ratio of lecithin:cholesterol (Chol) with 1:3 of the same showing the maximum resistive capacity at pH 3. However, the activity of the same molar ratio of lecithin:cholesterol was found to decrease with the decrease in pH [24]. In another study by Rao et al., it was reported that the β-galactosidase entrapped liposome retained the enzymatic activity for 20 days under refrigeration at neutral pH, while the activity was found to be decreased upon lowering than pH [22]. In a study by Chawan et al. liposomes containing fungal or bacterial β-galactosidase was found to be stable in pasteurized milk and buffer which is freshly prepared. Negligible hydrolysis was reported with bacterial β-galactosidase while the same was significant for fungal β-galactosidase during storage for 20 days at 4°C [25].

6.3.2 *LIPOSOMES AS A CARRIER FOR ESSENTIAL OILS (EOS)*

Essential oils (EOs) are characterized by strong aroma and are natural, volatile complex compounds [26]. EOs posses' properties like bactericidal, antifungal, and antioxidant activities [27–30]. Some EOs have been reported to have antiproliferative effects on tumor cells [31]. They are sensitive to temperature, light, and oxygen and thus are biologically unstable [32]. Having poor aqueous solubility, their distribution in the target site is limited [33]. Thus, to control the release of EOs and to improve their solubility and bioavailability different advanced drug delivery systems like nanoparticle, cyclodextrin inclusions, liposomes, and SLNs were developed [34–36]. Valenti et al. in their study showed that essential oil from *Santolina insularis* has been successfully incorporated in the liposomal system [37]. Different studies reported better oxidative stability of essential oil extracts from *Thymus* species, thymol, carvacrol when incorporated in the liposomal system as compared to their freely available forms [38]. In another study by Detoni et al. higher oxidation onset temperature for essential oil *Zanthoxylum tinguassuiba* incorporated in liposomes was reported [39]. Wen et al. in their study demonstrated that EOs from rose can be protected from thermal degradation if incorporated into liposomes [40]. Many studies have reported improved antimicrobial and biological activity of EOs incorporated into liposomes. In a study by Sinico et al. the essential oil extract of *Artemisia arborescens* L. incorporated liposomes showed higher antiherpetic and antiviral activity as compared to free oil [41]. The liposomal formulation of EOs of lemon, thymol, and carvacrol have shown better antimicrobial activity than their free forms [38]. Moghimipour et al. reported that the essential oil of *Eucalyptus camaldulensis* leaf when incorporated into a liposome showed enhanced antifungal activity [42]. Denton et al. also reported enhanced biological activity of *Zanthoxylum tinguassuiba* essential oil-based liposomes [39].

High entrapment efficiency of liposome-based essential oil has been reported by many researchers. Factors like lipid vesicle membrane may influence the entrapment efficiency of EOs. Ortan et al. demonstrated that entrapment efficiency of essential oil like linalool and lavandin was found to increase when incorporated in liposome by thin-film hydration method. However, the same was found to decrease if prepared by PGSS drying method [26]. Size of the developed liposomes also influence the entrapment efficiency. SUVs with unilamellar structure and smaller size have low entrapment efficiency than MLVs.

Liposomes may serve as a promising alternative for storing EOs for longer duration of time. In a study by Valenti et al. the essential oil of *S. insularis* incorporated into liposome by vesicle dispersion technique was found to be stable for a year. During the stability study, the researchers reported no alteration in vesicle size or oil leakage [37]. *A. arborescens* L. essential oil incorporated into MLV and SUV liposomes were reported to be stable for six months with intact size and retention of oil [43]. The developed liposomes of EOs of *A. graveolens* by Ortan was found to slightly leak the oil with a nominal increase in size during the six-month stability study [26]. The developed liposomes of *A. macrocephala Koidz* EOs by Wen et al. was found to be stable for a period of six months at a temperature of 25°C and relative humidity of 60% [40].

6.3.3   LIPOSOMES AS A CARRIER FOR VITAMINS

Vitamins are mostly unstable and vulnerable to degradation. Factors like vitamin structure, condition of processing and characteristics of food matrix, moisture, pH, light, oxygen content, and duration of storage and its conditions determine the rate of degradation of vitamins [43]. The presence of humidity, intense temperature, high oxygen content, pH nearing neutral and storage in containers made up of heavy metals makes ascorbic acid or vitamin C high unstable and susceptible to oxidation [44]. Maintaining pH, change in temperature, removal of oxygen is some of the classical approaches to overcome the degradation and instability issues of vitamins, however they were found to be inadequate. So, encapsulating vitamins in the liposomal system may provide promising results overcoming the instability issues of vitamins in food and nutraceutical industries. Liposomal product development of vitamins may protect both lipid and aqueous soluble vitamins from degradation. Kirby et al. successfully formulated and evaluated vitamin C encapsulated liposomes. The researchers reported that about 50% of the developed vitamin C based liposomal system survived after 50 days of study conducted at 15% relative humidity and 4°C temperature which is an extraordinary result in comparison to free vitamin C which gets completely disappeared within 18 days at the before mentioned temperature and within 6 days if left at room temperature. No effect of the external pH conditions was reported by the authors [45]. In a similar study by Yang et al., it was reported that about 70–80% of vitamin C was preserved inside the lipid bilayer after 24 hours which was just double from the preservation rate of free vitamin C (35%) [44]. Wechtersbach et al. confirmed in their study that the half-life of vitamin C based liposomal system is increased by 300-fold and rate of oxidation gets decreased in apple juice [46]. Tesoriere et al. successfully incorporated lipid-soluble vitamins like vitamin A and vitamin E into the liposomal system [47]. In another study by Banville et al. increased amount of vitamin D was recovered from the liposomal system in comparison to commercially available vitamin D [48]. Lee et al., in their study, demonstrated that multilamellar vesicle incorporated with vitamin A gets degraded to only 20% at 4°C and pH 7 which is five-fold better than free aqueous vitamin A which gets completely degraded within the same specified time and conditions [49].

Hydrophobic vitamin can be incorporated into MLV liposomal system. Lipid soluble vitamins can attain higher entrapment efficiency of about

80–99% in comparison to only 2–57% entrapment efficiency for water-soluble vitamins. Different liposomal methods, their constituents, and ratios influence the entrapment efficiency of vitamins into the liposomal system. The use of low phospholipid to buffer ratio, ratio of phosphatidyl-choline (PC) and Chol and the amount of core material by Wechtersbach et al. in preparing vitamin C encapsulated liposomal system resulted in the entrapment efficiency of only 2–8% for vitamin C. A much higher entrapment efficiency have been reported by other researchers for vitamin C [46]. The entrapment efficiency of retinol was found to be increased by increasing β-sitosterol and Chol content in the liposomal system [47]. Method opted for developing the liposomal system also effects the entrapment efficiency of the same vitamin. Freeze drying and dehydration/rehydration technique resulted in high entrapment efficiency.

To successfully evaluate the biological, chemical, and physical stability of liposome is a complex issue. Merging of the bilayer or phase transition causes physical instability in liposomes causing loss of entrapped vitamins. Oxidation and hydrolysis cause chemical instability in vitamins encapsulated liposomal system. Stability of bioactive liposomal system is evaluated in terms of EE, zeta potential and liposomal size. Low zeta potential causes particles to aggregate, resulting in instability [50]. Yang et al., in their study, demonstrated that vitamin C based nanoliposomes prepared by film evaporation and dynamic high pressure microfluidization technique exhibits stability for 24 hours at 37°C and for 60 days at 4°C. It was also demonstrated that the aged liposomes are more susceptible to leakage upon exposure to heat than the freshly prepared liposomes [46].

6.3.4   LIPOSOMES AS A CARRIER FOR ANTIMICROBIALS

Antimicrobials have known inhibitory effects against gram-positive bacteria in food and nutraceuticals and are also known to increase the shelf life [51]. However, direct addition of antimicrobial products in food and nutraceutical has its own limitations like significant loss of antimicrobial property and loss of starter culture as in the case of cheese, which is an important step. A classic example in this regard is nisin polypeptide which have the ability to inhibit the growth of gram-positive bacteria, however direct addition of nisin causes loss of activity and alteration in aroma, condition of food processing, food matrix, storage temperature, etc., [52].

Encapsulating nisin and other potential antimicrobials into the liposomal system may provide a potential solution to overcome the before mentioned limitations. It has reported that lysozyme/nisin has been encapsulated into a liposomal system to prevent the spoilage of cheese. In another result, it has been reported that encapsulating pediocin AcH into liposomes increased the antilisterial activity in diary-based slurries and muscle and tallow beef slurries [53]. In a study by Benech et al., it demonstrated that through the addition of nisin Z encapsulated liposomal system decrease the count of *Listeria innocua* by 1.5–3.0 log units during the period of six months of ripening and it was also reported that 90% of the nisin Z activity was still intact by the end of six months [54]. Nisin has also been incorporated into nanoliposomal system using distearoylphosphatidyl glycerol and disteraroylphosphotidyl choline was found to actively inhibit the growth of L. monocytogenes for 48 hours at 25°C and was not dependent on the milk fat content for its activity [55].

For nisin Z the entrapment efficiency was reported between 9.5% and 47% [56]. The entrapment efficiency was optimal for hydrogenated PC liposomes at about 34%, the increase in Chol concentration was found to inversely affect the entrapment efficiency of liposomes containing antimicrobials [57]. This is due to the reduced polypeptide affinity by the introduction of Chol, eventually decreasing the concentration of antimicrobials to be incorporated [56]. Were et al. co-encapsulated nisin with calcein into liposomal system using PC, Chol, PG at different molar ratios and reported high entrapment efficiency between 54% and 63%. Taylor et al. reported entrapment efficiency of about 70% to 90% for nisin bases liposomes when prepared by using distearoylphosphatidyl glycerol and disteraroylphosphotidyl choline [58].

Antimicrobials and the liposomal bi-layer member may negatively interact with each other leading to disrupting the structure [57]. Were et al. in their study reported the concentration-dependent effect leakage of liposomal content with PG and PC liposomes [57]. A slight change in the diameter of the developed liposomes were also reported by Were et al. [57]. Laridi et al. in its study reported that nisin Z based liposomes to be stable for 27 days at 4°C [56]. Long term stability study of nanoliposome has also been reported for 14 months at 4°C [8].

6.3.5   LIPOSOMES AS A CARRIER FOR PHENOLIC COMPOUNDS

Polyphenols has a number of biological functions and an integral part of animal and human diet. They are known to have anti-viral, anti-inflammatory, antibacterial, antioxidant effects [59]. They have potential to be used as anti-cancer and chemopreventive agents [60]. They are also known to inhibit the progress of neurodegenerative, cardiovascular, osteoporotic, and diabetic disorders. However, application of polyphenols as functional foods is limited owing to their stability concern, low solubility, and bioavailability [61]. Polyphenols have unpleasant taste and are capable of causing aggregation and precipitation by associating with proteins causing loss of functionality of polyphenols [62]. Encapsulating polyphenols into the liposomal system may be a positive approach to overcome these drawbacks and improve the biological functions of polyphenols as well as their shelf-life. The liposomal system has been reported to improve the aqueous solubility and bioavailability of phenolic compounds and helps in achieving sustained release. In a study by Caddeo et al. the developed liposomes were found to prevent the cytotoxicity of resveratrol (RESV) and avoids its immediate distribution and evokes its ability to proliferate the cancerous cells [50]. In a similar study by Isailovic et al. liposome entrapped RESV was reported to have low cytotoxicity and high antioxidant effect [63]. Takahashi et al. encapsulated CUR into liposomal system and observed improved absorption than the free form of CUR [64]. Anti-viral and antimicrobial effect of QUE enriched lecithin formulation was reported to be improved when incorporated into liposomal system [65].

Phenolic compounds entrapped into liposomal system have been reported with entrapment efficiency between 20% to more than 90%. Fan et al. reported low entrapment efficiency (20% to 57%) for salidroside [66]. High entrapment efficiency was also reported for vesicles entrapping CUR [62]. Fang et al., in their study, reported entrapment efficiency between 84% and 99% for EGCG, which was attributed to the presence of galloyl group in EGCG, which causes higher lipophilicity.

As per Fan et al. 10–15% leakage was observed after one month of storage at 4°C and 30°C, respectively, for salidroside based liposomal system [66]. In another study by Rashidinejad et al., the authors concluded that liposomes having EGCG or catechins are more stable as compared to empty liposomes in terms of oxidation [62]. A similar result was reported by Ramadan et al. where the researchers reported better oxidative stability of quercetin-enriched lecithin stability [67]. Gibis et al. reported improved oxidative and physical stability of poly-coated polyphenolic based liposomes [68].

6.4 CHALLENGES IN FORMULATING FUNCTIONAL FOOD AND NUTRACEUTICALS INTO LIPOSOMAL SYSTEMS

Liposomes efficiently encapsulates both lipophilic and hydrophilic molecules simultaneously. They had been extensively used as a flexible carrier in the cosmetic, pharmaceutical, and food industries. Liposomal supplement products can have several advantages over conventional supplements. When given orally, it can also indirectly enhance the distribution of food supplements after intestinal uptake. Liposomes have several advantages such as they improve the solubility of encapsulated molecules which leads to improved bioavailability, prevented chemical and enzymatic degradation of encapsulated molecules, reduced adverse reactions and toxicity [69], biocompatible, and biodegradable nature, and drug stability [70]. In addition, because of their particle size, they can easily flow in systematic circulation and sustain distortion, e.g., red blood cells [71]. However, the extensive usage of liposomes in food industries have been recently hampered because of different challenges which are as follows: (i) high production cost; (ii) Possibility of poor manufacturing including large particle size, poor quality of raw ingredients; (iii) Possibility of instability under the complex environments [1]; (iv) Low loading capacity for hydrophilic components [88]; (v) bioactive compound leakage; (vi) fast release; (vii) rapid elimination after administration [72]; (viii) It has been studied that charge liposomes are toxic. In liposome production methods, there may be a chance to present trace amounts of organic solvents in the final liposome preparation. Hence, commercialization is restricted by liposome associated cytotoxic effects [73].

The biggest challenge associated with liposomes is their manufacturing process such as scale-up problems, batch to batch irreproducibility, and absence of standard sterilization procedure. Among them, the greatest concern is the large-scale scalability of liposome technology [71, 74]. There are many classical and novel methods that have been developed to produce liposomes, such as transmembrane gradients, microfludization, supercritical reverse-phase evaporation, free thawing, and spray drying

[75]. These methods have been used to improve size reproducibility, EE, and production of freeze-dried liposomal powder for improving their stabilities. However, several other attempts also have been conducted to elevate the application of liposomes shown in Table 6.1.

Preliminary trials are conducted to improve the integrity and stability of liposomes by exploiting neutral long-chain saturated phospholipids and Chol, recently replacement of Chol with phytosterols (plant sterols) in liposomal structure has also been developed [76], since persons suffering from high Chol disorder should restrict its intake even in very low concentrations [77, 78]. Phytosterols can serve as a potential alternative for Chol in liposomal structure, its presence may also boost numerous bioactive properties with significant benefits for persons' health such as a reduction in the risk of cardiac diseases, antibacterial, anti-inflammatory, anti-atherosclerotic, anti-ulcerative, anti-oxidative, and anti-carcinogenic functions [79, 80]. Furthermore, the encapsulation of phytosterols in the liposomal structure can solve the issues with direct incorporation of phytosterols to food products because of their propensity to form insoluble crystals and very high melting point and [81]. Another study reported that 1-palmitoyl-2-oleoyl-sn-glycero-3-phosphocholine (POPC) or β-sitosterol based liposome structure shows significant lipid oxidation of lipophilic probe. In conjugation with the potential benefits of phytosterols on human health, smaller vesicle size, increase absolute zeta potential, better membrane fluidity, increased physical stability, and stabilizing with phospholipids in comparison to Chol are the encouraging results for the use of β-sitosterol in liposome formulation for employing in functional foods and nutraceutical sector [82].

The stability of liposomes was found to improve by modifying the same with numerous materials such as poloxamer, polyethylene glycol (PEG) [83], chitosan, carboxymethyl chitosan [84], and dextran derivatives [85]. Development of stimulus sensitive liposomes in the food industry is still an unmet need, although a variety of stimulus sensitive liposomes such as temperature-sensitive, pressure-sensitive have been developed in the pharmaceutical sector, which are capable of releasing their drug content in response to physical or chemical stimulus. For example, thermal sensitive liposomes have been developed by the incorporation of temperature-sensitive polymer, e.g., poly(N-isopropylacrylamide) in liposome [86]. In the food and nutraceutical sector, these types of carriers could be used in the baking industry, for illustrate, for essential oil or flavor agents release by enhancing the cooking temperature of the pre-cooked foods. pH-sensitive liposome [7] has also not been used in the encapsulation of nutraceutical ingredients to date [9]. However, they seem to have notable potential in the food and nutraceutical industries. A summarized table of advantages and challenges of liposomes in functional food and nutraceuticals is shown in Table 6.1.

TABLE 6.1 Advantages and Challenges of Liposomes in Functional Food and Nutraceuticals

Liposomes in Functional Food and Nutraceuticals		
Advantages	**Challenges**	**References**
Improved solubility	High production cost	[1, 69]
Improved bioavailability	Selection of GRAS or FDA approved substances	[87, 70]
Reduced chemical and enzymatic degradation	Instability due to complex environments	[1, 69]
Reduced side effects	Leakage of bioactive compounds	[69, 72]
Biocompatible and biodegradable formulation	Low loading of hydrophilic compounds	[70, 88]
Higher absorption	Rapid elimination	[72]
–	Cytotoxic effects	[73]
–	Development of stimulus sensitive liposomes	[86, 89]

Besides, the selection of food-grade or GRAS ingredients for its potential application in the food industry to formulate liposome formulation, since many polymers are approved for application in pharmaceutical or cosmeceutical industries are not permitted in food in large quantity, further experiments are required for the selection of appropriate substances for food applications. Furthermore, the substances should not also unpalatably influence the physicochemical properties of food or nutraceutical product that is encapsulated into liposome formulation [87]. Finally, food ingredients encapsulated in the liposomal formulation must exploit economical techniques for being able to be marketed. If researchers are able to develop more economical techniques, then this technology could significantly contribute to developing novel food and nutraceutical products.

6.5 GLOBAL MARKET OF LIPOSOMAL SYSTEM IN FUNCTIONAL FOOD AND NUTRACEUTICAL PRODUCTS

The global market of nutraceuticals as the food industry was $198.7 billion in 2016. It is expected to exceed $285 billion by 2021 with a compound annual growth rate (CAGR) of 8%. The market is estimated to reach $578.23 billion with a CAGR of around 9% in 2021 [90]. A recent market research report by Technavio on the nanoformulation-based food market estimates the market to grow at a CAGR of around 25% during the period 2019–2023 [91]. Around 50% of the population in the United States consumes functional food or nutraceuticals as a food supplement. The global demand for nutraceuticals is increased due to the following reasons: the awareness about healthy lifestyles, increased side effects associated with medicines, and hypercholesteremia [92]. In spite of the rapid global demand, the regulations of nutraceuticals are still unclear, and there is variation in the definitions for nutraceuticals or functional food around the globe. Thus, nutraceutical is not regulated as drug products in some countries but as food supplements [93]. This growth is usually directed by the increasing demand for fortified dairy products and energy drinks.

Asia is turning up to be a potential market for liposomal carriers in functional food and nutraceuticals. Among Asian countries, Japan stands as the largest consumer of functional food while China holds the second position as a consumer since the population is becoming more aware about their food habits. Furthermore, the European nutraceutical market has seen tremendous growth in the dietary supplement and functional food segments, which is driven by the growing average age of the population [94].

Various companies offer their innovative techniques to produce liposomes such as LipoCellTech™ and Zeal Technology™ [95] are innovative companies that supply the best liposomal powders worldwide. They have designed several liposome-based products in food and nutraceutical, e.g., liposomal vitamin C, liposomal multivitamin, liposomal CUR, liposomal magnesium, and liposomal glutathione (GSH) [96]. A summary of approved marketed products based on liposomal-based techniques in food and nutraceutical sectors enlisted in Table 6.2.

TABLE 6.2 A Summary of Approved Drug Product on Liposomal Based Techniques in Food and Nutraceuticals

Product Name	Manufacturer	Key Ingredients	Applications	References
R-Alpha Lipoic acid 250 mg	Doctor's Formulas	Lipoic Acid	Antioxidant, hypertension, heart failure, arteriosclerosis, strengthen immune system	[97]
Cogneo	Liposomal nutraceuticals	Docosahexaenoic acid (DHA)	Brain health	[98]
Foconeo	Liposomal nutraceuticals	DHA-Lutein	Vision health	[99]
Reposeo	Liposomal nutraceuticals	Cannabidiol and melatonin	Healthy sleep rhythm	[100]
Actinovo plus-liposomal resveratrol	Actinovo	Resveratrol	Food supplements	[101]
Actinovo plus-liposomal Vitamin C	Actinovo	Vitamin C	Antioxidant, tiredness, and fatigue	[102]
Actinovo plus-liposomal glutathione	Actinovo	Reduced glutathione	Antioxidant	[103]
Liposomal curcumin-resveratrol valimenta	Valimenta	Curcumin with resveratrol	Supplements	[104]
Liposomal CoQ_{10} Valimenta	Valimenta	Coenzyme Q10	Supplements	[105]
Liposomal Vitamin C valimenta	Valimenta	Vitamin C	Supplements	[106]
Liposomal Vitamin D3 valimenta	Valimenta	Vitamin D3	Supplements	[107]
Liposomal sleep formula	Core MedScience	Melatonin + GABA + Glutathione	Sleep	[108]

TABLE 6.2   *(Continued)*

Product Name	Manufacturer	Key Ingredients	Applications	References
Liposomal active b complex + mineral	Core Medscience	Vitamin B + mineral	Improved energy, healthy heart, clearer thinking, mood, sleep	[109]
Liposomal glutathione	Core Medscience	Glutathione	Clearer thinking, improved energy, physical performance, improved detoxification	[110]
Vitamin C liposomal	Goldman laboratories	Vitamin C	Collagen synthesis, antioxidant, anti-fatigue action	[111]
Vitamin D liposomal complex	Goldman laboratories	Vitamin D	Osteomalacia, rickets, osteoporosis, and other bone disorders	[112]
Glutathione liposomal	Goldman laboratories	Glutathione	Nutritional supplement	[113]
Phosal®	Lipoid GmbH	Curcuminoids	Gastrointestinal health, fat metabolism and liver health	[114]
PhytoSolve®	Lipoid GmbH	Coenzyme Q10, multivitamin, omega-3 fatty acids	Nutrient-enriched concentrate	[115]
Encapsome™	Encapsula NanoSciences LLC	Multivitamin, CoQ_{10}	Dietary supplements	[116]
Liposomal Vitamin C orange	Cure support	Vitamin C	Strengthen immune system, normal functioning of the nervous system, reduction of tiredness and fatigue	[117]

TABLE 6.3 A-List of Relevant Patents on Liposomal Technology in Food and Nutraceuticals

Publication/Patent No.	Title	Assignees/Applicant	Active Ingredients	Remarks	References
US10660853B2	Method for preparing liposome frozen powder capable of efficiently retaining its bilayer structure	Seoul National University	Microorganism, protein, enzyme.	Preparation of liposome frozen powder, thus advantageously preventing deterioration in stability of liposome particles.	[118]
US20160279073A1	Terpene and cannabinoid formulations	Full Spectrum Lab | Teewinot Technologies	Hemp oil and cannabinoid	Development of stable, fast-acting liposome formulation of hemp oil and cannabinoids that are suitable for nutraceutical applications.	[119]
TW202010488A	Method for preparing a stable controlled release propolis colloidal dispersion system for various uses	Yabifita	Propolis	A stable controlled release propolis colloidal dispersion system for use food as a component in functional foods.	[120]
KR2020-0005896A	Pharmaceutical composition and health functional food for preventing improving or treating inflammatory bowel disease including liposomes	Sejong University	Krill oil	A liposomal functional food for preventing, alleviating, or treating inflammatory intestinal diseases.	[121]
CN109463751A	Making method of health-care food of coenzyme Q10 liposome	Jiangsu Aland nourishment	Coenzyme Q10	A health-care food of a coenzyme Q10 liposome.	[122]

TABLE 6.3 (*Continued*)

Publication/Patent No.	Title	Assignees/ Applicant	Active Ingredients	Remarks	References
EP3417846A1	Food and/or nutraceutical composition	Gifar	Endocannabinoid	A food and/or nutraceutical composition, for example, in the form of liposomes.	[123]
CN108576779A	Konjac glucomannan-liposome composite nano food delivery system and preparation method and application thereof	Hebei University	Konjac glucomannan	A konjac glucomannan-liposome composite nano food delivery system has high entrapment rate, and synergy for a plurality of food components.	[124]
KR20180050954A	Method for preparing bioactive collagen peptide by using subcritical water, and liposome containing bioactive collagen peptide prepared thereby	Konkuk University	Collagen peptide	A stable liposome containing the bioactive collagen for the people who prefer high value-added functional product.	[125]
US10406117B2	Water-soluble lipophilic materials	Kemin	Lutein	Lutein-loaded liposomes are included in the dietary supplements industry, with potential nutritional bars, and functional foods.	[126]

TABLE 6.3 (Continued)

Publication/Patent No.	Title	Assignees/ Applicant	Active Ingredients	Remarks	References
CN106552255A	Preparation method of deer blood peptide liposome	Dalian Institute of Chemical Physics	Deer blood peptide	A deer blood peptide liposome improves bioavailability of oral deer blood peptide, enlarges a deer blood peptide eating method range, can be used in the fields of functional foods.	[127]
KR101740136B1	Natural liposome comprising red ginseng for improving blood circulation, process for the preparation thereof and food or pharmaceutical composition comprising the same	M&C Life Science	Red ginseng	A natural liposome comprising red ginseng for improving blood circulation, and a food composition comprising the same.	[128]

6.6 RECENT PATENTS RELATED TO LIPOSOMAL TECHNOLOGY IN FUNCTIONAL FOOD AND NUTRACEUTICALS

Liposomes can be used in the impending fields of nanotechnology such as cancer therapy, diagnosis, cosmetics, agriculture, gene therapy, and food technology. Authors have conducted detailed patent searches on-orbit intelligence, google patents, and espacenet on the liposomal formulation in functional food and nutraceuticals. This chapter will provide an overview of patented technologies and food applications of liposomes shown in Table 6.3. The search was performed between January 1, 2016 and June 30, 2020. A total number of 156 patents and 145 patents were nonrelevant based on their title, abstract, and claims information. Finally, 11 patents were considered in this chapter.

6.7 CONCLUSION

This chapter insights on the recent liposomal technology trends, challenges, and marketed products of liposomes in food and nutraceuticals industries. Immense efforts have been made in the last decades to enrich the liposomal delivery system in food industries. Many of them have shown considerable promise results for clinical applications, although stimuli sensitive polymer-based liposomes are still in the preclinical stage. The biggest limitation associated with liposomes is their scale-up manufacturing and it has not been addressed yet. Furthermore, there are many concerns such as a selection of food-grade polymers, the regulatory protocol of liposomes, along with the perspective of patent, which needs to be addressed before the commercialization of liposomal formulation for food or nutraceuticals sector. To optimize all these factors, the liposomal delivery system for the food industries can reach clinical relevance.

KEYWORDS

- large unilamellar vesicles
- multilamellar vesicles
- multivesicular vesicles
- oligolamellar vesicles
- small unilamellar vesicles
- unilamellar vesicles

REFERENCES

1. Shade, C. W., (2016). Liposomes as advanced delivery systems for nutraceuticals. *Integrative Medicine: A Clinician's Journal, 15*(1), 33–36.
2. Silva, A. C., Santos, D., et al., (2012). Lipid-based nanocarriers as an alternative for oral delivery of poorly water-soluble drugs: Peroral and mucosal routes. *Curr. Med. Chem., 19*(26), 4495–4510.
3. Alamri, A., Huang, Y., & Zhang, H., (2018). Liposomal nanoformulation of piperlongumine for improved aqueous solubility and enhanced anti-tumor activity *in-vitro. Int. J. Pharm. Pharmacol., 2*, 133. doi: 10.31531/2581–3080.1000133.
4. Samad, A., Sultana, Y., & Aqil, M., (2007). Liposomal drug delivery systems: An updated review. *Curr. Drug Deliv., 4*(4), 297–305.
5. Reineccius, G. A., (1995). *Liposomes for Controlled Release in the Food Industry* (pp. 113–131). ACS Symposium Series; American Chemical Society: Washington, DC.
6. Singh, H., Thompson, A., Liu, W., & Corredig, M., (2012). Liposomes as food ingredients and nutraceutical delivery systems. In: *Encapsulation Technologies and Delivery Systems for Food Ingredients and Nutraceuticals* (pp. 287–318). Woodhead Publishing.
7. Liu, W., Ye, A., Liu, W., Liu, C., & Singh, H., (2013). Liposomes as food ingredients and nutraceutical delivery systems. *Agro. Food Industry Hi-Tech., 24*(2), 68–71.
8. Colas, J. C., Shi, W., Rao, V. S. N. M., Omri, A., Mozafari, M. R., & Singh, H., (2007). *Microscopical Investigations of Nisin-Loaded Nanoliposomes Prepared by Mozafari Method and their Bacterial Targeting* (Vol. 38, pp. 841–847). Micron, New York.
9. Fathi, M., Mozafari, M. R., & Mohebbi, M., (2012). Nanoencapsulation of food ingredients using lipid-based delivery systems. *Trends Food Sci Technol., 23*, 13–27. https://doi.org/10.1016/j.tifs.2011.08.003.
10. Jahadi, M., Khosravi-Darani, K., Ehsani, M. R., et al., (2015). The encapsulation of flavorzyme in nanoliposome by heating method. *J. Food Sci. Technol., 52*, 2063–2072. https://doi.org/10. 1007/s13197-013-1243-0.
11. Khorasani, S., Danaei, M., & Mozafari, M. R., (2018). Nanoliposome technology for the food and nutraceutical industries. *Trends Food Sci. Technol., 79*, 106–115. https:// doi.org/10.1016/j.tifs. 2018.07.009.
12. Daud, M., Jalil, J. A., Madieha, I., et al., (2017). "Unsafe" nutraceuticals products on the internet: The need for stricter regulation in Malaysia. In: *Proceedings of the 5ᵗʰ International Conference on Information Technology for Cyber and IT Service Management at Bali*. Indonesia.
13. Gubernator, J., (2011). Active methods of drug loading into liposomes: Recent strategies for stable drug entrapment and increased *in vivo* activity. *Expert Opin. Drug Deliv., 8*, 565–580.
14. Sercombe, L., Veerati, T., Moheimani, F., et al., (2015). Advances and challenges of liposome assisted drug delivery. *Front Pharmacol., 6*, 1–13. https://doi.org/10.3389/fphar.2015.00286.
15. Kheadr, E. E., Vuillemard, J. C., & El Deeb, S. A., (2000). Accelerated cheddar cheese ripening with encapsulated proteinases. *Int. J. Food Sci. Technol., 35*, 483–495.

16. Thompson, A. K., (2003). *Liposomes: From Concepts to Applications* (Vol. 13, pp. S23–S32). Food New Zealand.
17. Wilkinson, M., & Kilcawley, K., (2005). Mechanisms of incorporation and release of enzymes into cheese during ripening. *Int. Dairy J., 15*, 817–830.
18. Picon, A., Serrano, C., Gaya, P., et al., (1996). The effect of liposome-encapsulated cyprosins on Manchego cheese ripening. *J. Dairy Sci., 79*, 1699–1705.
19. Walde, P., & Ichikawa, S., (2001). Enzymes inside lipid vesicles: Preparation, reactivity and applications. *Biomol. Eng., 18*, 143–177.
20. Laloy, E., Vuillemard, J. C., Dufour, P., et al., (1998). Release of enzymes from liposomes during cheese ripening. *J. Control Release, 54*, 213–222.
21. Kheadr, E. E., Vuillemard, J. C., & El Deeb, S., (2002). Acceleration of cheddar cheese lipolysis by using liposome entrapped lipases. *J. Food Sci., 67*, 485–492.
22. Rao, D., Chawan, C., & Veeramachaneni, R., (1995). Liposomal encapsulation of b-galactosidase: Comparison of two methods of encapsulation and *in vitro* lactose digestibility. *J. Food Biochem., 18*, 239–251.
23. Kirby, C., Brooker, B., & Law, B., (1987). Accelerated ripening of cheese using liposome-encapsulated enzyme. *Int. J. Food Sci. Technol., 22*, 355–375.
24. Matsuzaki, M., McCafferty, F., & Karel, M., (1989). The effect of cholesterol content of phospholipid vesicles on the encapsulation and acid resistance of bgalactosidase from *E. coli. Int. J. Food Sci. Technol., 24*, 451–460.
25. Chawan, C. B., Penmetsa, P. K., Veeramachaneni, R., et al., (1992). Liposomal encapsulation of b-galactosidase: Effect of buffer molarity, lipid composition and stability in milk. *J. Food Biochem., 16*, 349–357.
26. Ortan, A., Campeanu, G., DinuPirvu, C., et al., (2009). Studies concerning the entrapment of Anethum graveolens essential oil in liposomes. *Roum. Biotechnol. Lett., 14*, 4411–4417.
27. Mahapatra, S. K., Chakraborty, S. P., Das, S., et al., (2009). Methanol extract of *Ocimum gratissimum* protects murine peritoneal macrophages from nicotine toxicity by decreasing free radical generation, lipid and protein damage and enhances antioxidant protection. *Oxidative MedCell Longev, 2*, 222–230.
28. Boukhris, M., Bouaziz, M., Feki, I., et al., (2012). Hypoglycemic and antioxidant effects of leaf essential oil of pelargonium graveolens L'Her in alloxan-induced diabetic rats. *Lipids Health Dis., 11*, 81–90.
29. Esmaeili, A., (2012). Biological activities of eremostachys laevigata Bunge grown in Iran. *Pak. J. Pharm. Sci., 25,* 803–808.
30. Tian, J., Ban, X., Zeng, H., et al., (2011). Chemical composition and antifungal activity of essential oil from *Cicuta virosa* L. var. *Latisecta celak. Int. J. Food Microbiol., 145*, 464–470.
31. Jaganathan, S. K., Mazumdar, A., Mondhe, D., et al., (2011). Apoptotic effect of eugenol in human colon cancer cell lines. *Cell Biol. Int., 35*, 607–615.
32. Martin, A., Varona, S., Navarrete, A., et al., (2010). Encapsulation and co-precipitation processes with supercritical fluids: Applications with essential oils. *Open Chem. Eng. J., 4*, 31–41.
33. Shoji, Y., & Nakashima, H., (2004). Nutraceutics and delivery systems. *J. Drug Target., 12*, 385–391.

34. Woranuch, S., & Yoksan, R., (2013). Eugenol-loaded chitosan nanoparticles: I. Thermal stability improvement of eugenol through encapsulation. *Carbohydr. Polym., 96*, 578–585.

35. Hosseini, S. F., Zandi, M., Rezaei, M., et al., (2013). Two-step method for encapsulation of oregano essential oil in chitosan nanoparticles: Preparation, characterization and *in vitro* release study. *Carbohydr. Polym., 95*, 50–56.

36. De Oliveira, E. F., Paula, H. C., & Paula, R., (2014). Alginate/cashew gum nanoparticles for essential oil encapsulation. *Colloids Surf. B, 113*, 146–151.

37. Valenti, D., De Logu, A., Loy, G., et al., (2001). Liposome-incorporated *Santolina insularis* essential oil: Preparation, characterization and *in vitro* antiviral activity. *J. Liposome Res., 11*, 73–90.

38. Liolios, C., Gortzi, O., Lalas, S., et al., (2009). Liposomal incorporation of carvacrol and thymol isolated from the essential oil of *Origanum dictamnus* L. and *in vitro* antimicrobial activity. *Food Chem., 112*, 77–83.

39. Detoni, C. B., De Oliveira, D. M., Santo, I. E., et al., (2012). Evaluation of thermal-oxidative stability and anti-glioma activity of *Zanthoxylum tingoassuiba* essential oil entrapped into multi-and unilamellar liposomes. *J. Liposome Res., 22*, 1–7.

40. Wen, Z., You, X., Jiang, L., et al., (2011). Liposomal incorporation of rose essential oil by a supercritical process. *Flavor Fragr. J., 26*, 27–33.

41. Sinico, C., De Logu, A., Lai, F., et al., (2005). Liposomal incorporation of *Artemisia arborescens* L. essential oil and *in vitro* antiviral activity. *Eur J. Pharm. Biopharm., 59*, 161–168.

42. Moghimipour, E., Aghel, N., Mahmoudabadi, A. Z., et al., (2012). Preparation and characterization of liposomes containing essential oil of Eucalyptus camaldulensis leaf. *Jundishapur J. Nat. Pharm. Prod., 7*, 117–122.

43. Ball, G. F., (1998). *Bioavailability and Analysis of Vitamins in Foods.* London: Chapman and Hall.

44. Yang, S., Liu, W., Liu, C., et al., (2012). Characterization and bioavailability of vitamin C nanoliposomes prepared by film evaporation-dynamic high pressure micro fluidization. *J. Disperse Sci. Technol., 33*, 1608–1614.

45. Kirby, C., Whittle, C., Rigby, N., et al., (1991). Stabilization of ascorbic acid by microencapsulation in liposomes. *Int. J. Food Sci. Technol., 26*, 437–449.

46. Wechtersbach, L., Poklar, U. N., & Cigic, B., (2012). Liposomal stabilization of ascorbic acid in model systems and in food matrices. *LWT Food Sci. Technol., 45*, 43–49.

47. Tesoriere, L., Bongiorno, A., Pintaudi, A. M., et al., (1996). Synergistic interactions between vitamin A and vitamin E against lipid peroxidation in phosphatidylcholine liposomes. *Arch Biochem. Biophys., 326*, 57–63.

48. Banville, C., Vuillemard, J., & Lacroix, C., (2000). Comparison of different methods for fortifying cheddar cheese with vitamin D. *Int. Dairy J., 10*, 375–382.

49. Lee, S. C., Yuk, H. G., Lee, D. H., et al., (2002). Stabilization of retinol through incorporation into liposomes. *J. Biochem. Mol. Biol., 35*, 358–363.

50. Caddeo, C., Teskac, K., Sinico, C., et al., (2008). Effect of resveratrol incorporated in liposomes on proliferation and UV-B protection of cells. *Int. J. Pharm., 363*, 183–191.

51. Jay, J. M., Loessner, M. J., & Golden, D. A., (2005). *Modern Food Microbiology.* New York: Springer.

52. Thomas, L. V., & Delves-Broughton, J., (2005). Nisin. In: Davidson, P. M., Sofos, J. N., & Branen, A. L., (eds.), *Antimicrobials in Food* (pp. 237–274). New York: CRC Press.

53. Degnan, A., Buyong, N., & Luchansky, J. B., (1993). Antilisterial activity of pediocin AcH in model food systems in the presence of an emulsifier or encapsulated within liposomes. *Int. J. Food Microbiol., 18,* 127–138.

54. Benech, R. O., Kheadr, E., Laridi, R., et al., (2002). Inhibition of *Listeria innocua* in cheddar cheese by addition of nisin Z in liposomes or by *in situ* production in mixed culture. *Appl. Environ. Microbiol., 68,* 3683–3690.

55. Weiss, J., Gaysinsky, S., Davidson, M., et al., (2009). Nanostructured encapsulation systems: Food antimicrobials. In: Barbosa-Canovas, G., Mortimer, A., Lineback, D., et al., (eds.), *Global Issues in Food Science and Technology* (pp. 425–479). San Diego (CA): Academic Press.

56. Laridi, R., Kheadr, E., Benech, R. O., et al., (2003). Liposome encapsulated nisin Z: Optimization, stability and release during milk fermentation. *Int. Dairy J., 13,* 325–336.

57. Were, L. M., Bruce, B. D., Davidson, P. M., et al., (2003). Size, stability, and entrapment efficiency of phospholipid nanocapsules containing polypeptide antimicrobials. *J. Agric. Food Chem., 51,* 8073–8079.

58. Taylor, T. M., Gaysinsky, S., Davidson, P. M., et al., (2007). Characterization of antimicrobial-bearing liposomes by z-potential, vesicle size, and encapsulation efficiency. *Food Biophys., 2,* 1–9.

59. Bennick, A., (2002). Interaction of plant polyphenols with salivary proteins. *Crit. Rev. Oral Biol. Med., 13,* 184–196.

60. Arts, I. C., & Hollman, P. C., (2005). Polyphenols and disease risk in epidemiologic studies. *Am. J. Clin. Nutr., 81,* 317S–325S.

61. Fang, Z., & Bhandari, B., (2010). Encapsulation of polyphenols a review. *Trends Food Sci. Technol., 21,* 510–523.

62. Rashidinejad, A., Birch, E. J., Sun-Waterhouse, D., et al., (2014). Delivery of green tea catechin and epigallocatechin gallate in liposomes incorporated into low-fat hard cheese. *Food Chem., 156,* 176–183.

63. Isailovic, B. D., Kostic, I. T., Zvonar, A., et al., (2013). Resveratrol loaded liposomes produced by different techniques. *Innov. Food Sci. Emerging Technol., 19,* 181–189.

64. Takahashi, M., Uechi, S., Takara, K., et al., (2009). Evaluation of an oral carrier system in rats: Bioavailability and antioxidant properties of liposome-encapsulated curcumin. *J. Agric. Food Chem., 57,* 9141–9146.

65. Ramadan, M. F., Asker, S., & Mohamed, M., (2009). Antimicrobial and antiviral impact of novel quercetin-enriched lecithin. *J. Food Biochem., 33,* 557–571.

66. Fan, M., Xu, S., Xia, S., et al., (2007). Effect of different preparation methods on physicochemical properties of salidroside liposomes. *J. Agric. Food Chem., 55,* 3089–3095.

67. Ramadan, M. F., (2012). Antioxidant characteristics of phenolipids (quercetin-enriched lecithin) in lipid matrices. *Ind. Crops Prod., 36,* 363–369.

68. Gibis, M., Vogt, E., & Weiss, J., (2012). Encapsulation of polyphenolic grape seed extract in polymer-coated liposomes. *Food Funct., 3,* 246–254.

69. Semple, S. C., Chonn, A., & Cullis, P. R., (1996). Influence of cholesterol on the association of plasma proteins with liposomes. *Biochemistry, 35*, 2521–2525.

70. Pattni, B. S., Chupin, V. V., & Torchilin, V. P., (2015). New developments in liposomal drug delivery. *Chem. Rev., 115(19)*, 10938–10966.

71. Shukla, S., Haldorai, Y., Hwang, S. K., Bajpai, V. K., Huh, Y. S., & Han, Y. K., (2017). Current demands for food-approved liposome nanoparticles in food and safety sector. *Front Microbiol., 8*, 2398.

72. Mu, X., & Zhong, Z., (2006). Preparation and properties of poly(vinyl alcohol)-stabilized liposomes. *Int. J. Pharm., 318*, 55–61.

73. Alhajlan, M., Alhariri, M., & Omri, A., (2013). Efficacy and safety of liposomal clarithromycin and its effect on Pseudomonas aeruginosa virulence factors. *Antimicrob. Agents Chemother., 57*(6), 2694–2704.

74. Aishwarya, M., Subin, R. C. K. R., Quan, S. H., Laurent, B., & Chibuike, C. U., (2015). Encapsulation of food protein hydrolysates and peptides: A review. *RSC Adv., 5*, 79270–79278.

75. Peng, S., Zou, L., Liu, W., Li, Z., Liu, W., Hu, X., Chen, X., & Liu, C., (2017). Hybrid liposomes composed of amphiphilic chitosan and phospholipid: Preparation, stability and bioavailability as a carrier for curcumin. *Carbohydr. Polym., 156*, 322–332.

76. Alexander, M., Lopez, A. A., Fang, Y., et al., (2012). Incorporation of phytosterols in soy phospholipids nanoliposomes: Encapsulation efficiency and stability. *LWT Food Sci. Technol., 47*, 427–436.

77. Chan, Y. H., Chen, B. H., Chiu, C. P., et al., (2004). The influence of phytosterols on the encapsulation efficiency of cholesterol liposomes. *Int. J. Food Sci. Technol., 39*, 985–995.

78. Hwang, J. S., Tsai, Y. L., & Hsu, K. C., (2010). The feasibility of antihypertensive oligopeptides encapsulated in liposomes prepared with phytosterols-β-sitosterol or stigmasterol. *Food Res. Int., 43*, 133–139.

79. Cherif, A. O., (2012). Phytochemicals components as bioactive foods. In: Rasooli, I., (ed.), *Bioactive Compounds in Phytomedicine* (pp. 113–124). Rijeka: InTech.

80. Normén, L., Andersson, S., & Dutta, P., (2004). Does phytosterol intake affect the development of cancer? In: Dutta, P. C., (ed.), *Phytosterols as Functional Food Components and Nutraceuticals* (pp. 191–242). New York: Marcel Dekker.

81. McClements, D. J., Decker, E. A., & Weiss, J., (2007). Emulsion-based delivery systems for lipophilic bioactive components. *J. Food Sci., 72*, R109–R124.

82. Jovanović, A. A., Balanč, B. D., Ota, A., Grabnar, P. A., Djordjević, V. B., Šavikin, K. P., Bugarski, B. M., et al., (2018). comparative effects of cholesterol and β-sitosterol on the liposome membrane characteristics. *European J. Lipid Sci. and Tech., 120*, 9, 1–42.

83. Hattori, Y., Nakamura, M., Takeuchi, N., Tamaki, K., Ozaki, K., & Onishi, H., (2019). *Effect of Cationic Lipid Type in PEGylated Liposomes on siRNA Delivery Following the Intravenous Injection of siRNA Lipoplexes, 1*(2), 74–85.

84. Alomrani, A., Badran, M., Harisa, G. I., ALshehry, M., Alhariri, M., Alshamsan, A., & Alkholief, M., (2019). The use of chitosan-coated flexible liposomes as a remarkable carrier to enhance the antitumor efficacy of 5-fluorouracil against colorectal cancer. *Saudi Pharm. J., 27*(5), 603–611.

85. Elferink, M. G., De Wit, J. G., In't Veld, G., et al., (1992). The stability and functional properties of proteoliposomes mixed with dextran derivatives bearing hydrophobic anchor groups. *Biochim. Biophys. Acta., 1106,* 23–30.
86. Kono, K., Hayashi, H., & Takagishi, T., (1994). Temperature-sensitive liposomes: Liposomes bearing poly (N-isopropylacrylamide). *J. Control Release, 30,* 69–75.
87. Tamjidi, F., Shahedi, M., Varshosaz, J., et al., (2013). Nanostructured lipid carriers (NLC): A potential delivery system for bioactive food molecules. *Innov. Food Sci. Emerging Technol., 19,* 29–43.
88. McClements, D. J., (2014). *Nanoparticle- and Microparticle-Based Delivery Systems: Encapsulation, Protection and Release of Active Compounds.* Boca Raton (FL): CRC Press.
89. Liu, X., & Huang, G., (2013). Formation strategies, mechanism of intracellular delivery and potential clinical applications of pH-sensitive liposomes. *Asian J Pharm Sci., 8,* 319–328.
90. BCC Research, (2017). *Nutraceuticals: Global Markets.* Dublin, Ireland; 2017. Report No.: FOD013F.
91. https://www.businesswire.com/news/home/20190610005263/en/Global-Food-Nanotechnology-Market-2019-2023-Growing-Applications (accessed on 17 June 2021).
92. Fleisher, L. A., Roizen, M. F., & Roizen, J. D., (2018). *Essence of Anesthesia Practice* (4th edn., pp. 530, 531) Philadelphia, Pennsylvania, USA.: Elsevier and Saunders.
93. Jain, P. N., Rathod, M. H., Jain, V. C., & Vijayendraswamy, S. M., (2018). Current regulatory requirements for registration of nutraceuticals in India and USA. *Int. J. Drug Reg. Aff., 6*(2), 22–29.
94. https://www.annualreviews.org/doi/full/10.1146/annurev-food-022814-015507 (accessed on 17 June 2021).
95. https://aureabiolabs.com/zeal-technology/ (accessed on 17 June 2021).
96. https://www.lipocelltech.com/ (accessed on 17 June 2021).
97. https://www.doctorsformulas.com/en/product/r-alpha-lipoic-acid-250mg.htm (accessed on 17 June 2021).
98. https://liposomanutraceuticals.com/branded-liposomal-supplements/ (accessed on 17 June 2021).
99. https://liposomanutraceuticals.com/white-label-liposomal-supplements/liposomal-dha-lutein/ (accessed on 17 June 2021).
100. https://liposomanutraceuticals.com/branded-liposomal-supplements/reposeo/
101. https://www.actinovo.com/en/liposomal-resveratrol (accessed on 17 June 2021).
102. https://www.actinovo.com/en/liposomal-vitamin-c-orange-vanilla (accessed on 17 June 2021).
103. https://www.actinovo.com/en/liposomal-glutathione-dragonfruit-mango (accessed on 17 June 2021).
104. https://www.valimenta.com/product/liposomal-curcumin-resveratrol/ (accessed on 17 June 2021).
105. https://www.valimenta.com/product/liposomal-coq10/ (accessed on 17 June 2021).
106. https://www.valimenta.com/product/liposomal-vitamin-c/ (accessed on 17 June 2021).

107. https://www.valimenta.com/product/liposomal-vitamin-d3/ (accessed on 17 June 2021).
108. https://coremedscience.com/collections/all-products/products/liposomal-liquid-melatonin-sleep-aid-supplement-with-gaba-and-glutathione-10mg-spray-before-bed-get-natural-sound-sleep-without-drowsiness (accessed on 17 June 2021).
109. https://coremedscience.com/collections/all-products/products/copy-of-liposomal-glutathione-softgels-china-free-500-mg-serving-reduced-glutathione-gold-standard-setria-brand-glutathione-japan-30-servings-non-gmo-soy-free (accessed on 17 June 2021).
110. https://coremedscience.com/collections/all-products/products/liposomal-glutathione (accessed on 17 June 2021).
111. https://www.goldmanlaboratories.com/en/16-liposomal-vitamin-c (accessed on 17 June 2021).
112. https://www.goldmanlaboratories.com/en/32-liposomal-vitamin-d3-complex (accessed on 17 June 2021).
113. https://www.goldmanlaboratories.com/en/liposomal-glutathione/14-liposomal-glutathione.html (accessed on 17 June 2021).
114. https://www.lipoid.com/en/phosal (accessed on 17 June 2021).
115. https://www.lipoid.com/en/phytosolve (accessed on 17 June 2021).
116. https://clarencedesign.com/projects/encapsula/contract_manufacturing.html (accessed on 17 June 2021).
117. https://www.curesupport.com/product/curesupport-liposomal-vitamin-c-orange/ (accessed on 17 June 2021).
118. Chang, P. S., Park, K. M., Eun, H. Y. E. H., & Jung, H. S., (2020). *Method for Preparing Liposome Frozen Powder Capable of Efficiently Retaining its Bilayer Structure*. US patent 10660853B2.
119. Donsky, M., & Winnicki, R., (2016). *Terpene and Cannabinoid Formulations*. US patent 20160279073A1.
120. Panagiota, D., Konstantinos, G., Nikolaos, K., Sofia, L., Anna, P., & Anagnostis-Ioannis, T., (2020). *Method for Preparing a Stable Controlled-Release Propolis Colloidal Dispersion System for Various Uses*. TW patent 202010488A.
121. Hee, K. J., & Jeong, L. S., (2020). *Pharmaceutical Composition and Health Functional Food for Preventing Improving or Treating Inflammatory Bowel Disease Including Liposomes*. KR patent 20200005896A.
122. Haiying, C., Lin, L., Xiang, S., Xudong, X., Guodong, Z., & Xing, Z., (2019). *Making Method of Health-Care Food of Coenzyme Q10 Liposome*. CN patent 109463751A.
123. Garito, A., & Leo, E. G., (2018). *Food and/or Nutraceutical Composition*. EP patent 3417846A1.
124. Fatang, J., Ying, K., Jinling, L., Xuewen, N., Kao, W., & Man, X., (2018). *Konjac Glucomannan-Liposome Composite Nano Food Delivery System and Preparation Method and Application Thereof*. CN Patent 108576779A.
125. Ji, J. K., Gi, M. S., & Hee, P. S., (2018). *Method for Preparing Bioactive Collagen Peptide by Using Subcritical Water, and Liposome Containing Bioactive Collagen Peptide Prepared Thereby*. KR patent 20180050954A.
126. Martins, D. S., (2019). *Water Soluble Lipophilic Materials*. US patent 10406117B2.

127. Yan, J., & Hanfa, Z., (2017). *Preparation Method of Deer Blood Peptide Liposome.* CN patent 106552255A.
128. Jung, K. M., Kim, H., No-Hwan, P., Park, S., Park, J., Young, S. H., Okmin, et al., (2017). *Natural Liposome Comprising Red Ginseng for Improving Blood Circulation, Process for the Preparation Thereof and Food or Pharmaceutical Composition Comprising the Same.* KR Patent 101740136B1.
129. Fathi, M., Mozafari, M., & Mohebbi, M., (2012). Nanoencapsulation of food ingredients using lipid-based delivery systems. *Trends Food Sci. Technol., 23*, 13–27.
130. Mozafari, M. R., Reed, C. J., & Rostron, C., (2007). Prospects of anionic nanolipoplexes in nanotherapy: Transmission electron microscopy and light scattering studies. *Micron., 38*(8), 787–795.
131. Shiva, E., Azadmard-Damirchi, S., Seyed, H. P., Hadi, V., & Javad, H., (2016). Liposomes as carrier vehicles for functional compounds in the food sector. *Journal of Experimental Nanoscience, 11*(9), 737–759. doi: 10.1080/17458080.2016.1148273.

CHAPTER 7

Liposomes as Robust Anti-Tubercular and Anticancer Drugs Delivery System

AMEER KHUSRO and CHIROM AARTI

Research Department of Plant Biology and Biotechnology,
Loyola College, Chennai – 600034, Tamil Nadu, India

ABSTRACT

Considering the devastating threat and high mortalities due to the uncontrolled emergence of tuberculosis (TB) and cancer over the past few decades, the discovery of a new therapeutic approach is the desperate demand of this hour. Liposomes are a new drug delivery system that has created tremendous interest towards the treatment of TB and cancer. These microscopic spherical vesicles constitute one or few phospholipid bilayers representing polar groups of phospholipids towards the inner side and aqueous phase towards the outer side. These closed bilayer structures are generally preferred for encapsulation and delivery of potential drugs in order to entrap both hydrophilic and hydrophobic components because of the availability of aqueous core parts and lipid bilayers. Till date, several antitubercular and anticancer drugs have been formulated as liposomes for enhancing the therapeutic values. In addition, varieties of vesicular drug delivery systems viz. archeosomes, emulsomes, ethosomes, niosomes, transfersomes, vesosomes, and virosomes are developed as new generation liposomes for the treatment of various types of cancer. The prime focus of this chapter is to shed light on the antitubercular and anticancer aspects of the liposomes encapsulating diversified drugs, indicating its profound influence as effectual prophylactic and therapeutic alternatives to the existing other drug delivery systems in the future.

7.1   INTRODUCTION

Liposomes are small spherical vesicles that are mainly composed of aqueous compartment surrounded by phospholipid bilayer membranes. The dispersion of lipids or phospholipids in an aqueous medium through stirring forms liposomes spontaneously. Phospholipids are amphiphilic molecules with hydrophilic head and hydrophobic tail. The hydrophilic portion is generally phosphoric acid which bounds to water molecules and points towards the aqueous component, whereas the hydrophobic portion contains fatty acid or hydrocarbon chains and points in the opposite direction. Phospholipids such as phosphatidylcholine (PC), phosphatidylethanolamine (PE), phosphatidylglycerol, and phosphatidylserine (PS) are commonly used for liposome formulation [1].

The size of liposomes can vary from few nanometers to micrometers in diameter. Small unilamellar vesicles (SUV) range from 20 to 100 nm. Large unilamellar vesicles (LUV) and medium unilamellar vesicle (MUV) are of >100 nm, while multilamellar vesicles (MLVs) constitute many concentric lipid bilayers of 1–5 μm. On the other hand, unilamellar vesicle (UV) contains lipid bilayer of all sizes. Giant unilamellar vesicle (GUV) and multivesicular vesicle (MV) consist of lipid bilayer of more than 1 μm. Oligolamellar vesicle (OLV) contains a lipid bilayer of 0.1–1 μm. Based on the composition, they are categorized as cationic liposomes, conventional liposomes (CL), immuno-liposomes, long-circulating liposomes (LCL), and pH-sensitive liposomes. In view of the mode of preparations, it is categorized into reverse-phase evaporation vesicles (REV), multilamellar vesicle made by reverse-phase evaporation vesicles (MLV-REV), stable plurilamellar vesicle (SPLV), and frozen and thawed multilamellar vesicle (FATMLV) [2].

Liposomes are extensively utilized as carriers for bioactive molecules in pharmaceutical industries to entrap unstable components, particularly antioxidants and antimicrobial agents. The phospholipid layer of liposome forms a barrier, which is generally resistant to enzymes, pH, and free radicals within the body [3]. Liposomes are considered as one of the most successful drug carrier systems due to its high biodegradability nature, low toxicity, and potentiality to encapsulate hydrophilic and hydrophobic components [4]. Currently, several liposome-based formulations as anti-cancer, antifungal, and antiviral agents are in clinical phases. Additionally, liposomes are being used in vaccinology for developing therapeutic vaccines against viral and bacterial diseases [5].

In view of the disparate applications of liposomes in drug delivery such as site-specific targeting, sustained-release drug delivery, intraperitoneal administration, eye disorders, neurological disorders, microbial infections, and vaccines adjuvants, this chapter discusses the recent developments in liposome technology as drug carriers particularly for tuberculosis (TB) and cancer therapy.

7.2 CURRENT TB SITUATION AND END TB STRATEGY: AN OVERVIEW

TB is the leading contagious disease caused by *Mycobacterium tuberculosis*. In 2018, 10 million new TB cases were reported globally, and 1.5 million people died, constituting approximately 2,51,000 people with human immunodeficiency virus (HIV). However, the death rate due to TB infection reduced by 42% between 2000 and 2018. Globally in 2018, about 484,000 people developed TB that was resistant to rifampicin-resistant TB. Of them, 78% were diagnosed with multidrug-resistant TB (MDR-TB). In 2018, approximately 187,000 cases of MDR/rifampicin-resistant TB were observed, whereas 6.2% were reported with extensively drug-resistant TB (XDR-TB). On the other hand, in 2018, there were about 477,000 cases of TB which showed co-infection with HIV [6]. During 2000–2018, 1.6% of rate of reduction in TB cases was reported, whereas 2.0% of decline was observed during 2017–2018 worldwide. The total number of cases for TB deaths was reported 11.0% between 2015 and 2018, which was against the estimated rate of 35% reduction towards the End TB Strategy by 2020. There is an estimated target of 90.0 and 95.0% reduction in TB deaths by 2030 and 2035, respectively [7].

7.2.1 M. TUBERCULOSIS PATHOGENICITY

Initially, *M. tuberculosis* affects human pulmonary macrophage. Macrophages remove the invading microorganisms through phagocytosis process [8]. Generally, non-mycobacterium bacteria are trapped into the phagosomes, which fuse with lysosome and forms phagolysosome. Phagolysosomes inhibit the growth of invaded microbes. However, *M. tuberculosis* after phagocytized escapes from this defense mechanism by infecting macrophages and replicate in this drastic condition. The adaptive mechanism of bacteria depends on varied survival processes. *M. tuberculosis* arrests the maturation process of phagosomes, prevents the acidification of phagosomal compartments, and avoids the formation of phagolysosome. Furthermore, bacteria impair the apoptosis of macrophages and suppress the antimicrobial responses, thus helping the bacterium to escape from the phagosomes [8]. The bacterium becomes undetectable to the innate immune system because major histocompatibility complex class II antigen keeps away its presentation. Thus, *M. tuberculosis* is able to manipulate and survive in the adverse condition. The unique composition of the cell wall is a leading factor for bacterial pathogenicity [9].

7.2.2   RECENTLY APPROVED ANTI-TB DRUGS

Over the past five decades, only three new drugs have shown effectiveness in the treatment of TB. Bedaquiline (a diarylquinoline compound) exhibited antitubercular role against drug-sensitive and drug-resistant *M. tuberculosis* with a novel mode of action including the inhibition of bacterial adenosine triphosphate (ATP) synthase (by binding to subunit c of the enzyme), the enzyme used by bacteria to generate energy [10]. Hards et al. [11] clearly showed the uncoupling process induced by bedaquiline. Bedaquiline along with other antitubercular drugs is used for treating MDR-TB. It should only be used when an effective TB treatment cannot otherwise be provided and other drugs are showing severe toxicity to patients. Bedaquiline is not used for treating extra-pulmonary TB, drug-sensitive TB, latent TB infections, and infection caused by mycobacteria other than *M. tuberculosis*. This drug was approved based on the phase IIb clinical trial results. However, phase III trials are still ongoing and its final approval is pending [6].

Delamanid (another approved anti-TB drug) belongs to nitroimidazole class [12] and targets bacterial cell wall components, particularly methoxy mycolic acid and ketomycolic acid [12]. In fact, it is a pro-drug which shows its property through the action of deazaflavin dependent nitroreductase. The intermediate metabolite produced between delamanid and desnitro-imidazooxazole derivative blocks the mycolic acid production [13]. This drug is also under trial a phase (phase III) which is used along with moxifloxacin. In addition to this, other ongoing trials include assessing the utilization of this drug for treating pediatric MDR-TB. It should be noteworthy that bedaquiline and delamanid should be administrated considering the WHO-suggested regimen for MDR-TB infected patients for avoiding the development of drug resistance *M. tuberculosis* [13]. Sometimes both drugs will be given to the same patient in the case of difficult-to-treat patients.

The U.S. Food and Drug Administration (FDA) approved Pretomanid (an oral drug; nitroimidazopyran derivatives) in 2019 for the treatment of a specific type of MDR-TB of the lungs. It has also shown activity against both latent and active TB. It is given to the patients along with bedaquiline and linezolid. Vomiting, nausea, acne, headache, peripheral neuropathy, anemia, enhanced liver enzymes, dyspepsia, rash, hyperamylasemia, visual impairment, hypoglycemia, and diarrhea are the most common adverse effect of this drug. The use of this drug should be avoided for patients with hypersensitivity towards bedaquiline and linezolid.

7.3 LIPOSOMES AS DRUG DELIVERY SYSTEM FOR TB THERAPY

Pharmacodynamic and pharmacokinetic properties are generally improved due to the implementation of liposomes as drug delivery system. Liposomes as a drug delivery system reduce toxicity and enhance the antimicrobial activities of drugs against intracellular and extracellular pathogens [14]. Liposomes have gained immense interest due to its unique properties such as selective passive drug targeting and promising therapeutic efficiency. Liposomes are engulfed by macrophages and release the antitubercular drugs immediately in order to inhibit the growth of *M. tuberculosis*. Thus, it is an emerging potential drug delivery system as antimicrobials. Additionally, the rate of elimination of *M. tuberculosis* from liver and spleen using liposomes-encapsulated drug was reported higher than that of free drugs, indicating it as a highly effective therapeutic strategy for TB [15].

Both the pulmonary and extra-pulmonary TB can be treated using liposomes. The route of administration plays a paramount role in the success of liposome-based TB therapy [16]. Various routes except the oral route are generally considered for the administration of liposomes because liposomes are sensitive to intestinal lipases [17]. The intravenous administration of liposomes containing antitubercular drugs shows outflow of liposomal constituents in plasma before reaching the target tissue. Sometimes it leads to the rapid clearance from the bloodstream [18]. After intravenous administration, accumulation of liposomes containing antitubercular drugs occurs at the liver and spleen, instead of being accumulated at the lung, thereby suggesting a promising strategy for extra-pulmonary TB therapy [19]. The intravenous administration route may be of utmost interest in pulmonary TB treatment [20]. In addition, non-invasive routes, particularly inhalation route is another attractive strategy as antitubercular therapy [21]. The inhalatory administration of liposomes for TB treatment depends on the size of liposomes [22]. Liposomes of about 1 nm size often deposit in the upper respiratory tract (nose, pharynx, and larynx). Tracheal and bronchi regions entrap liposomes of 5 nm in size, while liposomes of 20 nm are the most favorable for accumulation into the deep alveolar region of the lung. Liposomes of size >15 μm should be avoided as they retain in the throat and swallowed [22]. After the inhalation of liposomes, it translocates to extra-pulmonary sites and reaches other target organs by crossing the respiratory tract epithelium into the interstitium, accessing the bloodstream, and distributing through the lymphatic pathways [23].

Liposomes are utilized for targeting specified organs by passive and active methods [24]. In passive targeting, liposomes are mainly composed of phospholipids and sterols [25]. Combining passive targeting in pulmonary TB with the inhalation route is of great interest, since the particles possess an adequate size. This passive targeting is thus possible because of particles engulfing tendency of macrophages. Combining intravenous and inhalation routes with passive targeting in extra-pulmonary TB is also of interest. After reaching into the blood, liposomes are easily engulfed by phagocytic cells, thereby reaching to the lysosomes [26]. Gaspar et al. [19] formulated liposomes by encapsulating rifabutin and administrated intravenously every 3 days over a 2-week period in mice and compared with the free drug administrated by the similar route. Findings showed that the levels of drug in the lungs were lower than that of the liver or spleen, suggesting its role in the treatment of extra-pulmonary TB, particularly in patients co-infected with HIV.

Amikacin was encapsulated into SUVs made of PC/Chol/distearoylphosphatidylglycerol by passive method and administered intravenously in mice for 3 days a week over 4 weeks. Findings showed increased antitubercular activity and higher half-life time towards the free drug [27]. Pyrazinamide was encapsulated in MLVs made of dipalmitoylphosphatidylcholine/Chol and administered intravenously. Results showed a higher reduction in *M. tuberculosis* counts in the lungs when compared with the free drug for 6 days/week, thereby suggesting that the drug-encapsulated liposomes were more effective than the free drug [26]. Rifampicin and isoniazid were encapsulated in MLVs composed of PC/Chol and evaluated for antitubercular roles in guinea pigs via the inhalation route. Findings showed the presence of both the drugs in the lungs until day 5 post-nebulization. Results suggested a possible reduction in bacterial counts by daily administration frequency of the two first-line antitubercular drugs [28].

Active targeting corresponds to the alteration in the composition and structure of liposomes [25]. These alterations involve the applications of charged lipids or the attachment of any ligand for delivering the drug to the targeted sites. Additionally, coating of liposomal surface with PEG is also an integral part of the active targeting method [18]. The variation in the liposomal surface using negatively charged lipids such as dicetylphosphate (DCP) increases the affinity for capturing anionic liposomes [18]. The tetrapeptide tuftsin (Thr-Lys-Pro-Arg) is a liposomal ligand which is used as a natural macrophage activator. This peptide binds specifically to macrophages, monocytes, and polymorphonuclear leukocytes, thereby possessing a broad-spectrum bactericidal activity. This method stimulates the non-specific cells against the bacterial infection [29]. Mannose-targeted based drug delivery system for the treatment of macrophage-related disease is also being implied. Mannose ligands in liposomes (mannosylated liposomes) can also be used for the treatment of TB [30]. Deol et al. [20] depicted the co-encapsulation of isoniazid and rifampicin in MLVs liposomes containing PC, Chol, O-steroyl amylopectin (O-SAP), DCP and distearoylphosphatidylethanolamine-polyethylene glycol (PEG) 2000 (DSPE-PEG2000). This formulation was administrated intravenously, twice a week for 6 weeks, which revealed 40% accumulation in the lungs of normal and TB infected mice. The use of PEG 2000, O-SAP, and DCP prolonged the blood circulation time and reduced the opsonization by avoiding the accumulation in the liver and spleen macrophages, and preferentially accumulating at the lungs. The encapsulation of these two drugs were observed less toxic, estimating a lower level of total bilirubin and hepatic enzymes when compared with the free drug. Labana et al. [31] formulated the encapsulation of rifampicin and isoniazid in MLVs made of PC, Chol, DCP, and DSPE-PEG 2000, and administered once a week for 6 weeks in mice. Results showed a significant reduction in the total count of bacilli in the lungs, liver, and spleen with respect to the free drugs. The formulation showed no toxicity to the liver with reduced bilirubin, alanine transaminase, and alkaline phosphatase seric levels. Agarwal et al. [32] formulated tuftsin-bearing liposome for the encapsulation of rifampicin. The formulation was administrated intravenously in mice twice a week for 2 weeks. Results showed the higher effectiveness of formulation in terms of reducing the total count of bacilli in lungs compared with the free drug. Zaru et al. [33] studied several MLV formulations composed of PC, dipalmitoylphosphatidylcholine or distearoyloglycerophosphatidylcholine with or without Chol encapsulating rifampicin. The toxicity and mucoadhesive traits of this formulation were examined in lung cancer cell line (A549), which showed reduced toxicity as compared to the free drug. However, the mucoadhesive properties were reported to be reduced in the presence of formulation. The LUVs liposomes made up of PC/Chol were encapsulated with several antitubercular drugs and formulated for administration via inhalation route. Results exhibited an effective drug loading for isoniazid and pyrazinamide, but the encapsulation of ethionamide, streptomycin, and rifampicin was unsuccessful. Likewise, MLVs, and SUV liposomes were used to encapsulate amikacin which revealed that the negatively charged MLV formulation made of phosphatidylcholine/DCP/Chol was effective [34].

Liposomes as drug delivery systems have also shown interest in designing vaccines for TB. The vaccine RUTI exhibited efficacy towards TB treatment following chemotherapy in mice and guinea pigs [35, 36]. Vaccines containing cytokines, interleukin-2 (IL-2), granulocyte macrophage-colony stimulating factor, and tumor necrosis factor can be very useful to elicit humoral and cellular immune responses. DNA vaccine combination expressing mycobacterial heat shock protein 6 and IL-12 using the hemagglutinating virus of Japan and TB subunit vaccine are vaccines of liposomes-based technology which showed its potential as promising candidate for TB treatment [37].

Liposomes encapsulating clofazimine, resorcinomycin A, and PD117558 exhibited complete inhibition of bacterial growth at concentrations ranging from 8 to 31 µg/mL [38]. In another study, the treatment of TB using liposomal clofazimine showed a significant reduction in the bacterial counts of tissues without any toxicity. Findings suggested that liposomal clofazimine can be used as an efficacious antitubercular agent [39].

Kaul et al. [40] investigated the effect of rifampicin and ofloxacin-loaded PEGylated liposome for therapeutic roles in mycobacterium infection. Results indicated satisfactory EE of 66.89 and 40.61% for rifampicin and ofloxacin, respectively with promising anti-mycobacterial activity. Therapeutic efficiency investigations further suggested that liposomes-based delivery of the antitubercular drugs is effective in the murine model of infection.

Effect of the encapsulation of usnic acid into liposomes and its combination with rifampicin and isoniazid against MDR-TB clinical isolates was investigated using microdilution and checkerboard method. Results showed a synergistic interaction between rifampicin and usnic acid or usnic acid-loaded liposomes. The combination of usnic acid or usnic acid-loaded liposomes exhibited no potential effect. Findings suggested that usnic acid-loaded liposomes may be utilized to enhance the antitubercular potency of rifampicin [41].

Nkanga and Krause [42] demonstrated efficient encapsulation of a complex system, containing isoniazid-hydrazone-phthalocyanine conjugate (Pc-INH) in gamma-cyclodextrin (γ-CD), in liposome using crude soybean lecithin through organic solvent-free technique, heating process. Findings indicated Pc-INH/γ-CD-liposomes a promising system for site-specific antitubercular drugs delivery and liposomal encapsulation of large hydrophobic components.

The potency of crude soybean lecithin (CL) to co-encapsulate rifampicin and isoniazid for liposomal dual delivery was evaluated. Rifampicin was encapsulated in CL-liposomes with/without Chol. Isoniazid was incorporated using a freeze-thawing technique. Liposomes constituting CL alone (CLL) exhibited 90% EE for rifampicin and 59% for isoniazid. Findings suggested CLL as an ideal vehicle for macrophage-targeting drug delivery [43].

Niu et al. [44] developed nanoliposomes for facilitating the delivery of antitubercular products to THP-1-derived human macrophages as *Mycobacterium* host cells, and evaluated drug potency as well as the impact of a transforming growth factor-β1-specific short interfering RNA (siRNA) delivery system employing nanoliposomes. Nanoliposomes were reported weakly cytotoxic towards human macrophages. Nanoliposomal short interfering transforming growth factor-β1 (siTGF-β1) could significantly down-regulate TGF-β1 in THP-1-derived human macrophages *in vitro*. Findings concluded that isoniazid/rifampin/pyrazinamide-loaded nanoliposomes with siTGF-β1 have the potency to improve spinal TB chemotherapy via nano-encapsulation of antitubercular drugs. Some of the liposomes as nanocarriers for delivery of antitubercular drugs by active and passive targeting are summarized in Table 7.1.

TABLE 7.1 Liposomes as Anti-Tubercular Drug Delivery System

Liposomes Types	Drug	Animal	Types of Targeting	Mechanism of Action	References
SUVs PC and tuftsin	Rifampicin	Mice	Active	Enhanced efficiency compared to free drug.	[32]
MLVs PC/Chol, and O-SAP, MBSA or DCP	Rifampicin	Rats	Active	Preferential lung accumulation.	[18]
MLVs PC/Chol/DCP/ DSPE-PEG 2000, and O-SAP	Rifampicin and isoniazid	Mice	Active	Improved efficacy of the formulation. Reduced bacilli at the lungs, liver, and spleen. No toxic effects on the liver. Controlled bilirubin, serum alanine transaminase, and alkaline phosphatase levels.	[31]
MLVs PC/ Chol/O-SAP/DCP/ DSPC-PEG 2000	Rifampicin and isoniazid	Mice	Active	High efficacy of the formulation. Reduced liver, kidneys, and lungs CFUs and normal lung weight. Hepatotoxicity reduction and normal lung morphology.	[20]
MLVs PC/Chol	Rifampicin and isoniazid	Guinea pigs	Passive	Increased therapeutic drug levels in plasma with a single dose.	[28]
MLVs DPPC/Chol	Pyrazinamide	Mice	Passive	Increased efficiency compared with free drug. Reduced *M. tuberculosis* bacilli in the lungs.	[26]
SUVs PC/Chol/DSPG	Amikacin	Mice	Passive	Enhanced antitubercular activity and improved half-life of the drug encapsulated liposomes compared with the free drug.	[27]

TABLE 7.1 *(Continued)*

Liposomes Types	Drug	Animal	Types of Targeting	Mechanism of Action	References
MLVs PC/DPPG	Rifabutin	Mice	Passive	Reduced bacilli present in the spleen and liver compared with the free drug. No significant differences in the lungs	[19]
MLVs DSPC/Chol	Rifampicin	*In vitro*	Not available	Inferior toxicity compared with the free drug.	[33]

Abbreviations: Chol: cholesterol; DCP: dicetylphosphate; DPPC: dipalmitoylphosphatidylcholine; DPPG: dipalmitoylphosphatidylglycerol; DSPC: distearoyloglycerophosphatidylcholine; DSPE: distearoylphosphatidylethanolamine; DSPG: distearoylphosphatidylglycerol; MBSA: maleylated bovine serum albumin; MLV: multilamellar vesicle; O-SAP: O-steroyl amylopectin; PC: phosphatidylcholine; PEG: polyethylene glycol; SUV: small unilamellar vesicle.

7.4 CANCER: AN OVERVIEW AND ITS CURRENT SITUATION

Cancer is a multi-cellular and multi-genic non-infectious disease which shows uncontrolled growth of abnormal cells. It is in fact, the colossal public burden of this century, causing millions of mortalities per year. This life-threatening disease causes irregular or uncontrolled growth of malignant cells, which invade normal tissues or organs and destroy them. Cancer cells grow by forming new blood vessels via angiogenesis process. Tumor cells spread from the primary site to secondary locations in the body through blood vessels and lymphatic streams, causing metastasis [45]. The interaction between carcinogens and a person's DNA causes mutation which leads to cancer. These carcinogens may be physical (ultraviolet and ionizing radiation), chemical (components of tobacco, asbestos, and aflatoxins), and infectious (certain viruses, bacteria, and parasites). Surprisingly, about 50% of all cancer cases are due to unknown risk factors or carcinogens. Consumption of tobacco causes about 20 different types of cancer. Currently, about 1.1 billion people are regular consumers of tobacco products which might be responsible for the death of over 1 billion deaths this century [46]. Alcohol consumption is another huge risk factor for cancer which is generally associated with 10 different types of cancers [47]. In 2016, alcohol drinking was responsible for 4–5% of all cancer deaths. According to WHO, there is no minimum level of alcohol consumption at which no risks are involved. Likewise, obesity is responsible for 6% of cases of breast cancer, 8% of colon cancer, and 34% of corpus uteri cancer [48]. On the other hand, air pollution is also responsible for the global burden of cancer.

Millions of people around the globe had cancer, and 9.6 million died from the disease in 2018. By 2040, those counts are expected to be double. The most frequently diagnosed cancer is lung cancer (11.6% of all cases), followed by female breast (11.6%) and colorectal cancers (10.2%). Lung cancer is the foremost cause of death from cancer (18.4% of all deaths), followed by colorectal (9.2%) and gastric cancer (8.2%). The most common cancer type varies among countries, with certain cancers, such as cervical cancer and Kaposi sarcoma, much more common in countries with low-human development index (HDI) than in high-HDI countries. The economic effect of cancer is escalating. The total annual economical cost of cancer in 2010 was calculated about US$ 1.16 trillion [49].

At present, chemotherapy is the most commonly used therapeutic approach among varied strategies for the treatment of cancer. However, it possesses several limitations and severe side effects. On the other hand, most of the anticancer drugs exhibit poor bioavailability and pharmaco-kinetics attributes. Therefore, there is a prerequisite for developing ideal drug delivery systems, which can not only enhance the bioavailability and pharmacokinetic traits but also deliver the potential drug candidates to the site of action, without affecting the normal cells. A plethora of studies showed the delivery of liposomes encapsulated anticancer drugs at the site of solid tumors with minimal toxicity as compared to free drugs.

7.5 LIPOSOMES AS DRUG DELIVERY SYSTEM FOR CANCER THERAPY

At present, several conventional products are available as anticancer drugs delivery vehicle. Doxil (a PEGylated liposomal formulation) was the first liposomal product encapsulating doxorubicin which was approved by the FDA to treat Kaposi's sarcoma in acquired immunodeficiency syndrome (AIDS) patients [50]. LipoDox is another liposomal formulation made in India by Sun Pharma. The first generic version of Doxil was approved by the FDA and prepared by Sun Pharma in 2013 [51]. Doxil was found effective against refractory ovarian cancer, breast cancer, and multiple myeloma [50]. DaunoXome is the liposomal formulation of daunorubicin to treat AIDS-related Kaposi's sarcoma [52]. Myocet (a non-PEGylated liposomal formulation of doxorubicin) in combination with cyclophospha-mide was approved to treat metastatic breast cancer in Europe but was not yet approved by the FDA for use in the United States [53]. Myocet in combination with Herceptin (trastuzumab) and Taxol (paclitaxel) is under trial phase for the treatment of metastatic breast cancer [54]. Marqibo (liposomal formulation of vincristine) was approved by the FDA to treat acute lymphoblastic leukemia. The CPX-351 (a liposomal formulation of cytarabine and daunorubicin) exhibited promising effect against secondary acute myeloid leukemia [55]. The CPX-1 (liposomal formulation of irino-tecan HCl and floxuridine) completed phase II clinical trial on patients with advanced colorectal cancer [56]. MM-398 (liposomal formulation encapsulating irinotecan) was evaluated to treat various types of cancers [57, 58]. MM-302 encapsulating doxorubicin is currently being tested in phase I clinical trials against metastatic HER2-positive breast cancer [59]. MBP-426 (liposomal formulation of oxaliplatin) is being used in phase II clinical trials to treat gastric cancer [60]. Lipoplatin (liposomal formulation of cisplatin) is being evaluated in phase III clinical trial against non-small cell lung cancer [61]. Stimuvax is an anti-MUC1 cancer vaccine for preventing non-small cell lung cancer and currently is in phase III clinical trial [62]. Another liposomal formulation of doxorubicin is under phase III clinical trial against primary hepatocellular carcinoma and in phase II trial against breast cancer and colorectal liver metastasis [63, 64].

Cisplatin (CDDP) is a chemotherapeutic agent used to treat testicle, ovary, head, lung, and neck carcinomas [65–70]. It is also administrated by intraperitoneal route for the treatment of peritoneal carcinomatosis. In view of its nephrotoxicity, liposomes composed of PC/PS/Chol encapsulating CDDP were assessed in IgM immunocytoma bearing rats. Results depicted reduced incidence and severity of renal lesions after the administration of liposomal formulation injection as compared to the free CDDP [71]. A long-circulating formulation composed of hydrogenated soy phosphatidylcholine/DSPE-PEG2000/Chol (SPI-077) was developed and tested *in vivo* against tumor models [72]. SPI-077 showed enhanced tumor platinum uptake and increased anticancer activity with respect to free CDDP [73]. Another long-circulating liposomal formulation containing CDDP (Lipoplatin) showed toxicity effect against tumor cells lines and revealed reduced toxicity in normal bone marrow stem cells as compared to free CDDP [74]. Lipoplatin reduced tumor size in xenografts of human breast, prostate, and pancreatic cancer. Further, the histopathological analyzes of the tumors exhibited apoptosis in the tumor cells similar to free CDDP [75].

Mice and rats treated with CDDP caused tubular damage, but mice treated with the same dose of lipoplatin were completely free of kidney injury [76]. Stathopoulos et al. [77] investigated the pharmacokinetics and toxicity of lipoplatin in patients with pre-treated advanced malignant tumors. Results showed that the elimination of urine was about 40% of the infused concentration in 3 days. The treatment with higher doses showed grades 1 and 2 gastrointestinal tract and hematological toxicity. Lipoplatin showed the potentiality to preferentially concentrate on the malignant tissue of both primary and metastatic origin [78]. In a phase I trial study, the administration of lipoplatin revealed lack of nephrotoxicity at doses of 100 and 125 mg/m^2. However, grade 2 neutropenia and grade 1 nausea, fatigue, diarrhea, and neurotoxicities were observed after the administration of lipoplatin at the similar doses [79]. In another study, lipoplatin showed nephrotoxicity, gastrointestinal toxicity, and myelotoxicity as the major adverse reactions [80]. A phase II study for metastatic breast cancer, lipoplatin, and vinorelbine revealed absolute response in 9.4%, fair response in 43.8%, stable disease in 37.5%, and progressive infection in 9.4% of patients. Additionally, no grade 3/4 nephrotoxicity and neurotoxicity were reported [81]. In another phase II trial, lipoplatin (120 mg/m^2), administered with gemcitabine (1000 mg/m^2) for non-small cell lung cancer patients, showed high efficiency. Further, lipoplatin treatment showed reduced nephrotoxicity, as compared to CDDP treatment [82]. The pharmacokinetics of lipoplatin along with 5-fluorouracil revealed that the liposomal formulation had high elimination rate and a short half-life than free CDDP, which confirmed decreased nephrotoxicity [83]. CDDP is also widely used for the treatment of peritoneal carcinomatosis by the intraperitoneal route. However, it shows severe toxicity by administrating through the intraperitoneal route [84]. The failure of intraperitoneal therapy is mainly because of low penetration of CDDP. Hence, in order to obtain maximum penetration within the tumor, higher doses of drug are suggested. Araújo et al. [85] evaluated the tissue distribution of SpHL-CDDP after their intraperitoneal administration in Ehrlich ascitic tumor-bearing mice. The CDDP AUC obtained for ascitic fluid and blood after SpHL-CDDP administration was 3.3-fold higher and 1.3-fold lower, respectively, as compared to the control treatment. Likewise, SpHL-CDDP treatment showed no variations in the blood urea and creatinine levels of the mice [86, 87]. These findings suggested the application of SpHL-CDDP as a potent drug delivery system for improving the therapeutic effect of the CDDP-based therapy. In another report, the survival of Ehrlich ascitic tumor-bearing mice treated with SpH-CDDP was higher as compared to free CDDP after intraperitoneal administration, thereby indicating the potency of SpHL-CDDP in clinical studies [88]. In spite of the potential anticancer trait of oxaliplatin (an analog of CDDP), it is associated with severe adverse effects such as neurotoxicity, tubular necrosis, hematologic toxicity, and gastrointestinal tract toxicity [89]. Lipoxal is a liposomal formulation of oxaliplatin constituting hydrogenated soy phosphatidylcholine (HSPC)/DPPG/cholesterol/DSPE-PEG. It showed removal of breast cancers in mice at 16 mg/kg of concentration. Lipoxal was well-tolerated and showed a reduction in the adverse impact of free oxaliplatin [90].

Doxorubicin, daunorubicin, and their derivatives belong to anthracyclines and are used as first-line drugs for cancer therapy [91]. In spite of their potential impact as anticancer drugs, their toxicities are a significant concern for us, mainly when administrated along with other chemotherapeutic drugs [92]. Liposomes based drug delivery systems have shown a prominent role in decreasing adverse effects of anthracyclines groups of drugs. Forssen et al. [93] reported the potency of liposomes encapsulating daunorubicin for accumulating in P-1798 murine lymphosarcoma and MA16C mammary adenocarcinoma tumor models. Results showed maximum liposomal uptake with respect to the free drug. Similar reports were observed by Forssen and Ross [94]; and Pea et al. [95] too who demonstrated improved efficacy and reduced toxicity of liposomes encapsulating daunorubicin (DaunoXome), as compared to free drug. In another study of phase I/II clinical trials, the administration of DaunoXome administration exhibited enhancement in plasma AUC, increased plasma levels, reduced distribution volume, and decreased elimination rate, with respect to the free drug [96]. A randomized phase III trial was conducted to analyze the safety of DaunoXome in the treatment of advanced AIDS-related Kaposi's sarcoma. DaunoXome showed significantly less alopecia and neuropathy, and reported no evidence of cardiac toxicity as compared to doxorubicin, bleomycin, and vincristine [97]. In 1996, the FDA approved DaunoXome as a first-line therapy for HIV-related Kaposi's sarcoma.

Myocet (a liposomal product) along with cyclophosphamide is approved in Europe for treating breast cancer. It contains egg phosphatidylcholine (EPC)/cholesterol and encapsulated doxorubicin. Preclinical toxicity studies revealed significant toxicity of Myocet against cancer cells, as compared to free doxorubicin [98]. For the ascitic model, the maximum concentration in tumors drugs exposure was about 10-fold increment for liposomal doxorubicin, as compared to free doxorubicin, suggesting Myocet as ideal choice for therapeutics [99]. Study also reported a significant reduction in the cardiotoxicity of Myocet when combined with cyclophosphamide [100]. A phase III trial study revealed a significant reduction in cardiac events and congestive heart failure by administrating Myocet, as compared to doxorubicin [101]. Cowens et al. [102] observed reduced toxicities due to the administration of Myocet, as compared to the administration of free doxorubicin at the similar doses. Batist et al. [53]

demonstrated that the combination of Myocet (60 mg/m^2) with cyclophosphamide (600 mg/m^2) for the treatment of metastatic breast cancer reduced the cardiotoxicity. Doxorubicin encapsulated in a long-circulating liposomal formulation made up of HSPC/DSPE-PEG2000/cholesterol (Doxil/Caelyx) demonstrated that Doxil/Caelyx was significantly more active than free doxorubicin [103] and showed reduced cardiac toxicity [104]. In association with Velcade (Bortezomib), Doxil/Caelyx is approved by the FDA to treat multiple myeloma. Plethora of anticancer studies is currently in progress using Doxil/Caelix for treating other malignancies [105, 106].

Paclitaxel is another pivotal anticancer drug that inhibits endothelial cell proliferation, motility, and tube formation [107]. Zhang et al. [108] prepared a liposomal formulation of paclitaxel constituting 1,2-dioleyl-sn-glycero-3-phosphocholine/cholesterol/cardiolipin (LEPETU) and demonstrated better growth inhibition of human ovarian (OVCAR-3), human lung (A549), breast (MX-1), and prostate (PC-3) cancer cells than that of free drugs. Liposomal paclitaxel has been found less toxic than that of free paclitaxel. Yang et al. [109] developed a PEGylated liposomal formulation of paclitaxel and demonstrated that cytotoxicity of paclitaxel formulation against human breast cancer cell lines (MDA-MB-231 and SK-BR-3) was similar to Taxol. Biodistribution analyzes in the breast cancer xenograft nude mouse model depicted that the uptake of PEGylated liposomes was significantly improved in tumor tissues with high inhibition rate of tumor growth after its administration. Paclitaxel was encapsulated in cationic liposomes made of dioleytrimethylammoniumpropane/dioleoylphosphatidylcholine (EndoTAG-1) as for treating solid tumors [110]. Findings revealed that the vascular targeting with EndoTAG-1 improved tumor microvessel leakage because of the vascular damage.

DepoCyte (a sustained-release formulation of cytarabine) was used to evaluate the potency and safety in 28 patients with lymphomatous meningitis. The response rate was reported significantly higher in DepoCyte [111]. In another study, a greater median time to neurological progression was obtained due to the administration of DepoCyte for treating patients with solid tumor neoplastic meningitis [112]. Likewise, the administration of liposomal cytarabine showed a reduced number of lumbar injections, with respect to the standard schedule, improving the patients' quality of life [113]. On the other hand, Marqibo (a DSPC/cholesterol encapsulation of vincristine sulfate) has been proved to be a potential anticancer agent in patients with heavily pretreated aggressive non-Hodgkin lymphoma [114].

Wang-Gallam et al. [115] demonstrated the impact of nanoliposomal formulation of irinotecan, alone, and along with fluorouracil and folinic acid against metastatic pancreatic ductal adenocarcinoma. Findings showed that the nanoliposomal irinotecan formulated with the fluorouracil and folinic acid combination extended the survival of patients with the disease. Metformin hydrochloride (Met) encapsulated liposomal vesicles were developed for enhanced anticancer therapeutics. Liposomal Met was developed using thin-film hydration (TFH) via passive loading, active loading, and drug-loaded lipid film. Findings depicted high activity of positively charged liposomes with lower IC_{50} (50% inhibitory concentration) value, reduced cell migration trait, decreased colonies, and promising apoptosis-stimulated property in breast cancer cells [116]. In another study, Met-encapsulating liposome (LP-Met) and Herceptin-conjugated LP-Met (Her-LP-Met) were assessed for their anticancer properties against breast cancer stem cells *in vitro* and *in vivo*. Her-LP-Met showed greater inhibition of breast cancer stem cells proliferation *in vitro* than free Met or LP-Met. In addition, the anti-migration effect of Her-LP-Met on breast cancer stem cells was higher than that of free Met or LP-Met. The activity was increased along with doxorubicin. Activity of Her-LP-Met along with free doxorubicin was higher than that of free Met, free doxorubicin, or Her-LP-Met in mouse model, thereby suggesting Her-LP-Met a new therapeutic modality to target breast cancer stem cells [117].

Cancer cells are targeted by liposomes using antibody-based approach by adding certain antibodies to the liposomal surface (immunoliposomes; ILP). These IPL are specific to the cancer cells [118]. Maruyama et al. [119] prepared pendant type ILP (34A-PEG-ILP) that showed target specific action (lung endothelial cells and tumor tissue) as compared to ordinary liposomes. Unfortunately, this technique is limited to the antigen-antibody specifications.

Enzyme-responsive liposomes are another effective approach for the delivery of anticancer drugs because the presence of certain enzymes such as secreted phospholipase A2 (sPLA2), matrix metalloproteinases, urokinase plasminogen activator, and elastase in cancer patients [120]. Overexpression of these enzymes could be an effective target for controlling drug release from liposomes [121]. Mock et al. [122] developed doxorubicin-loaded sPLA2-responsive liposomes for the treatment of prostate cancer. *In vivo* results showed that liposomes were approximately two times more effective than sterically stabilized liposomes at reducing the growth of cancer cells. Allison and Gregoriadis [123] reported the potency of liposomes to stimulate immunity response of entrapped antigens. Hydrophilic antigens (nucleic acids, peptides, carbohydrates, proteins, and hapten) are encapsulated within the aqueous phase of liposomes, while hydrophobic components (antigens and lipopeptides) are intercalated into the outer lipid bilayers [5].

Folate receptors (tumor-associated antigens) have the potentiality to bind with folic acid and its conjugates and ingest the bound molecules the endocytosis process. Several folates attached liposomes have shown effectiveness as the target-based delivery system. Paclitaxel and imatinib co-encapsulated liposomes were formulated [124]. Folate-conjugated liposomes showed significantly higher anticancer activities against MCF-7 and PC-3 cells compared with CL. In addition, targeted liposomes in MCF-7 and PC-3 cells revealed a decrease in vascular endothelial growth factor gene expression compared with CL. Handali et al. [125] developed 5-fluorouracil-loaded folate-targeted liposomes by thin-film hydration method using DPPC, Chol, and folic acid-PEG-DSPE for improving the safety and effectiveness of 5-fluorouracil. Based on *in vitro* tests findings, the folate-targeted liposomes had higher toxicity against Caco-2, HT-29, CT26, MCF-7, and HeLa cell lines than CL and free 5-fluorouracil. *In vivo* findings showed that targeted liposomal formulation significantly decreased tumor volume (169 mm^3) as compared to free 5-fluorouracil (326.4 mm^3). Results summarized that folic acid-targeted liposomes may be used for delivery of 5-fluorouracil to cancer cells. Folic acid-conjugated liposomes containing celastrol and irinotecan for targeted therapy against breast cancer was developed by Soe et al. [126]. Liposomes were developed using thin-film hydration technique. Results exhibited high *in vitro* cell uptake and improved apoptosis in folate receptor-positive breast cancer cells (MDA-MB-231 and MCF-7), but not in folate receptor-negative lung cancer cells (A549). *In vivo* experiment with MDA-MB-231 xenografts in a mouse tumor model showed potent drug delivery system for folic acid-conjugated liposomes. Gazzano et al. [127] developed folate-conjugated liposomal formulations of nitrooxy-doxorubicin for improving active targeting against the Pgp-expressing tumors. The liposomal formulation not only revealed significant anticancer attributes against Pgp-positive and folate receptor-positive cells but also triggered mitochondria-dependent apoptosis, thereby suggesting its role in the treatment of folate receptor/Pgp-positive tumors. Moghimipour et al. [128] developed folate-targeted liposomes for the delivery of 5-fluorouracil to target colon cancer cells that revealed and increased reactive oxygen species (ROS) production with respect to free 5-fluorouracil. In addition, 5-fluorouracil-encapsulated folate-targeted liposomes showed higher inhibition of cancer cells with respect to free 5-fluorouracil without any adverse effect. Qiu et al. [129] demonstrated the treatment of lymphoma by developing a vincristine-loaded folate decorated liposomes system. Vincristine-loaded folate-conjugated liposomes exhibited higher anticancer trait in the murine model bearing lymphoma xenografts and showed a targeted impact in vincristine delivery to the B-cell lymphoma cells. Monteiro et al. [130] developed paclitaxel-loaded pH-sensitive long-circulating folate-coated liposomes radio-labeled with technetium-99 m (99 mTc), which showed higher tumor-to-muscle ratio than CL and free paclitaxel, suggesting its feasibility as an efficient paclitaxel delivery system to folate-receptor-positive cancer cells.

7.6   THERAPEUTIC ROLES OF NEW GENERATION LIPOSOME

Despite the tremendous role of liposomes as therapeutics and the success of its formulations in clinical trials phases, none of them showed Physico-chemical stability at the target sites [131]. Hence, it is essential to over-come liposomes associated with these issues in terms of finding new drug delivery system as a new generation of liposomes.

Archaeosomes are liposomal formulations that are prepared with diether and/or tetraether linkages, present in archaebacterial membrane [132]. These archaebacterial lipids are stable towards varied abiotic factors [133]. Archaeosomes are generally prepared using conventional process at any temperature and are considered a better carrier than CL, particularly for protein and peptide delivery because of their high stability. Virosomes are another liposomal formulation which contain fusogenic viral envelope protein [134]. They are generally used for the intracellular delivery of drugs and DNA [135]. Virosome-based vaccines are immunogenic and well-tolerated in children. It can be used as an efficient delivery system for varied antigens and drugs, including nucleic acids, cytotoxic drugs, and toxoids [136]. Niosomes are the vesicles made up of non-ionic surfactants (alkyl ethers and alkyl esters) and Chol. It is known to increase the stability of the encapsulated drugs. It not only improves the oral bioavailability of poorly absorbed drugs (due to the presence of non-ionic surfactants) but also enhances penetration of drugs in the skin. Niosomes based delivery system is biodegradable, biocompatible, nonimmunogenic, and improves the therapeutic performance of drug molecules [137]. Similar to liposomes, niosomes are unsuitable for transdermal delivery because of low skin permeability [138]. Transfersome is used for the delivery of proteins and peptides. This formulation is a non-invasive method for transdermal delivery. Transfersomes are lipid-based supramolecular aggregates and penetrate easily [139]. Ethosomes are modified liposomes which contain phospholipids and alcohol. It is considered a unique formulation due to the availability of high concentrations of ethanol, which causes modification in the skin lipid bilayer organization, thereby improving drug distribution potency in stratum corneum lipids [140]. In a like manner, novasomes are the modified forms of liposomes which are formulated by mixing monoester of polyoxyethylene fatty acid, Chol, and free fatty acid [141]. Novasomes are more effective when administrated under non-occluded circumstances from a restricted dose. Novasome-based vaccines against various infections have been developed [142]. Cryptosomes is a liposomal composition comprising poloxamer molecule as well as liposome encapsulating few delivery agents [137]. Emulsome represents the formulation containing both liposomes and emulsions. Unlike liposomes, the internal core of emulsomes contain lipid. Sonication helps produce emulsomes of smaller size [143]. In fact, it is a new emerging delivery system which provides sustainable and controlled release of drugs, thereby suggested to play a pivotal role in the treatment of life-threatening diseases [143]. Chol is an important constituent of emulsomes that affects the stability of emulsomal systems and contributes in the encapsulation of drugs [144]. Emulsomes protects the encapsulated drug from adverse conditions of the stomach. Vesosomes are heterogeneous, aggregated, and large lipid bilayer enclosing liposomes that fix active targeting components. As a matter of fact, vesosomes are liposomes within liposomes that are used as drug delivery systems [145]. Genosomes are artificially functional complexes for functional gene delivery to the cell. It is generally made up of cationic phospholipids [146].

In view of the pivotal applications of liposomal derivatives or new generation liposomes as drug delivery systems for the treatment of disparate diseases, its applications as promising anticancer agents have gained tremendous interest that would open a new door for the medical sciences in the future. Table 7.2 shows applications of new generation liposomes towards the treatment of varied cancer types.

TABLE 7.2 Anticancer Traits of New Generation Liposomes

Liposomes	Encapsulating Agents	Cancer Type	Mechanism of Action	References
Archaeosomes	Ovalbumin	Lung	Induced strong CD4 (+) as well as CD8 (+) CTL responses to entrapped soluble antigens.	[147]
Archaeosomes	Paclitaxel	Breast	Toxicity of archaeosomal was more than the standard paclitaxel.	[148]
Archaeosomes	Bacterial and Archaeal lipid	Colon	Interacted with cell membranes predominantly by fusion and endocytosis.	[149]
Archaeosomes	Ovalbumin	Melanoma	Induced strong CD8+ T cell responses and protection from a B16-Ovalbumin melanoma tumor.	[150]
Archaeosomes	Triptorelin acetate	Prostate	Showed toxicity of the PC3 prostate cancer cell line.	[151]
Archaeosomes	Papillomavirus L1/E6/E7 genes	Cervical	Exhibited potent anticancer trait.	[152]
Emulsomes	Piceatannol	Colon	Enhanced apoptotic trait.	[153]
Emulsomes	Curcumin- and Piperine	Colorectal	Cell cycle arrest at G2/M phase. Induced apoptosis. Increase in Caspase 3 level.	[154]
Ethosomes	Curcumin	Melanoma	Greater drug deposition on the skin.	[155]
Ethosomes	5-fluorouracil	Melanoma	Improved intracellular localization of ethosomes and facilitated endocytosis.	[156]
Ethosomes	5-fluorouracil	Melanoma	High entrapment efficiency.	[157]
Ethosomes	Itraconazole	Skin	High entrapment efficiency.	[158]
Ethosomes	Cryptotanshinone	Acne	High entrapment efficiency.	[159]
Ethosomes	Fisetin	Skin	High entrapment efficiency. Decrease in the levels of TNF-α and IL-1α.	[160]
Ethosomes	Thymoquinone	Breast	High entrapment efficiency. Cytotoxic activity of the ethosomic thymoquinone against MCF-7 cell lines was greater than free thymoquinone.	[161]

TABLE 7.2 *(Continued)*

Liposomes	Encapsulating Agents	Cancer Type	Mechanism of Action	References
Ethosomes	Sulforaphane	Skin	Increased percutaneous permeation of sulforaphane.	[162]
Niosomes	Balanocarpol	Breast and ovary	Strong *in vitro* cytotoxicity than over free drug.	[163]
Niosomes	Curcumin	Ovary	Enhanced cytotoxic activity and apoptotic rate.	[164]
Niosomes	Doxycycline	Prostate	Regulated alterations in PC3 cells cycle.	[165]
Niosomes	Cisplatin	Breast	Increased cytotoxicity.	[166]
Niosomes	Lawsone	Breast	Increased anticancer trait.	[167]
Niosomes	Capecitabine	Breast	Enhanced release rate and toxicity.	[168]
Niosomes	5-fluorouracil and Leucovorin	Colon	High entrapment efficiency.	[169]
Niosomes	Doxorubicin and Quercetin	Gastric	Potent cytotoxicity.	[170]
Niosomes	D-limonene	Lung and liver	Enhanced cytotoxic activity.	[171]
Niosomes	Paclitaxel and Curcumin	Breast	High entrapment efficiency and cytotoxicity.	[172]
Niosomes	Thymoquinone	Breast	G2/M arrest and decreased the migration of MCF7 cells.	[173]
Transfersomes	5-fluorouracil	Skin	Greater permeation rate and depth.	[174]
Transfersomes	5-fluorouracil	Skin	Improved penetration in skin.	[175]
Vesosomes	Doxorubicin and 5-fluorouracil	Cervical	Enhanced antitumor efficacy. No obvious toxicity at the treatment dose.	[176]
Virosomes	Her-2/neu multi-peptide	Breast	Enhanced cellular immune responses.	[177]

7.7 CONCLUSIONS AND FUTURE PERSPECTIVES

Liposomes have revealed its tremendous potential as ideal drug delivery system with enhanced pharmacokinetics and pharmacodynamics. Liposomes have offered a plethora of advantages in designing novel pathogens- or host-directed treatments and vaccines due to its pronounced anti-pathogenic, anticancer, and immunomodulatory attributes. In view of the severe adverse effects of leading antitubercular drugs, these lipid-based formulations have emerged as promising drug carrier system for target-specific treatment of TB. It has the potency to reduce the duration of TB treatment, increase treatment compliance, and reduce the administration of more than one antitubercular drug to a single formulation. On the other hand, liposomes have revolutionized the current era of cancer treatment by overcoming the limitations of conventional chemotherapy. These nanocarriers overcome the undesirable side effects on non-infected cells and improve the bioavailability. In addition, new generation liposomes have exhibited their paramount role in the treatment of different kinds of cancer with high entrapment efficiency and toxicity. These new generation liposomes can be ideal drug delivery systems for treating TB and cancer in the next few years. Some of liposome-based drug delivery systems are available in the market and many more are under clinical trial phases for the successful treatment of TB and cancer. However, research is still desperately required to understand in depth about the absorption inducing abilities of lipids and find better predictive tools for determining *in vivo* response of varied lipids with various types of ligands. In a nutshell, liposomes, and new generation liposomes-based drug delivery system would be leading approach as therapeutics in the future.

CONFLICT OF INTEREST

None declared.

KEYWORDS

- conventional liposomes
- drug delivery system
- liposomes

- **medium unilamellar vesicle**
- **new generation liposomes**
- **stable plurilamellar vesicle**
- **tuberculosis**
- **cancer therapy**

REFERENCES

1. Batista, C. M., Carvalho, C. M. B., & Magalhães, N. S. S., (2007). Liposomes and their therapeutic: State of art applications. *Braz. J. Pharm. Sci., 43*, 167–179.
2. Samad, A., Sultana, Y., & Aqil, M., (2007). Liposomal drug delivery systems: An updated review. *Curr. Drug Deliv., 4,* 297–305.
3. Nisini, R., Poerio, N., Mariotti, S., De Santis, F., & Fraziano, M., (2018). The multirole of liposomes in therapy and prevention of infectious diseases. *Front. Immunol., 9,* 155. doi.org/10.3389/fimmu.2018.00155.
4. Bozzuto, G., & Molinari, A., (2015). Liposomes as nanomedical devices. *Int. J. Nanomedicine, 10*, 975–999.
5. Schwendener, R. A., (2014). Liposomes as vaccine delivery systems: A review of the recent advances. *Ther. Adv. Vaccines, 2,* 159–182.
6. Khusro, A., Aarti, C., Barbabosa-Pliego, A., & Salem, A. Z. M., (2018a). Neoteric advancement in TB drugs and an overview on the antitubercular role of peptides through computational approaches. *Microb. Pathog., 114,* 80–89.
7. WHO Global Tuberculosis Report, (2019). Geneva: World Health Organization, 2019 WHO. Available online: https://www.who.int/tb/publications/global_report/en/ (accessed on 17 June 2021).
8. Filion, L. G., Al-Ahdal, M. N., & Tetro, J. A., (2013). Vaccines against *Mycobacterium tuberculosis*: Exploring alternate strategies to combat a near-perfect pathogen. *J. Mycobac. Dis., S1,* 003. doi.org/10.4172/2161-1068.S1-003.
9. Welin, A., Raffetseder, J., Eklund, D., Stendahl, O., & Lerm, M., (2011). Importance of phagosomal functionality for growth restriction of *Mycobacterium tuberculosis* in primary human macrophages. *J. Innate Immun., 3,* 508–518.
10. Field, S. K., (2015). Bedaquiline for the treatment of multidrug-resistant tuberculosis: Great promise or disappointment? Ther. Adv. Chronic Dis., 6, 170–184.
11. Hards, K., Robson, J. R., Berney, M., Shaw, L., Bald, D., Koul, A., et al., (2015). Bactericidal mode of action of bedaquiline. *J. Antimicrob. Chemother., 70,* 2028–2037.
12. Gualano, G., Capone, S., Matteelli, A., & Palmieri, F., (2016). New antituberculosis drugs: From clinical trial to programmatic use. *Infect. Dis. Rep., 8,* 6569.
13. Xavier, A. S., & Lakshmanan, M., (2014). Delamanid: A new armor in combating drug-resistant tuberculosis. *J. Pharmacol. Pharmacother., 5,* 222–224.

14. Zou, Y., Lee, H. Y., Seo, Y. C., & Ahn, J., (2012). Enhanced antimicrobial activity of nisin-loaded liposomal nanoparticles against foodborne pathogens. *J. Food Sci.*, 77, 165–170.

15. Bibhas, C. M., Gitanjali, M., Subas, C. D., & Narahari, N. P., (2017). Exploring the use of lipid-based nano-formulations for the management of tuberculosis. *J. Nanosci. Curr. Res.*, 2, 112, doi.org/10.4172/2572-0813.1000112.

16. Hoet, P., Legiest, B., Geys, J., & Nemery, B., (2009). Do nanomedicines require novel safety assessments to ensure their safety for long-term human use? *Drug Safety*, 32, 625–636.

17. Pandey, R., & Khuller, G. K., (2006). Nanotechnology-based drug delivery system(s) for the management of tuberculosis. *Indian J. Exp. Biol.*, 44, 357–366.

18. Vyas, S. P., Kannan, M. E., Jain, S., Mishra, V., & Singh, P., (2004). Design of liposomal aerosols for improved delivery of rifampicin to alveolar macrophages. *Int. J. Pharm.*, 269, 37–49.

19. Gaspar, M. M., Cruz, A., Penha, A. F., Reymao, J., Sousa, A. C., Eleuterio, C. V., et al., (2008). Rifabutin encapsulated in liposomes exhibits increased therapeutic activity in a model of disseminated tuberculosis. *Int. J. Antimicrob. Agents*, 31, 37–45.

20. Deol, P., & Khuller, G. K., (1997). Lung specific stealth liposomes: Stability, biodistribution and toxicity of liposomal antitubercular drugs in mice. *Biochim. Biophys. Acta*, 1334, 161–172.

21. Pison, U., Welte, T., Giersing, M., & Groneberg, D. A., (2006). Nanomedicine for respiratory diseases. *Eur. J. Pharmacol.*, 533, 341–350.

22. Gill, S., Löbenberg, R., Ku, T., Azarmi, S., Roa, W., & Prenner, E. J., (2007). Nanoparticles: Characteristics, mechanisms of action, and toxicity in pulmonary drug delivery: A review. *J. Biomed. Nanotechnol.*, 3, 107–119.

23. Andrade, F., Videira, M., Ferreira, D., & Sarmento, B., (2011). Nanocarriers for pulmonary administration of peptides and therapeutic proteins. *Nanomedicine*, 6, 123–141.

24. Surendiran, A., Sandhiya, S., Pradhan, S. C., & Adithan, C., (2009). Novel applications of nanotechnology in medicine. *Indian J. Med. Res.*, 130, 689–701.

25. Santos, N., & Castanho, M., (2002). Liposomes: Has the magic bullet hit the target? *Quim. Nova*, 25, 1181–1185.

26. El-Ridy, M. S., Mostafa, D. M., Shehab, A., Nasr, E. A., & Abd, E. S., (2007). Biological evaluation of pyrazinamide liposomes for treatment of *Mycobacterium tuberculosis. Int. J. Pharm.*, 330, 82–88.

27. Dhillon, J., Fielding, R., Adler-Moore, J., Goodael, R. L., & Mitchison, D., (2001). The activity of low-clearance liposomal amikacin in experimental murine tuberculosis. *J. Antimicrob. Chemoth.*, 48, 869–876.

28. Pandey, R., Sharma, S., & Khuller, G. K., (2004). Nebulization of liposome-encapsulated antitubercular drugs in guinea pigs. *Int. J. Antimicrob. Agents*, 24, 93–94.

29. Agrawal, A. K., & Gupta, C. M., (2000). Tuftsin-bearing liposomes in treatment of macrophage-based infections. *Adv. Drug Deliver. Rev.*, 41, 135–146.

30. Irache, J. M., Salman, H. H., Gamazo, C., & Espuelas, S., (2008). Mannose-targeted systems for the delivery of therapeutics. *Exp. Opin. Drug Deliv.*, 5, 703–724.

31. Labana, S., Pandey, R., Sharma, S., & Khuller, G. K., (2002). Chemotherapeutic activity against murine tuberculosis of once-weekly administered drugs (isoniazid and rifampicin) encapsulated in liposomes. *Int. J. Antimicrob. Agents, 20*, 301–304.

32. Agarwal, A., Kandpal, H., Gupta, H., Singh, N., & Gupta, C., (1994). Tuftsin-bearing liposomes as rifampin vehicles in treatment of tuberculosis in mice. *Antimicrob. Agents Chemother., 38*, 588–593.

33. Zaru, M., Mourtas, S., Klepetsanis, P., Fadda, A. M., & Antimisiaris, S. G., (2007). Liposomes for drug delivery to the lungs by nebulization. *Eur. J. Pharm. Biopharm., 67*, 655–666.

34. Ravaoarinoro, M., Toma, E., Agbaba, O., & Morisset, R., (1993). Efficient entrapment of amikacin and teicoplanin in liposomes. *J. Drug Target, 1*, 191–195.

35. Dascher, C. C., Hiromatsu, K., Xiong, X., Morehouse, C., Watts, G., Liu, G., et al., (2003). Immunization with a mycobacterial lipid vaccine improves pulmonary pathology in the guinea pig model of tuberculosis. *Int. Immunol., 15*, 915–925.

36. Cardona, P. J., & Amat, I., (2006). Origin and development of RUTI, a new therapeutic vaccine against *Mycobacterium tuberculosis* infection. *Arch Bronconeumol., 42*, 25–32.

37. Christensen, D., Agger, E. M., Andreasen, L. V., Kirby, D., Andersen, P., & Perrie, Y., (2009). Liposome-based cationic adjuvant formulations (CAF): Past, present, and future. *J. Liposome Res., 19*, 2–11.

38. Mehta, R. T., Keyhani, A., McQueen, T. J., Rosenbaum, B., Rolston, K. V., et al., (1993). *In vitro* activities of free and liposomal drugs against *M. avium*-M. intracellular complex and *M. tuberculosis. Antimicrob. Agents Chemother., 37*, 2584–2587.

39. Adams, L. B., Sinha, I., Franzblau, S. G., Krahenbuhl, J. L., & Mehta, R. T., (1999). Effective treatment of acute and chronic murine tuberculosis with liposome-encapsulated clofazimine. *Antimicrob. Agents Chemother., 43*, 1638–1643.

40. Kaul, A., Chaturvedi, S., Attri, A., Kalra, M., & Mishra, A. K., (2016). Targeted theranostic liposomes: Rifampicin and ofloxacin loaded pegylated liposomes for theranostic application in mycobacterial infections. *RSC Adv., 6*, 28919. doi. org/10.1039/c6ra01135g.

41. Ferraz-Carvalho, R. S., Pereira, M. A., Linhares, L. A., Lira-Nogueira, M. C., Cavalcanti, I. M., Santos-Magalhães, N. S., et al., (2016). Effects of the encapsulation of usnic acid into liposomes and interactions with antituberculous agents against multidrug-resistant tuberculosis clinical isolates. *Memorias do Instituto Oswaldo Cruz, 111*, 330–334.

42. Nkanga, C. I., & Krause, R. W. M., (2019). Encapsulation of isoniazid-conjugated phthalocyanine-in-cyclodextrin-in-liposomes using heating method. *Sci. Rep., 9*, 11485. doi.org/10.1038/s41598–019–47991-y.

43. Nkanga, C. I., Noundou, X. S., Walker, R. B., & Krause, R. W. M., (2019). Co-encapsulation of rifampicin and isoniazid in crude soybean lecithin liposomes. *S. Afr. J. Chem., 72*, 80–87.

44. Niu, N. K., Shi, Z. Y., Yang, Z. Q., Li, Y. A., Shi, J. D., & Wang, Z. L., (2020). *Preparation, Characterization, and In-Vitro Cytotoxicity of Nanoliposomes Loaded with Anti-Tuberculous Drugs and TGF-β1 siRNA for Improving Spinal Tuberculosis Therapy*. Preprint (Version 1) available at Research Square [+https://doi.org/10.21203/rs.3.rs-36126/v1+].

45. Khusro, A., Aarti, C., Barbabosa-Pliego, A., Rivas-Cáceres, R. R., & Cipriano-Salazar, M., (2018b). Venom as therapeutic weapon to combat dreadful diseases of 21st century: A systematic review on cancer, TB, and HIV/AIDS. *Microb. Pathog., 125*, 96–107.

46. WHO Report on the Global Tobacco Epidemic, (2019). *Offer Help to Quit Tobacco Use*. Geneva: World Health Organization; 2019. https://apps.who.int/iris/bitstream/handle/10665/274603/9789241565639-eng.pdf?ua=1 (accessed on 17 June 2021).

47. Bagnardi, V., Rota, M., Botteri, E., Tramacere, I., Islami, F., Fedirko, V., et al., (2015). Alcohol consumption and site-specific cancer risk: A comprehensive dose-response meta-analysis. *Br. J. Cancer, 112*, 580 -593.

48. Ferlay, J., Ervik, M., Lam, F., Colombet, M., Mery, L., Piñeros, M., et al., (2020). *Global Cancer Observatory: Cancer Today*. Lyon: International Agency for Research on Cancer. https://gco.iarc.fr/today (accessed on 3 July 2021).

49. WHO Report on Cancer, (2020). Geneva: World Health Organization; 2020 WHO. Available online: https://www.who.int/healthinfo/statistics/en/ (accessed on 17 June 2021).

50. Barenholz, Y., (2012). Doxil--the first FDA-approved nano-drug: lessons learned. *J. Control Release, 160*, 117–134.

51. Chou, H., Lin, H., & Liu, J. M., (2015). A tale of the two PEGylated liposomal doxorubicin. *Onco. Targets Ther., 8*, 1719–1720.

52. Petre, C. E., & Dittmer, D. P., (2007). Liposomal daunorubicin as treatment for Kaposi's sarcoma. *Int. J. Nanomedicine, 2*, 277–288.

53. Batist, G., Ramakrishnan, G., Rao, C. S., Chandrasekharan, A., Gutheil, J., Guthrie, T., et al., (2001). Reduced cardiotoxicity and preserved antitumor efficacy of liposome-encapsulated doxorubicin and cyclophosphamide compared with conventional doxorubicin and cyclophosphamide in a randomized, multicenter trial of metastatic breast cancer. *J. Clin. Oncol., 19*, 1444–1454.

54. Baselga, J., Manikhas, A., Cortes, J., Llombart, A., Roman, L., Semiglazov, V. F., et al., (2014). Phase III trial of nonpegylated liposomal doxorubicin in combination with trastuzumab and paclitaxel in HER2-positive metastatic breast cancer. *Ann. Oncol., 25*, 592–598.

55. Cortes, J. E., Goldberg, S. L., Feldman, E. J., Rizzeri, D. A., Hogge, D. E., Larson, M., et al., (2015). Phase II, multicenter, randomized trial of CPX-351 (cytarabine:daunorubicin) liposome injection versus intensive salvage therapy in adults with first relapsed AML. *Cancer, 121*, 234–242.

56. Batist, G., Gelmon, K. A., Chi, K. N., Miller, W. H. Jr., Chia, S. K., Mayer, L. D., et al., (2009). Safety, pharmacokinetics, and efficacy of CPX-1 liposome injection in patients with advanced solid tumors. *Clin. Cancer Res., 15*, 692–700.

57. Ko, A. H., Tempero, M. A., Shan, Y. S., Su, W. C., Lin, Y. L., Dito, E., et al., (2013). A multinational phase 2 study of nanoliposomal irinotecan sucrosofate (PEP02, MM-398) for patients with gemcitabine-refractory metastatic pancreatic cancer. *Br. J. Cancer., 109*, 920–925.

58. Saif, M. W., (2014). MM-398 achieves primary endpoint of overall survival in phase III study in patients with gemcitabine refractory metastatic pancreatic cancer. *J. Pancreas, 15*, 278–279.

59. Geretti, E., Leonard, S. C., Dumont, N., Lee, H., Zheng, J., De, S. R., et al., (2015). Cyclophosphamide-mediated tumor priming for enhanced delivery and anti-tumor

activity of HER2-targeted liposomal doxorubicin (MM-302). *Mol. Cancer Ther., 14,* 2060–2071.

60. Goldberg, M. S., Hook, S. S., Wang, A. Z., Bulte, J. W., Patri, A. K., Uckun, F. M., et al., (2013). Biotargeted nanomedicines for cancer: Six tenets before you begin. *Nanomedicine, 8,* 299–308.

61. Fantini, M., Gianni, L., Santelmo, C., Drudi, F., Castellani, C., Affatato, A., et al., (2011). Lipoplatin treatment in lung and breast cancer. *Chemother. Res. Pract., 2011,* 125192. doi.org/10.1155/2011/125192.

62. Broglio, K. R., Stivers, D. N., & Berry, D. A., (2014). Predicting clinical trial results based on announcements of interim analyses. *Trials, 15,* 73. doi. org/10.1186/1745-6215-15-73.

63. Poon, R. T., & Borys, N., (2011). Lyso-thermosensitive liposomal doxorubicin: An adjuvant to increase the cure rate of radiofrequency ablation in liver cancer. *Future Oncol., 7,* 937–945.

64. Staruch, R., Chopra, R., & Hynynen, K., (2011). Localized drug release using MRI-controlled focused ultrasound hyperthermia. *Int. J. Hyperthermia, 27,* 156–171.

65. Guillot, T., Spielmann, M., Kac, J., Luboinski, B., Tellez-Bernal, E., Munck, J. N., et al., (1992). Neoadjuvant chemotherapy in multiple synchronous head and neck and esophagus squamous cell carcinomas. *Laryngoscope, 102,* 311–319.

66. Le Chevalier, T., Brisgand, D., Douillard, J. Y., Pujol, J. L., Alberola, V., Monnier, A., et al., (1994). Randomized study of vinorelbine and cisplatin versus vindesine and cisplatin versus vinorelbine alone in advanced non-small-cell lung cancer: Results of a European multicenter trial including 612 patients. *J. Clin. Oncol., 12,* 360–367.

67. Shirazi, F. H., Molepo, J. M., Stewart, D. J., Ng, C. E., Raaphorst, G. P., & Goel, R., (1996). Cytotoxicity, accumulation, and efflux of cisplatin and its metabolites in human ovarian carcinoma cells. *Toxicol. Appl. Pharmacol., 140,* 211–218.

68. Kondagunta, G. V., Bacik, J., Donadio, A., Bajorin, D., Marion, S., Sheinfeld, J., et al., (2005). Combination of paclitaxel, ifosfamide, and cisplatin is an effective second-line therapy for patients with relapsed testicular germ cell tumors. *J. Clin. Oncol., 23,* 6549–6555.

69. Hirai, M., Minematsu, H., Hiramatsu, Y., Kitagawa, H., Otani, T., Iwashita, S., et al., (2010). Novel and simple loading procedure of cisplatin into liposomes and targeting tumor endothelial cells. *Int. J. Pharmaceutics, 391,* 274–283.

70. Krieger, M., Eckstein, N., Schneider, V., Koch, M., Royer, H. D., Jaehde, U., et al., (2010). Overcoming cisplatin resistance of ovarian cancer cells by targeted liposomes *in vitro. Int. J. Pharmaceutics, 389,* 10–17.

71. Steerenberg, P. A., Storm, G., De Groot, G., Claessen, A., Bergers, J. J., Franken, M. A., et al., (1988). Liposomes as drug carrier system for cis-diamminedichloroplatinum (II). II. Antitumor activity *in vivo,* induction of drug resistance, nephrotoxicity and Pt distribution. *Cancer Chemother. Pharmacol., 21,* 299–307.

72. Newman, M. S., Colbern, G. T., Working, P. K., Engbers, C., & Amantea, M. A., (1999). Comparative pharmacokinetics, tissue distribution, and therapeutic effectiveness of cisplatin encapsulated in long-circulating, pegylated liposomes (SPI-077) in tumor-bearing mice. *Cancer Chemother. Pharmacol., 43,* 1–7.

73. Vaage, J., Donovan, D., Wipff, E., Abra, R., Colbern, G., Uster, P., et al., (1999). Therapy of a xenografted human colonic carcinoma using cisplatin or doxorubicin

encapsulated in long-circulating pegylated stealth liposomes. *Int. J. Cancer, 80,* 134–137.

74. Arienti, C., Tesei, A., Ravaioli, A., Ratta, M., Carloni, S., Mangianti, S., et al., (2008). Activity of lipoplatin in tumor and in normal cells *in vitro*. *Anticancer Drugs, 19,* 983–990.

75. Boulikas, T., (2004). Low toxicity and anticancer activity of a novel liposomal cisplatin (Lipoplatin) in mouse xenografts. *Oncol. Rep., 12,* 3–12.

76. Devarajan, P., Tarabishi, R., Mishra, J., Ma, K., Kourvetaris, A., Vougiouka, M., et al., (2004). Low renal toxicity of Lipoplatin compared to cisplatin in animals. *Anticancer Res., 24,* 2193–2200.

77. Stathopoulos, G. P., Boulikas, T., Vougiouka, M., Deliconstantinos, G., Rigatos, S., Darli, E., et al., (2005). Pharmacokinetics and adverse reactions of a new liposomal cisplatin (Lipoplatin): Phase I study. *Oncol. Rep., 13,* 589–595.

78. Boulikas, T., Stathopoulos, G. P., Volakakis, N., & Vougiouka, M., (2005). Systemic Lipoplatin infusion results in preferential tumor uptake in human studies. *Anticancer Res., 25,* 3031–3039.

79. Stathopoulos, G. P., Boulikas, T., Kourvetaris, A., & Stathopoulos, J., (2006). Liposomal oxaliplatin in the treatment of advanced cancer: A phase I study. *Anticancer Res., 26,* 1489–1493.

80. Stathopoulos, G. P., Rigatos, S. K., & Stathopoulos, J., (2010). Liposomal cisplatin dose escalation for determining the maximum tolerated dose and dose-limiting toxicity: A phase I study. *Anticancer Res., 30,* 1317–1321.

81. Farhat, F. S., Temraz, S., Kattan, J., Ibrahim, K., Bitar, N., Haddad, N., et al., (2011). A phase II study of lipoplatin (liposomal cisplatin)/vinorelbine combination in HER-2/neu-negative metastatic breast cancer. *Clin. Breast Cancer, 11,* 384–389.

82. Mylonakis, N., Athanasiou, A., Ziras, N., Angel, J., Rapti, A., Lampaki, S., et al., (2010). Phase II study of liposomal cisplatin (Lipoplatin) plus gemcitabine versus cisplatin plus gemcitabine as first-line treatment in inoperable (stage IIIB/IV) non-small cell lung cancer. *Lung Cancer, 68,* 240–247.

83. Jehn, C. F., Boulikas, T., Kourvetaris, A., Possinger, K., & Lüftner, D., (2007). Pharmacokinetics of liposomal cisplatin (lipoplatin) in combination with 5-FU in patients with advanced head and neck cancer: First results of a phase III study. *Anticancer Res., 27,* 471–475.

84. Chauffert, B., Favoulet, P., Polycarpe, E., Duvillard, C., Beltramo, J. L., Bichat, F., et al., (2003). Rationale supporting the use of vasoconstrictors for intraperitoneal chemotherapy with platinum derivatives. *Surg. Oncol. Clin. North America, 12,* 835–848.

85. Araújo, J. G., Mota, L. G., Leite, E. A., Maroni, L. C., Wainstein, A. J., Coelho, L. G., et al., (2011). Biodistribution and antitumoral effect of long-circulating and pH-sensitive liposomal cisplatin administered in Ehrlich tumor-bearing mice. *Exp. Biol. Med., 236,* 808–815.

86. Leite, E. A., Giuberti, C. S., Wainstein, A. J., Wainstein, A. P., Coelho, L. G., Lana, A. M., et al., (2009). Acute toxicity of long-circulating and pH-sensitive liposomes containing cisplatin in mice after intraperitoneal administration. *Life Sci., 84,* 641–649.

87. Leite, E. A., Lana, A. M., Junior, A. D., Coelho, L. G., & De Oliveira, M. C., (2012). Acute toxicity study of cisplatin loaded long-circulating and pH-sensitive liposomes administered in mice. *J. Biomed. Nanotechnol., 8*, 229–239.
88. Maroni, L. C., Silveira, A. C. O., Leite, E. A., Melo, M. M., Ribeiro, A. F. C., Cassali, G. D., et al., (2012). Antitumor effectiveness and toxicity of cisplatin loaded long-circulating and pH-sensitive liposomes against Ehrlich ascitic tumor. *Exp. Biol. Med., 237*, 973–984.
89. Stathopoulos, G. P., Boulikas, T., Vougiouka, M., Rigatos, S. K., & Stathopoulos, J. G., (2006). Liposomal cisplatin combined with gemcitabine in pretreated advanced pancreatic cancer patients: A phase I-II study. *Oncol. Rep., 15*, 1201–1204.
90. Boulikas, T., Pantos, A., Bellis, E., & Christofis, P., (2007). Designing platinum compounds in cancer: Structures and mechanisms. *Cancer Ther., 5*, 537–583.
91. Leonard, R. C. F., Williams, S., Tulpule, A., Levine, A. M., & Oliveros, S., (2009). Improving the therapeutic index of anthracycline chemotherapy: Focus on liposomal doxorubicin (Myocet®). *The Breast, 18*, 218–224.
92. Safra, T., (2003). Cardiac safety of liposomal anthracyclines. *The Oncologist, 8*, 17–24.
93. Forssen, E. A., Coulter, D. M., & Proffitt, R. T., (1992). Selective *in vivo* localization of daunorubicin small unilamellar vesicles in solid tumors. *Cancer Res., 52*, 3255–3261.
94. Forssen, E. A., & Ross, M. E., (1994). Daunoxome treatment of solid tumors: Preclinical and clinical investigations. *J. Liposome Res., 4*, 481–512.
95. Pea, F., Russo, D., Michieli, M., Baraldo, M., Ermacora, A., Damiani, D., et al., (2000). Liposomal daunorubicin plasmatic and renal disposition in patients with acute leukemia. *Cancer Chemother. Pharmacol., 46*, 279–286.
96. Gill, P. S., Espina, B. M., Muggia, F., Cabriales, S., Tulpule, A., Esplin, J. A., et al., (1995). Phase I/II clinical and pharmacokinetic evaluation of liposomal daunorubicin. *J. Clin. Oncol., 13*, 996–1003.
97. Gill, P. S., Wernz, J., Scadden, D. T., Cohen, P., Mukwaya, G. M., Von, R. J. H., et al., (1996). Randomized phase III trial of liposomal daunorubicin versus doxorubicin, bleomycin, and vincristine in AIDS-related Kaposi's sarcoma. *J. Clin. Oncol., 14*, 2353–2364.
98. Kanter, P. M., Bullard, G. A., Pilkiewicz, F. G., Mayer, L. D., Cullis, P. R., & Pavelic, Z. P., (1993). Preclinical toxicology study of liposome-encapsulated doxorubicin (TLC D-99): Comparison with doxorubicin and empty liposomes in mice and dogs. *In Vivo, 7*, 85–95.
99. Harasym, T. O., Cullis, P. R., & Bally, M. B., (1997). Intratumor distribution of doxorubicin following i.v. administration of drug encapsulated in egg phosphatidylcholine/cholesterol liposomes. *Cancer Chemother. Pharmacol., 40*, 309–317.
100. Hofheinz, R. D., Gnad-Vogt, S. U., Beyer, U., & Hochhaus, A., (2005). Liposomal encapsulated anticancer drugs. *Anticancer Drugs, 16*, 691–707.
101. Harris, L., Batist, G., Belt, R., Rovira, D., Navari, R., Azarnia, N., et al., (2002). Liposome- encapsulated doxorubicin compared with conventional doxorubicin in a randomized multicenter trial as first-line therapy of metastatic breast carcinoma. *Cancer, 94*, 25–36.

102. Cowens, J. W., Creaven, P. J., Greco, W. R., Brenner, D. E., Tung, Y., Ostro, M., et al., (1993). Initial clinical (phase I) trial of TLC D-99 (doxorubicin encapsulated in liposomes). *Cancer Res., 53*, 2796–2802.

103. Vaage, J., Donovan, D., Mayhew, E., Uster, P., & Woodle, M., (1993). Therapy of mouse mammary carcinomas with vincristine and doxorubicin encapsulated in sterically stabilized liposomes. *Int. J. Cancer, 54*, 959–964.

104. Gabizon, A., Shmeeda, H., & Barenholz, Y., (2003). Pharmacokinetics of pegylated liposomal doxorubicin. *Clin. Pharmacokin, 42*, 419–436.

105. Robert, N. J., Vogel, C. L., Henderson, I. C., Sparano, J. A., Moore, M. R., Silverman, P., et al., (2004). The role of the liposomal anthracyclines and other systemic therapies in the management of advanced breast cancer. *Semin. Oncol., 31*, 106–146.

106. Hau, P., Fabel, K., Baumgart, U., Rümmele, P., Grauer, O., Bock, A., et al., (2004). Pegylated liposomal doxorubicin efficacy in patients with recurrent high-grade glioma. *Cancer, 100*, 1199–1207.

107. Zhang, Q., Huang, X. E., & Gao, L. L., (2009). A clinical study on the premedication of paclitaxel liposome in the treatment of solid tumors. *Biomed. Pharmacoth., 63*, 603–607.

108. Zhang, J. A., Anyarambhatla, G., Ma, L., Ugwu, S., Xuan, T., Sardone, T., et al., (2005). Development and characterization of a novel Cremophor EL free liposome-based paclitaxel (LEP-ETU) formulation. *Eur. J. Pharm. Biopharm., 59*, 177–187.

109. Yang, T., Cui, F. D., Choi, M. K., Cho, J. W., Chung, S. J., Shim, C. K., et al., (2007). Enhanced solubility and stability of PEGylated liposomal paclitaxel: *In vitro* and *in vivo* evaluation. *Int. J. Pharmaceutics, 338*, 317–326.

110. Strieth, S., Eichhirn, M. E., Werner, A., Sauer, B., Teifeil, M., Michaelis, U., et al., (2008). Paclitaxel encapsulated in cationic liposomes increases tumor microvessel leakiness and improves therapeutic efficacy in combination with cisplatin. *Clin. Cancer Res., 14*, 4603–4611.

111. Glantz, M. J., Lafollette, S., Jaeckle, K. A., Shapiro, W., Swinnen, L., Rozental, J. R., et al., (1999a). Randomized trial of a slow-release versus a standard formulation of cytarabine for the intrathecal treatment of lymphomatous meningitis. *J. Clin. Oncol., 17*, 3110–3116.

112. Glantz, M. J., Jaeckle, K. A., Chamberlain, M. C., Phuphanich, S., Recht, L., Swinnen, L. J., et al., (1999b). A randomized controlled trial comparing intrathecal sustained-release cytarabine (DepoCyte) to intrathecal methotrexate in patients with neoplastic meningitis from solid tumors. *Clin. Cancer Res., 5*, 3394–3402.

113. Spina, M., Chimienti, E., Martellotta, F., Vaccher, E., Berretta, M., Zanet, E., et al., (2010). Phase 2 study of intrathecal, long-acting liposomal cytarabine in the prophylaxis of lymphomatous meningitis in human immunodeficiency virus-related non-Hodgkin lymphoma. *Cancer, 116*, 1495–1501.

114. Rodriguez, M. A., Pytlik, R., Kozak, T., Chhanabhai, M., Gascoyne, R., Lu, B., et al., (2009). Vincristine sulfate liposomes injection (Marqibo) in heavily pretreated patients with refractory aggressive non-Hodgkin lymphoma: Report of the pivotal phase 2 study. *Cancer, 115*, 3475–3482.

115. Wang-Gillam, A., Li, C. P., Bodoky, G., Dean, A., Shan, Y. S., Jameson, G., et al., (2016). Nanoliposomal irinotecan with fluorouracil and folinic acid in metastatic

pancreatic cancer after previous gemcitabine-based therapy (NAPOLI-1): A global, randomized, open-label, phase 3 trial. *Lancet, 387*, 545–557.

116. Shukla, S. K., Kulkarni, N. S., Chan, A., Parvathaneni, V., Farrales, P., Muth, A., et al., (2019). Metformin-encapsulated liposome delivery system: An effective treatment approach against breast cancer. *Pharmaceutics, 11*, 559. doi.org/10.3390/pharmaceutics11110559.

117. Lee, J. Y., Shin, D. H., & Kim, J. S., (2020). Anticancer effect of metformin in Herceptin-conjugated liposome for breast cancer. *Pharmaceutics, 12*, 11. doi.org/10.3390/pharmaceutics12010011.

118. Kunjachan, S., Ehling, J., Storm, G., Kiessling, F., & Lammers, T., (2015). Noninvasive imaging of nanomedicines and nanotheranostics: Principles, progress, and prospects. *Chem. Rev., 115*, 10907–10937.

119. Maruyama, K. L., Ishida, O., Takizawa, T., & Moribe, K., (1999). Possibility of active targeting to tumor tissues with liposomes. *Adv. Drug Deliv. Rev., 40*, 89–102.

120. Olusanya, T. O. B., Ahmad, R. R. H., Ibegbu, D. M., Smith, J. R., & Elkordy, A. A., (2018). Liposomal drug delivery systems and anticancer drugs. *Molecules, 23*, 907. doi.org/10.3390/molecules23040907.

121. Yamashita, S., Yamashita, J., & Ogawa, M., (1994). Overexpression of group II phospholipase A2 in human breast cancer tissues is closely associated with their malignant potency. *Br. J. Cancer, 69*, 1166–1170.

122. Mock, J. N., Costyn, L. J., Wilding, S. L., Arnold, R. D., & Cummings, B. S., (2013). Evidence for distinct mechanisms of uptake and antitumor activity of secretory phospholipase A2 responsive liposome in prostate cancer. *Integr. Biol., 5*, 172–182.

123. Allison, A. C., & Gregoriadis, G., (1974). Liposomes as immunological adjuvants. *Nature, 252*, 252.

124. Peres-Filho, M. J., Santos, A. P. D., Nascimento, T. L., Ávila, R. I. D., Ferreira, F. S., Valadares, M. C., et al., (2017). Antiproliferative activity and VEGF expression reduction in MCF7 and PC-3 cancer cells by Paclitaxel and Imatinib co-encapsulation in folate-targeted liposomes. *AAPS PharmSciTech, 19*, 1–12.

125. Handali, S., Moghimipour, E., Rezaei, M., Ramezani, Z., Kouchak, M., Amini, M., et al., (2018). A novel 5-fluorouracil targeted delivery to colon cancer using folic acid conjugated liposomes. *Biomed. Pharmacother., 108*, 1259–1273.

126. Soe, Z. C., Thapa, R. K., Ou, W., Gautam, M., Nguyen, H. T., Jin, S. G., et al., (2018). Folate receptor-mediated celastrol and irinotecan combination delivery using liposomes for effective chemotherapy. *Colloids Surf. B Biointerfaces, 170*, 718–728.

127. Gazzano, E., Rolando, B., Chegaev, K., Salaroglio, I. C., Kopecka, J., Pedrini, I., et al., (2017). Folate-targeted liposomal nitrooxy-doxorubicin: An effective tool against P-glycoprotein-positive and folate receptor-positive tumors. *J. Control. Release, 270*, 37–52.

128. Moghimipour, E., Rezaei, M., Ramezani, Z., Kouchak, M., Amini, M., Angali, K. A., et al., (2017). Folic acid-modified liposomal drug delivery strategy for tumor targeting of 5-fluorouracil. *Eur. J. Pharm. Sci., 114*, 166–174.

129. Qiu, L., Dong, C., & Kan, X., (2018). Lymphoma-targeted treatment using a folic acid-decorated vincristine-loaded drug delivery system. *Drug Des. Dev. Ther., 12*, 863–872.

130. Monteiro, L. F., Malachias, Â., Poundlana, G., Magalhãespaniago, R., Mosqueira, V. C. F., Oliveira, M. C., et al., (2018). Paclitaxel-loaded pH-sensitive liposome: New insights on structural and physicochemical characterization. *Langmuir, 34*, 5728–5737.

131. Torchilin, V. P., (2005). Recent advances with liposomes as pharmaceutical carriers. *Nature Rev. Drug Discov., 4*, 145–160.

132. Patel, G. B., Agnew, B. J., Deschatelets, L., Fleming, L. P., & Sprott, G. D., (2000). *In vitro* assessment of archaeosome stability for developing oral delivery systems. *Int. J. Pharmaceutics, 194*, 39–49.

133. Sprott, G. D., (1992). Structures of archaeobacterial membrane lipids. *J. Bioenerg. Biomembranes, 24*, 555–566.

134. Kaneda, Y., (2000). Virosomes: Evolution of the liposome as a targeted drug delivery system. *Adv. Drug Deliv. Rev., 43*, 197–205.

135. Cusi, M. G., Terrosi, C., Savellini, G. G., Genova, G. D., Zurbriggen, R., & Correale, P., (2004). Efficient delivery of DNA to dendritic cells mediated by influenza virosomes. *Vaccine, 22*, 735–739.

136. Gluck, R., Moser, C., & Metcalfe, I. C., (2004). Influenza virosomes as an efficient system for adjuvanted vaccine delivery. *Exp. Opin. Biol. Ther., 4*, 1139–1145.

137. Gangwar, M., Singh, R., Goel, R. K., & Nath, G., (2012). Recent advances in various emerging vesicular systems: An overview. *Asian Pac. J. Trop. Biomed., 2*, 1176–1188.

138. Garg, B. J., Saraswat, A., Bhatia, A., & Katare, O. P., (2010). Topical treatment in vitiligo and the potential uses of new drug delivery systems. *Indian J. Dermatol. Venereol. Leprol., 76*, 231–238.

139. Jain, S., Jaio, N., Bhadra, D., Tiwary, A. K., & Jain, N. K., (2005). Delivery of nonsteroidal anti-inflammatory agents like diclofenac. *Curr. Drug Deliv., 2*, 223–227.

140. Patel, S., (2007). *Ethosomes: A Promising Tool for Transdermal Delivery of Drug* (Vol. 5, pp. 1–5). Pharma Info. Net.

141. Lasic, D. D., (1997). Liposomes and niosomes. In: Rieger, M. M., & Rhein, L. D., (eds.), *Surfactants in Cosmetics* (pp. 263–283). Marcel Dekker: New York.

142. Frézard, F., (1999). Liposomes: From biophysics to the design of peptide vaccines. *Braz. J. Med. Biol. Res., 32*, 181–189.

143. Vyas, S. P., Subhedar, R., & Jain, S., (2006). Development and characterization of emulsomes for sustained and targeted delivery of an antiviral agent to liver. *J. Pharmacol. Pharmacotherapy., 58*, 321–326.

144. Barry, B. W., (2001). Novel Mechanisms and devices to enable successful transdermal drug delivery. *Eur. J. Pharm. Sci., 14*, 101–114.

145. Kisak, E., Coldren, B., & Zasadzinski, J. A., (2002). Nanocompartments enclosing vesicles, colloids, and macromolecules via interdigitated lipid bilayers. *Langmuir, 18*, 284–288.

146. Alatorre-Meda, M., Gonzales-Perez, A., & Rodriguez, J. R., (2010). DNA Metafectene pro-complexation: A physical chemistry study. *Phys. Chem. Chem. Phys., 12*, 7464–7472.

147. Krishnan, L., Sad, S., Patel, G. B., & Sprott, G. D., (2003). Archaeosomes induce enhanced cytotoxic T lymphocyte responses to entrapped soluble protein in the absence of interleukin 12 and protect against tumor challenge. *Cancer Res., 63*, 2526–2534.

148. Alavi, S. E., Mansouri, H., Esfahani, M. K. M., Movahedi, F., Akbarzadeh, A., & Chiani, M., (2014). Archaeosome: As new drug carrier for delivery of paclitaxel to breast cancer. *Indian J. Clin. Biochem., 29*, 150–153.

149. Ameri, A., Moghimipour, E., Ramezani, Z., Kargar, M., Hashemitabar, M., Saremy, S., et al., (2016). Formulation of a new generation of liposomes from bacterial and archeal lipids. *Trop. J. Pharm. Res., 15*, 215–220.

150. Stark, F. C., Agbayani, G., Sandhu, J. K., Akache, B., McPherson, C., & Deschatelets, L., (2019). Simplified admix archaeal glycolipid adjuvanted vaccine and checkpoint inhibitor therapy combination enhances protection from murine melanoma. *Biomedicines, 7*, 91. doi.org/10.3390/biomedicines7040091.

151. Mohammadi, Z. H., Abolmaali, S., & Akbarzadeh, A., (2018). Preparing nanoarchaeosome containing triptorelin acetate and evaluation of its cellular toxic effect on PC3 prostate cancer cell line. *Jorjani Biomed. J., 6*, 21–31.

152. Karimi, H., Soleimanjahi, H., Abdoli, A., & Banijamali, R. S., (2020). Combination therapy using human papillomavirus L1/E6/E7 genes and archaeosome: A nanovaccine confer immune adjuvanting effects to fight cervical cancer. *Sci. Rep., 10*, 5787.

153. Alhakamy, N. A., Badr-Eldin, S. M., Ahmed, O. A. A., Asfour, H. Z., Aldawsari, H. M., Algandaby, M. M., et al., (2020). Piceatannol-loaded emulsomes exhibit enhanced cytostatic and apoptotic activities in colon cancer cells. *Antioxidants, 13*, 419, doi.org/10.3390/antiox9050419.

154. Bolat, Z. B., Islek, Z., Demir, B. N., Yilmaz, E. N., Sahin, F., & Ucisik, M. H., (2020). Curcumin- and piperine-loaded emulsomes as combinational treatment approach enhance the anticancer activity of curcumin on HCT116 colorectal cancer model. *Front. Bioeng. Biotechnol., 8*, 50, doi.org/10.3389/fbioe.2020.00050.

155. Kollipara, R. K., Tallapaneni, V., Sanapalli, B. K. R., Vinoth, K. G., & Karri, V. V. S. R., (2019). Curcumin loaded ethosomal vesicular drug delivery system for the treatment of melanoma skin cancer. *Res. J. Pharm. Tech., 12*, 1783–1792.

156. Khan, N. R., & Wong, T. W., (2018). 5-Fluorouracil ethosomes - skin deposition and melanoma permeation synergism with microwave. *Artif. Cells Nanomed. Biotechnol., 46*, 568–577.

157. Shaji, J., & Bajaj, R., (2017). Formulation development of 5-fluorouracil transethosomes for skin cancer therapy. *Int. J. Pharm. Pharm. Res., 11*, 453–464.

158. Saraf, S., & Gupta, M. K., (2018). Itraconazole loaded ethosomal gel system for efficient treatment of skin cancer. *Int. J. Drug Deliv., 10*, 12–19.

159. Yu, Z., Lv, H., Han, G., & Ma, K., (2016). Ethosomes loaded with cryptotanshinone for acne treatment through topical gel formulation. *PLoS One, 11*, e0159967. doi: 10.1371/journal.pone.0159967.

160. Moolakkadath, T., Aqil, M., Ahad, A., Imam, S. S., Praveen, A., Sultana, Y., et al., (2019). Fisetin loaded binary ethosomes for management of skin cancer by dermal application on UV exposed mice. *Int. J. Pharm., 560*, 78–91.

161. Nasri, S., Ebrahimi-Hosseinzadeh, B., Rahaie, M., Hatamian-Zarmi, A., & Sahraeian, R., (2020). Thymoquinone-loaded ethosome with breast cancer potential: Optimization, *in vitro* and biological assessment. *J. Nanostructure Chem., 10*, 19–31.

162. Cristiano, M. C., Froiio, F., Spaccapelo, R., Mancuso, A., Nisticò, S. P., Udongo, B. P., et al., (2019). Sulforaphane-loaded ultra-deformable vesicles as a potential natural

nanomedicine for the treatment of skin cancer diseases. *Pharmaceutics, 12.* doi. org/10.3390/pharmaceutics12010006.

163. Obeid, M. A., Gany, S. A. S., Gray, A. I., Young, L., Igoli, J. O., & Ferro, V. A., (2020). Niosome encapsulated balanocarpol: Compound isolation, characterization, and cytotoxicity evaluation against human breast and ovarian cancer cell lines. *Nanotechnology, 31,* 195101.

164. Xu, Y. Q., Chen, W. R., Tsosie, J. K., Xie, X., Li, P., Wan, J. B., et al., (2016). Niosome encapsulation of curcumin: Characterization and cytotoxic effect on ovarian cancer cells. *J. Nanomaterials, 2016,* Article ID 6365295.

165. Akbarzadeh, I., Yaraki, M. T., Bourbour, M., Noorbazargan, H., Lajevardi, A., Shilsar, S. M. S., et al., (2020). Optimized doxycycline-loaded niosomal formulation for treatment of infection-associated prostate cancer: An *in-vitro* investigation. *J. Drug Deliver. Sci. Technol., 57,* 101715.

166. Kanaani, L., Javadi, I., Ebrahimifar, M., Shahmabadi, H. M., Khiyavi, A. A., & Mehrdiba, T., (2017). Effects of cisplatin-loaded niosomal nanoparticles on BT-20 human breast carcinoma cells. *Asian Pac. J. Cancer Prev., 18,* 365–368.

167. Barani, M., Mirzaei, M., Torkzadeh-Mahani, M., & Nematollahi, M. H., (2018). Lawsone-loaded niosome and its antitumor activity in MCF-7 breast cancer cell line: A nano-herbal treatment for cancer. *DARU J. Pharm. Sci., 26,* 11–17.

168. Nazari-Vanani, R., Karimian, K., Azarpira, N., & Heli, H., (2019). Capecitabine loaded nanoniosomes and evaluation of anticancer efficacy. *Artif. Cells Nanomed. Biotechnol., 47,* 420–426.

169. Karthick, K., & Arul, K. K. S. G., (2016). Formulation and evaluation of niosomes co-loaded with 5-fluorouracil and leucovorin: Characterization and *in vitro* release study. *Int. J. Res. Pharm. Nano Sci., 5,* 239–250.

170. Hemati, M., Haghiralsadat, F., Jafary, F., Moosavizadeh, S., & Moradi, A., (2019). Targeting cell cycle protein in gastric cancer with CDC20siRNA and anticancer drugs (doxorubicin and quercetin) co-loaded cationic PEGylated nanoniosomes. *Int. J. Nanomedicine, 14,* 6575–6585.

171. Hajizadeh, M. R., Maleki, H., Barani, M., Fahmidehkar, M. A., Mahmoodi, M., & Torkzadeh-Mahani, M., (2019). *In vitro* cytotoxicity assay of D-limonene niosomes: An efficient nano-carrier for enhancing solubility of plant-extracted agents. *Res. Pharm. Sci., 14,* 448–458.

172. Alemi, A., Reza, J. Z., Haghiralsadat, F., Jaliani, H. Z., Karamallah, M. H., Hosseini, S. A., et al., (2018). Paclitaxel and curcumin coadministration in novel cationic PEGylated niosomal formulations exhibit enhanced synergistic antitumor efficacy. *J. Nanobiotechnol., 16,* 28, doi.org/10.1186/s12951-018-0351-4.

173. Barani, M., Mirzaei, M., Torkzadeh-Mahani, M., & Adeli-Sardou, M., (2019). Evaluation of Carum-loaded niosomes on breast cancer cells: Physicochemical properties, *in vitro* cytotoxicity, flow cytometric, DNA fragmentation and cell migration assay. *Sci. Rep., 9,* 7139.

174. Zhang, Z., Wang, X., Chen, X., Wo, Y., Zhang, Y., & Biskup, E., (2015). 5-fluorouracil-loaded transfersome as theranostics in dermal tumor of hypertrophic scar tissue. *J. Nanomater., 253712,* 1–9.

175. Khan, M. A., Pandit, J., Sultana, Y., Sultana, S., Ali, A., Aqil, M., et al., (2015). Novel carbopol-based transfersomal gel of 5-fluorouracil for skin cancer treatment: *In vitro* characterization and *in vivo* study. *Drug Deliv., 22*, 795–802.

176. Zhang, X., Zong, W., Wang, J., Dong, M., Cheng, W., Sun, T., et al., (2019). Multi-compartmentalized vesosomes containing DOX loaded liposomes and 5FU loaded liposomes for synergistic tumor treatment. *New J. Chem., 43*, 4895–4899.

177. Wiedermann, U., Wiltschke, C., Jasinska, J., Kundi, M., Zurbriggen, R., Garner-Spitzer, E., et al., (2010). A virosomal formulated Her-2/neu multi-peptide vaccine induces Her-2/neu-specific immune responses in patients with metastatic breast cancer: A phase I study. *Breast Cancer Res. Treat., 119*, 673–683.

CHAPTER 8

Liposomal Supplements

NILOUFAR MAGHSOUDNIA,[1] MARYAM EDALAT,[2]
MOHAMMAD HAJIMOLAALI,[3] SHEIDA IRANPOUR,[1]
PARSA KHOSHKHAT,[4] PARINAZ INANLOO,[5]
REZA BARADARAN EFTEKHARI,[1] SOPHIA G. ANTIMISIARIS,[3] and
FARID ABEDIN DORKOOSH[1]

[1]Department of Pharmaceutics, Faculty of Pharmacy, Tehran University of Medical Sciences, Tehran, Iran

[2]Pharmaceutical Sciences Research Center, Islamic Azad University, Tehran Medical Sciences University (IAUTMU), Tehran, Iran

[3]Laboratory of Pharmaceutical Technology, Dept. of Pharmacy, School of Health Sciences, University of Patras, Rio 26504, Greece

[4]International Campus, School of Pharmacy, Tehran University of Medical Sciences, Tehran, Iran

[5]Islamic Azad University, Tehran Medical Branch, Tehran, Iran

ABSTRACT

Liposomes, enclosed phospholipid vesicles with bilayered membrane structures, are effective and widely used encapsulation systems in the pharmaceutical and dietary supplement industries such as nutritional supplements. These encapsulating materials have attracted great attention due to their bioavailability, biodegradability, absence of toxicity, ability to deliver a wide variety of bioactive compounds, increasing ingredients solubility, and enhanced stability against a range of environmental, enzymatic, and chemical stresses. There has been rapid development of liposomes to carry different functional compounds known as supplements needed for a healthy life. In this chapter, the application of liposomes as emerging carrier vehicles of supplements such as minerals, vitamins, herbal extracts, essential fatty acids, and antioxidant compounds is discussed in detail.

8.1 INTRODUCTION

Nutritional supplements are any dietary products that provide nutrients that may not be consumed in sufficient quantities, such as vitamins, minerals, herbal extracts, antioxidants, and other nutritional elements. There are broad ranges of supplements that demonstrated health benefits. As the dietary supplement industry has been developed in an incredible way recently, researchers are paying attention to improving bioavailability and product formulation technology. Among various technologies for achieving mentioned issues, liposomes, usually formed from phospholipids, have been applied to encapsulate not only drugs but even nutritional supplements [1]. It is worth mentioning that liposomes were first developed for medical and cosmetics purposes. The idea of oral liposomal formulation dated back to 1970's when phospholipid-based liposomes attracted scientists' interest due to their stable chemical structure, safety profile, protection of therapeutically active compounds from metabolic degradation, and ability to control the rate of drug dissolution and release [2]. Then studies revealed that compared to free drug administration, liposome-encapsulated drugs showed better absorption [3]. Moreover, labile molecules such as peptides and proteins were able to be protected against chemical and enzymatic degradation in the gastrointestinal tract with the use of liposomal formulations. Thus, oral liposomal delivery of therapeutic agents became a new trend in pharmaceutical researches. Regarding dietary supplements, liposomes are a very effective method of oral supplement delivery [1]. Liposomes possess precious characteristics such as the possibility of large-scale production, biocompatibility, non-toxic, and non-immunogenic properties, targetability, ability to carry a wide variety of bioactive agents (entrapping both hydrophilic and lipophilic ones) and health benefits of liposomal ingredients for human that have made these efficient carriers take enormous attention in food and supplement industry [4]. In comparison to conventional methods used in the production of supplements, liposomes can enhance stability of products against a wide range of environmental, enzymatic, and chemical issues such as exposure to extreme pH, temperature, and high ion concentrations [4] and they could also improve absorption efficiency due to the similar structure to that of cellular membranes [5]. This matter could result in modifying the distribution of supplements and increasing their efficacy *in vivo* [6]. As an example of liposomes' benefits in nutritional supplements, off-flavors could be avoided, and flow properties as well as stability could be increased by encapsulating both water- and fat-soluble vitamins [7]. Another noticeable property of liposomes is related to reduction of normal degradation of active agents during processing or storage conditions in many food systems such as vitamins that avoids the need for using a larger than functionally necessary amount of the active materials. This issue can improve the economic aspects of product formulation and decrease potential toxicity problems [8].

An ideal liposomal formulation in supplement industry should exhibit the following characteristics: 1) components of liposomes structure should be non-reactive with the active ingredients being delivered; 2) the active agents should be protected by liposomes against environmental conditions; 3) liposomes materials and the active ingredients should be food-grade and inexpensive due to the economical explanation [9].

In this chapter, the application of liposomes in encapsulating nutritional supplements such as minerals, vitamins, herbal extracts, essential fatty acids, and antioxidant compounds and their characteristics as well as challenges and opportunities have been discussed in detail (Figure 8.1).

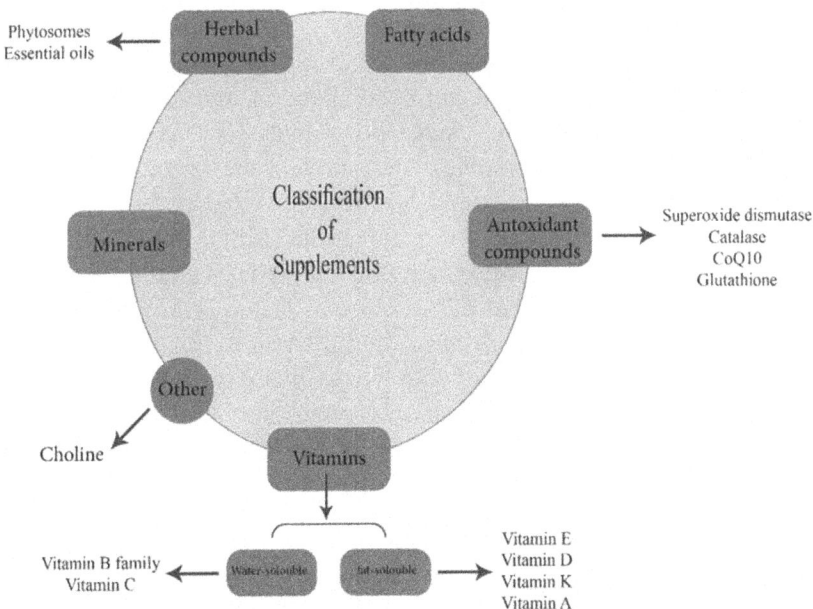

FIGURE 8.1 Nutritional supplements.

8.2 MINERALS

Liposomes have been applied widely to carry minerals fortifying the food matrix. They could also be employed in dairy products to control the release of minerals during heat treatment and minimize unwanted aggregation [8]. Additionally, application of liposomes for minerals delivery in supplements has taken scientists' attention recently that are discussed in the following.

Iron, one of the most abundant metals on Earth, plays an essential role in many important physiological functions such as oxygen transport, energy metabolism and DNA synthesis [5]. Iron (Fe) deficiency is highly identified as the nutritional deficiency worldwide due to its insufficiency in dietary intake and lack of bioavailability in foods. This essential trace element plays a vital role in many metabolic processes as iron deficiency may lead to serious diseases such as anemia [10]. Iron fortification of food is one of the effective ways to prevent iron deficiency. Liposomes are potential carriers for iron fortification of food products. As an antagonism relation exists between iron and milk calcium, there is a low amount of iron in milk. Therefore, the addition of enough iron into milk as microencapsulated by liposomes has been proposed to supply the milk with sufficient iron protected against iron-calcium antagonism. Highly soluble iron salts such as ferrous sulfate have been applied widely in this issue due to their cheapness, higher bioavailability, and prevention of the negative effects of free iron. Using antioxidants such as ascorbic acid in ferrous encapsulated liposomes could protect the ferrous ion against oxidation [11].

In a study, milk was fortified with ferrous sulfate liposomes (FLs) made by reverse-phase evaporation method (encapsulation efficiency (EE) of liposomes was about 67%) and was stable at low temperature for 1 week [12]. Ding et al. [9] also used the reverse-phase evaporation method to encapsulate liposomes with ferrous glycinate that demonstrated an acceptable stability in simulated gastrointestinal juice for 5 h [9]. In another study, ferrous sulfate was encapsulated in a lecithin liposome for milk enrichment and was stable for 6 months without shown any decrease of Fe bioavailability [13].

Oral iron supplements such as ferrous sulfate, ferrous gluconate, and ferrous fumarate, used to treat iron deficiency-related diseases, are often absorbed ineffectively, and can cause adverse effects in the gastrointestinal tract due to the oxidative toxicity of ferrous iron. Liposomal iron supplements with high absorption efficiency and minimal side effects could be considered as possible candidates in this matter [10]. Iron liposomes could increase iron levels and hemoglobin concentrations in serum significantly while not aggravating oxidative stress at the same time in comparison with traditional iron supplements. They could also maintain physical stability and provide gradual release properties [14]. It has been studied that absorption of non-heme iron, such as ferrous sulfate, could be influenced by several factors like divalent metal ions (such as Zn^{2+}) and phytic acid in diets based on cereals and legumes. Regarding this issue, application of liposomes could improve the bioavailability of encapsulated iron ions [14].

In a study, iron liposomes encapsulated with ferric ammonium citrate (FAC) or heme iron were used to treat exercise-associated anemia in rats. Serum iron and liver iron contents were significantly increased in anemic rats treated with iron liposomes in comparison with those of control groups. The results indicated iron liposomes could improve the efficiency of iron delivery as a supplement for the treatment of exercise-associated iron deficiency anemia with minimal side effects [10]. In another research, heme liposomes (HEME-LIP) and ferric citrate liposomes (FAC-LIP) were developed by rotary-evaporated film-ultrasonication method and both types of iron liposomes exhibited stable physical properties. FAC-LIP and HEME-LIP increased serum iron levels by 119% and 54% higher than ferric citrate (FAC) and heme alone, respectively, in rat models of AI (anemia of inflammation). Hepcidin, the key regulator of iron homeostasis was also up-regulated by iron liposomes. Therefore, it was concluded the absorption of iron liposomes was improved over that of unencapsulated iron agents as they showed better efficacy and they also induced no toxicity problems [5].

In a study, fortification of soya milk with Calcium (Ca) salts was applied by encapsulating the Ca salt (calcium lactate) in a lecithin liposome resulted in providing 100 g soya milk with up to 110 mg Ca; reaching levels equivalent to those in normal cow's milk [13]. Encapsulating calcium in liposome form withstand the gastric environment for optimal assimilation in the system blood and increase the absorption compared to conventional calcium supplements.

8.3   VITAMINS

Vitamins are essential organic micronutrients responsible for the natural function of hundreds of enzymes and various cellular pathways. Vitamins as essential nutrients, cannot be synthesized within the human body, and thus require to be supplied through diet [15]. From structural view, most vitamins are not single molecules with a distinct structure but rather a group of related molecules, commonly known as vitamers [16]. Considering vitamin E as an instance, this vitamin consists of a group of four tocopherols and four tocotrienols [17]. The exact number of vitamins has been a subject of debate as to whether choline, a nutrient that is required to produce acetylcholine, belongs to this category of nutrients [18]. Nevertheless, in this chapter, choline is explained as a vitamin that is not classified as either fat-soluble or water-soluble and is discussed in a separate section at the end. As a common approach, vitamins are categorized according to their water-solubility. This classification divides vitamins into two main groups of water-soluble and fat-soluble molecules. Vitamins B family and vitamin C are water-soluble, while vitamin A, D, E, and K belong to the fat-soluble group. Each vitamin possesses unique role in human body functions. Many problems encountered in the case of oral digestion of free supplements will be overcome with the use of a liposomal delivery system since pharmacokinetic properties of the liposomal system override the usual pharmacokinetic parameters of the free supplement [1]. This phenomenon is especially more important for supplements with a regulated pattern of absorption as if vitamin C, that is better and more readily absorbed within a liposome [19]. Liposomal delivery of vitamins along with a brief introduction to their functions in maintaining overall health, is provided as in the following subsections.

8.3.1   VITAMIN B FAMILY

There are eight members in this family; each is either a cofactor for crucial metabolic processes or a precursor needed to generate one. B vitamins are abundantly found in meat and also, in small amounts, in whole carbohydrate-based foods. Vitamin B family has numerous roles in the normal functioning of human cells [20]. Table 8.1. summarizes some of the natural functions of vitamin B family members.

TABLE 8.1   Natural Functions of Vitamin B Family

Vitamin	Also Known as	Function	References
Vitamin B1	Thiamine	Functions as a coenzyme in the catabolism of amino acids and sugars.	[21]
Vitamin B2	Riboflavin	Precursor of FAD and FMN coenzymes needed for flavoprotein enzyme reactions	[22]
Vitamin B3	Niacin, nicotinic acid, and nicotinamide	Precursor of NAD and NADP coenzymes, two main components of metabolic processes.	[23]
Vitamin B5	Pantothenic acid	Precursor of coenzyme A needed to metabolize a considerable number of molecules.	[24]
Vitamin B6	Pyridoxine, pyridoxal, and pyridoxamine	Involves in many enzymatic reactions in cell metabolism.	[25]
Vitamin B7	Biotin	Coenzyme for carboxylase enzymes, involved in gluconeogenesis.	[26]
Vitamin B9	Folate	Involves in cell growth and a precursor needed to generate and repair DNA.	[27]
Vitamin B12	Cobalamin	Coenzyme involved in the cellular metabolism as well as DNA synthesis and regulation.	[28]

Among the vitamin B family, vitamin B_{12} has devoted considerable share of researches mainly due to its significant role in the production of red blood cells. Deficiency of vitamin B_{12} is connected with the development of various diseases, including Pernicious anemia, hyper-homocysteinemia which itself is linked with the development of osteoporosis, and neurologic disorders [29]. Epidemiological studies also suggest a high prevalence of vitamin B_{12} deficiency in American adults with approximate 39% of the population between 26 and 83 in the low normal (below 258 picomoles per liter) range of vitamin B_{12} levels [30]. At the first step of vitamin B_{12} absorption from diet, the vitamin binds to a salivary vitamin B_{12}-binding protein called haptocorrin, which then will be subjected to proteolysis by pancreatic proteases within the duodenum. The free vitamin B_{12} binds to intrinsic factor to be further absorbed through receptor-mediated endocytosis [31]. Therefore, bioavailability of vitamin B_{12} is affected by several factors and highly dependent on the healthy functioning of the gastrointestinal system. In a clinical study conducted by Vitetta et al. a liposome oral spray formulation of B_{12} consisting of 1000 µg B_{12} per dose, was shown to increase the baseline serum levels of B_{12} by 14% compared to 1000 µg B_{12} tablet in 6 h in healthy adults [32]. A similar study used a sublingual formulation of the liposomal B_{12} supplement which delivered 1000 µg of methylcobalamin per day to patients. In total 53 participants in the study, vitamin B_{12} was increased by about 54.5%, 105.51%, and 269.66% after 1 week, 1 month and 2 months, respectively. Moreover, metformin did not induce any decrease in B_{12} serum levels in those patients taking the drug for blood sugar control which highlights the role of liposomal formulation in maintaining the B_{12} absorption [33]. Previous studies showed that those treated with metformin might experience decreased vitamin B_{12} absorption from intestine [34]. Additionally, liposomal formulations showed faster absorption than the free vitamin B_{12} [35]. Currently, in the market, there are several liposomal formulations of vitamin B family supplements, some of which also contain other vitamins from different groups. Table 8.2 introduces some of the common brands in this category.

8.3.2 VITAMIN C

Vitamin C, also known as ascorbic acid, is an enolic derivative of alpha-ketolactone and has a similar structure with glucose. As a reversible biologic reducing agent, vitamin C plays a critical role in the normal functioning of several enzymes possessing copper or iron in their structure. It is abundantly found in various food sources such as citrus fruits, spinach, potatoes, sprouts, tomatoes, strawberries, cabbage, and brussels sprouts [36]. The distal small intestine is the main site of absorption for vitamin C, where it is transported through an energy-dependent active transport process [37]. Absorption capacity of human intestine for vitamin C allows doses up to 100 mg/day to be almost completely absorbed; however, as ingested concentrations increase, smaller amounts are absorbed; doses more than 1000 mg/day, which are available as supplements, are not able to be absorbed more than 50%. Liposomal delivery systems of vitamin C provide a faster and greater blood concentration following oral administration. In a study done by Davis et al., 4 g of vitamin C, either encapsulated within phosphatidylcholine (PC) liposomes or unencapsulated, was orally administered to 11 adults within the range of 45–70 years. Plasma vitamin C concentrations of participants after two, three, and four hours, were greater following oral administration of vitamin C encapsulated within liposomes than placebo and unencapsulated vitamin C, indicating greater bioavailability for liposome-encapsulated formulation [19]. As blood is the main medium for vitamin C to be distributed among the tissues, the study highlights the role of liposomal formulations in clinical applications. As if vitamin B, liposomal formulations of ascorbic acid are available in the market which are introduced in brief in Table 8.3.

TABLE 8.2   Common Products of Liposomal Vitamin B Family in the Market

Brand Name	Components	Manufacturer
Lypo-Spheric® Vitamin B	Vitamin B complex + phosphatidylcholines + essential phospholipids from soy lecithin	LivOn Labs
Liposomal active methylated B-complex	Vitamin A, vitamin C, vitamin D, vitamin E, vitamin K1, vitamin K2, and vitamin B complex	Core MedScience
B-complex	Vitamin B complex + para-aminobenzoic acid (PABA)	Doctor's Formulas
Biomax® activated VIT B.CO	Vitamin B complex + 5-methyltetrahydrofolate (the active form of folate)	Coyne Healthcare
Liposomal B-complex	Vitamin B complex + PABA	CureSupport
Liposomal methyl B-complex	Vitamin B complex + milk thistle seed extract	Quicksilver Scientific
Altrient liposomal Vitamin B complex	Vitamin B complex + cinnamon and phospholipids	LivOn Labs
Liposomal liquid Vitamin B complex	Vitamin B complex + phospholipids from sunflower	Yoga Nutrition
Liposomal bioactive B complex	Vitamin B complex	BioBalance

TABLE 8.3   Common Products of Liposomal Vitamin C in the Market

Brand Name	Components	Manufacturer
Lypo-Spheric® Vitamin C	Vitamin C+ phosphatidylcholines + essential phospholipids from soy lecithin	LivOn labs
Mega liposomal Vitamin C™	Vitamin C + Stevia leaf extract	Aurora nutrascience
Liposomal Vitamin C	Vitamin C + Paprika extract	Dr. Mercola
Liposomal Vitamin C liquid	Vitamin C	Pure encapsulations
Optimal liposomal Vitamin C	Vitamin C	Seeking health
Liposomal Vitamin C	Vitamin C	Quicksilver scientific
Altrient liposomal Vitamin C	Vitamin C + Lecithin phospholipids	LivOn Labs
Liposomal C	Vitamin C	Bioceuticals

8.3.3   VITAMIN A

Vitamin A is a family of lipid-soluble compounds commonly known as retinoic acids. The family consists of four isoprenoid molecules connected in a head-to-tail manner. Intestinal absorption of vitamin A is dependent on the vitamin source as plant-based foods are rich in provitamin A which needs to be cleaved to retinol in order to be absorbed through intestine wall, while vitamin A from animal sources is in the form of preformed vitamin A (i.e., retinol, retinal, retinoic acid, and retinyl esters) which is subject to hydrolysis into retinol within the small intestine lumen. Without regard to the primary source of digested vitamin A, formed retinols are re-esterified joining the chylomicrons and transfer within the blood plasma. Chylomicrons are afterwards crushed into several remnants that contain retinyl esters such as apolipoproteins B and E [38]. Vitamin A plays multiple roles in human physiological processes, including growth, development, immune system, and retina [39]. Rhodopsin, a molecule needed for color vision, is generated by a combination of vitamin A with the protein opsin [40]. Additionally, vitamin A is involved in maintaining surface tissues such as lungs, bladder, intestine, and skin [41]. Sachaniya et al. reported a liposomal formulation of vitamin A with an enhanced pharmacokinetic profile compared to free vitamin for the treatment of osteoporosis. They used PC liposomes to encapsulate vitamin A and evaluated the prepared nanoformulation in terms of release kinetics and cell proliferation in SaOS-2 osteoblast cell line. Nano-liposomes containing vitamin A were able to control the release rate of the entrapped material within the 24 h following oral administration in a way that only 50% of vitamin A was released while free vitamin showed about 95% release during the same period. Results of proliferation of osteoblast cells caused by treatment with either free or encapsulated vitamin A revealed that only liposomal formulations were able to induce an increase in cell proliferation (around 1.2-fold) after 24 h, which was further raised to 1.5-fold increase after 72 h [42]. Long-term stability remains one of the challenges in the development of many liposomal formulations; however, vitamin A-encapsulated liposomes from the previous study demonstrated no physicochemical changes during 1 month of storage at 4°C, which indicated the feasibility of liposomal formulation of vitamin A to be commercially available in the market. Moreover, liposomal formulations of retinol cause less lytic effects on RBC in the blood, which in the case of free retinol is a major drawback resulting in significant toxicity [43].

8.3.4   VITAMIN D

Vitamin D is a subclass of fat-soluble vitamins consisting of a group of secosteroids. In humans, the most important members of this family are vitamin D2, also known as ergocalciferol, and vitamin D3, cholecalciferol. Although vitamin D is naturally synthesized through a UV radiation-dependent reaction in the skin epidermis, consumption rate often overrides the production capacity (that is highly variable in individuals) and thus, dietary supplementation of vitamin D is recommended to avoid subclinical vitamin D deficiency. Vitamin D provided either from diet or skin synthesis needs to be activated through a two-step process involving the liver and kidneys. Adequate levels of Vitamin D in the blood plasma are necessary due to the significant clinical role the vitamin has in both calcium homeostasis and bone metabolism [44]. Various forms of vitamin D supplements are available for the treatment of vitamin D deficiency among which cholecalciferol and ergocalciferol are more commonly prescribed to patients since the cost is lesser than activated forms of vitamin D. The data upon benefits of liposomal vitamin D supplementation is not consistent enough to draw a firm conclusion; however, multiple companies worldwide manufacture liposomal vitamin D supplementations which are allegedly superior to the conventional vitamin D supplements mostly in terms of pharmacokinetic properties. Products claims are mostly based on pharmacological concepts as nano-liposomes are could be readily absorbed via passive absorption across transmucosal membranes. These theoretically assumptions, albeit lack clinical and empirical basis, has made a fertile ground for pharmaceutical companies to develop lipid-based vitamin D products, some of which are introduced in Table 8.4.

TABLE 8.4   Common Products of Liposomal Vitamin D in the Market

Manufacturer/Product	Ingredients
Liquid health naturals/liposomal vegan Vitamin D3	Vitamin D3, ACTIVAloe aloe vera juice, vegetable glycerin (USP), natural flavoring, potassium sorbate, citricidal grapefruit seed extract.
Dr. Mercola/liposomal Vitamin D3	Vitamin D3, sunflower lecithin, hydroxypropyl methylcellulose, medium-chain triglycerides, sunflower oil.
Doctor's formula/VITAMIN D3 2500 IU and K2 200 mcg	Vitamin D3, vitamin K2 and a complex of phosphatidylcholines
ACTINOVO/liposomal Vitamin D3 + K2	Purified Water, phospholipids, maltodextrin, Vitamin D3, Vitamin K2-MK7 and natural sea buckthorn-extract
Bioceuticals/liposomal D3	Colecalciferol, lecithin, tocopherols concentrate-mixed (high-alpha type), purified water, glycerol, ethanol, tocofersolan (vitamin E polyethylene glycol succinate), orange oil, potassium sorbate.
NOW/Vitamin D-3 and K-2 liposomal spray	Vitamin D3 (as Cholecalciferol) (from Lanolin), Vitamin K2 (as Menaquinone), de-ionized water, vegetable glycerin, xylitol, MCT oil (medium-chain triglycerides), natural fruit flavors, phospholipids, rosemary extract, potassium sorbate and citric acid.
Henry blooms/VitaQIK™ liposomal D3 and K2	Colecalciferol, menaquinone, sugar, lactose, gluten, wheat, milk derivatives, animal products and artificial colors and flavors, ethanol, potassium sorbate, sodium benzoate.

8.3.5 VITAMIN K

Vitamin K family is composed of structurally similar compounds sharing 2-methyl-1,4-naphthoquinone nucleus with differing lengths of carbon side chains. The side chain in vitamin K1 is phytyl, designating the vitamin the synonymous name "phylloquinone," while vitamin K2 has several forms, differing in the isoprenoid side chain, indicated as MK-4 through MK-13 based on the length of the side chain. However, the most common form of vitamin K2 is MK-4 which is alternatively known as menatetrenone. Vitamin K absorption is a multi-step process requiring normal functions of pancreatic enzymes, bile salts, and intestinal lymphatics. Firstly, protein-bound vitamin K from diet is released by the proteolytic effect of pancreatic enzymes. Secondly, in the presence of bile salts, mixed micelles are formed, which then are incorporated into chylomicrons assisting intestinal absorption into the lymphatic system. The free vitamin K is eventually available bound to very-low-density lipoprotein as it circulates within the body [45]. Vitamin K plays a crucial role in coagulation pathways as well as bone formation [46]. Dalmoro et al. reported a polymer-coated liposomal formulation of vitamin K2 as a novel micronutrients delivery model for nutraceutical aims. They used L-α-PC from soybean as the main component of the vesicles and chitosan with medium molecular weight as the coating agent. As per the study results, vitamin K2 was efficiently encapsulated within the liposomes with an entrapment efficiency of 94.7%. Addition of chitosan at 0.005% (w/v) concentration to the surface of nano-liposomes resulted in a fairly high mucoadhesiveness of around 80%, which is in agreement with the previous studies suggesting mucoadhesive properties for chitosan polymer and chitosan-coated nanoparticles. Oral drug delivery systems equipped with mucoadhesive agent were demonstrated to show promising potentials in the delivery of various therapeutic/supplementary compounds. Moreover, stability study of the nano-liposomes showed no significant change in physicochemical properties of vitamin-loaded nano-liposomes over the 1 month of storage [47]. Unlike vitamin B and C, there is not much data available upon oral liposomal delivery of vitamin K either in simulated studies or those using animal models. Nevertheless, current evidences as well as pharmacological perspectives suggest a better pharmacokinetic profile with enhanced clinical outcomes for liposomal formulations of vitamin K as compared to free vitamin K conventional oral dosage forms.

8.3.6   VITAMIN E

A total of eight compounds, including four tocopherols and four tocotrienols, compromises the vitamin E family. However, the primary form of vitamin E that is active in the body is alpha-tocopherol which is itself a group of eight isomers with three chiral centers in their structures. The RRR-isomer is commonly known as the natural source of vitamin E, yet dietary supplements possess all eight forms with approximately one-half of the activity of natural-source vitamin E. As an antioxidant, vitamin E plays an important role in scavenging free radicals [48]. Numerous clinical studies have established the role of antioxidants, including vitamin E in the prevention of chronic diseases; thus, vitamin E supplementation has a notable role in clinical practice [17(b), 49]. Oral supplements of vitamin E are commonly in the form of tocopheryl esters which are resistant to oxidation; however, required to be unesterified prior to absorption. Thus, absorption of supplemental vitamin E is dependent on the function of intestinal esterases, the fact that limits their absorption in terms of both rate and capacity. On the other side, the active form of α-tocopherol is too viscous to be soluble in water and thus is prone to oxidization by atmospheric oxygen [50]. Hence, liposomal formulation of vitamin E offers advantages to modify the pharmacokinetic profile of the dietary supplement. Nacka et al. used n-3 polyunsaturated fatty acids (PUFA)-rich liposomes to overpass parameters affecting absorption of vitamin E. In their study, vitamin E-contained liposomes were administered to rats through a gastric feeding tube followed by collection of the lymph for the next 24 h. Vitamin E recovery from rats' lymph showed two-times higher recovery in rats treated with liposomal formulations compared with those treated with sardine oil, both containing 0.01 mol% α-tocopherol [51]. Currently, a number of liposomal vitamin E supplements are available in the market, most of which are claimed to have superior bioavailability to the conventional α-tocopherol supplements.

8.3.7   CHOLINE

Choline is an essential nutrient naturally occurs, in the form of PC, in various food sources such as egg, salmon, broccoli, soybean oil and chicken liver. From structural view, choline is basically a family of quaternary ammonium compounds with choline hydroxide commonly known as the choline base. In humans, choline is produced through the phosphatidyleth-anolamine N-methyltransferase (PEMT) pathway, however, the pathway production capacity does not meet the body requirements, and thus adequate amounts must be obtained from the diet [52]. The compound has been a subject of debate as to whether choline is classified in the vitamins family or be treated as an independent compound. From some viewpoints, it is similar to B family vitamins as along with folate, riboflavin, and vitamin B_{12}, choline is needed for methyl-group transfer [53]. Bioavail-ability of choline is highly dependent on both individual factors as well as the structure of choline in different sources. A recent study evaluates the plasma kinetic profile of choline following oral administration of two different choline sources: choline bitartrate salt and choline in the form of PC present in SuperbaBoost™ krill oil. Blood samples of 18 healthy adults were collected before and after intake of the study products at distinct time points to draw concentration-time curves of free choline and its metabolites in plasma. Compared to free choline bitartrate salt, higher choline levels were observed in those administrated with SuperbaBoost™. Area under curve (AUC) during the first 12 h after administration was recorded 26.07 (μmol/L)*h for SuperbaBoost™ while choline bitartrate salt resulted in 23.28 (μmol/L)*h for the same parameter. Maximum concentration was also higher in SuperbaBoost™ group with 4.21 μmol/L compared to 3.67 μmol/L in the choline salt group. Moreover, blood samples within the SuperbaBoost™ group showed greater concentrations of betaine and dimethylglycine, two of the main metabolites of choline, than choline salt subjects [54]. Considering the lack of evidence upon direct comparison of liposomal choline formulation versus free choline supplements, the results of the mentioned clinical study might suggest potential advantages brought in by liposomal formulations of choline as well as PC.

8.4 HERBAL EXTRACTS

For thousands of years, herbal extracts have been used because of their dietary, cosmetic, and therapeutic properties in all over the world. Extracts are multi-component mixtures obtained from herbs. There are different acquisition methods for obtaining the phytochemicals from the whole plant or its parts, such as extraction, expression distillation, fractionation, purification, etc., [55]. The main advantage of using extracts over raw herbs is improvement of oral bioavailability. Low bioavailability, low aqueous solubility, and substantial first-pass metabolism are the limiting factors for clinical application of phyto-components with proven therapeutic potential [56]. Novel drug delivery system (NDDS) for herbal constituents is a step beyond plant extracts for overcoming phytochemical's clinical limitations. NDDSs include: encapsulated or phospholipid combined forms (liposomal or phytosomal forms), polymeric nanoparticles, nanoemulsions, micelles, etc. The main categories of phytochemicals and their sub-classified groups are described in Figure 8.2. Among all phytochemicals, the health-promoting ones are flavonoids and other phenolic compounds. Thus, it seems fundamental to design suitable delivery systems for the delivery of sufficient amounts of such active ingredients to the site of action.

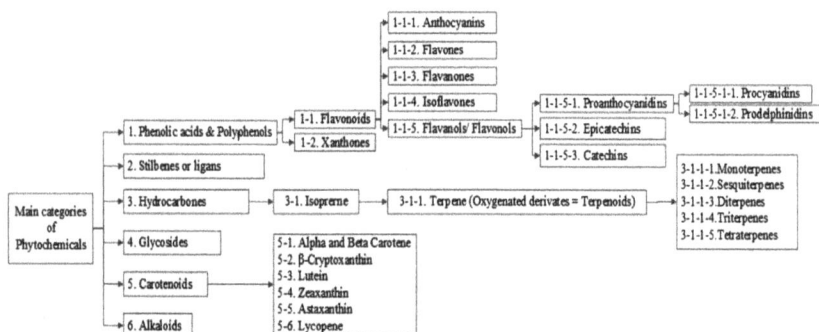

FIGURE 8.2   Classification of phytochemicals.

There are three major obstacles in the way of food industry to gain the nutritional benefit of freshly harvested plants from the oral route of administration. First: the destructive effect of processing techniques such as cooking on phytochemicals, second: the huge multiple-ring molecular structure of most of the phytochemicals that limits its transportation across the lipoid cell membranes, third: low lipid solubility and the inability of these phyto-constituents to be actively absorbed through intestinal cells and their strong tendency to self-aggregate [55, 57]. Due to the mentioned limitations, the clinical use of many potentially effective phytochemi-cals is restricted. However, clarification of these hurdles is achievable by means of encapsulation methods. In the phytopharmaceutical area, encapsulated phytochemicals (liposomal or phytosomal forms) have gained unprecedented popularity in the market, and it is mainly owing to the increased bioavailability as well as the straightforward preparation techniques.

8.4.1   LIPOSOMES AS PHYTOCHEMICALS CARRIERS

The application of liposomes in herbal formulations is wide, and it is evident from the studies that liposomes release herbal phytochemicals in a sustained manner with reducing peak-valley fluctuations. Targeted delivery is another advantage of using liposomes by improving the solubility and bioavailability. The maintenance of hydrophilic-hydrophobic structure in liposomes made them a great choice for wide-range phytochemical targeting, whereby inner aqueous core is the appropriate location for hydrophilic components and the phospholipid bilayer supplies the second lipophilic space. By means of manipulation of the phospholipid bilayers of liposomal structures such as surface modification, it is possible to change its properties such as surface charge, composition, and its size [57(b), 58].

In one study using *Santolina insularis*, liposomes were able to contain more than 95% of the compound over a year at 4–5°C [59]. Chemical stability of *Anethum graveolens* was significantly increased by encapsulation within the multilamellar vesicles (MLV), indicating the role of liposomal membranes in protection of entrapped materials [60]. Thermal-induced stability of *Zanthoxylum tinguassuiba* was enhanced by the use of liposomes as the drug carrier [61]. Salidroside, a glucoside found in the plant *Rhodiola rosea*, is reported to increase stress resistance in several models. The compound also possesses antitumor activity through tumor-suppressive kinases. Treatment with 0.8% salidroside is connected with enhanced activity of T cells and overall immunity. Nevertheless, clinical application of salidroside is limited due to its poor low bioavailability. Fan et al. reported a liposomal formulation of salidroside prepared by the melting method with improved physicochemical properties which might cause a meaningful increase in salidroside bioavailability [62]. In another study conducted by Takahashi et al. curcumin (CUR) was encapsulated within liposomes in order to increase the CUR absorption following oral administration. *In vivo* studies in rats revealed that CUR -loaded liposomes showed a faster rate absorption with higher C_{max} achieved at earlier time compared to free CUR or a mixture of CUR and lipids. Data from

concentration-time curve indicated higher area-under-curve value for
CUR-encapsulated liposomes at all points. Antioxidant activity of CUR
as a pharmacodynamical marker, was significantly greater following oral
administration of liposomes containing CUR than free drug, the fact that
proposes liposomal formulation of CUR as a promising nutrient delivery
system [63]. Liposomes might also be employed to reduce the toxicity
of therapeutic compounds. Resveratrol (RESV), a type of natural phenol
with multiple clinical benefits in various diseases, induces cytotoxicity
when higher total dosages are used, while studies suggest that accumulation of the drug in the site of disease is required for a suitable clinical
outcome. High local concentrations of RESV in tissues causes notable
cytotoxicity in animal models. Besides, RESV as if many other natural
phenols, is sacredly soluble in water and highly labile to oxidation. Thus,
formulations with better solubility as well as more biocompatibility are
of interest. Isailović et al. prepared RESV-incorporated liposomes via
the thin-film method and evaluated their physicochemical properties as
well as the toxicity profile using morphological changes of keratinocytes
treated with liposomes. As per study results, compared to free RESV
solution, liposomal formulations with the same concentration of RESV
induce significantly lower cytotoxicity, whereas antioxidant effects of the
drug were preserved and equal in either free RESV solution or liposomes
[64]. Table 8.5 introduces some liposomal herbal compounds and the main
advantages of their formulations.

TABLE 8.5 Liposomal Phytochemicals

Phytochemical	Advantages of Liposomal Formulation	References
Curcumin	Cationic liposome-PEG-PEI complex (LPPC) of Curcumin increased the accumulation of active component in the cells, as well as supplying the prolonged circulation time.	[65]
Ginsenoside	The entrapment efficiency (EE) of positively charged liposomes encapsulating Ginsenoside (stearyl amine as a positively charge inducer) were superior over neutral and negatively charged ones.	[66]
Tripterygium wilfordii extract	Heat stability was improved along with a reduction in side effects.	[67]
Quercetin	Therapeutic efficacy was improved along with a reduction in side effects.	[68]

TABLE 8.5   *(Continued)*

Phytochemical	Advantages of Liposomal Formulation	References
Silymarin	Oral bioavailability was improved by means of increasing gastrointestinal solubility aggregation of herbal compound was prevented by maintaining liposomal stability through encapsulation as hybrid liposomes.	[69]

8.4.2   *PHYTOSOMES AS PHYTOCHEMICAL CARRIERS*

Phytosomes are liposome-like vesicles with unique features (physico-chemical and spectroscopic) that could deliver the phytochemicals, especially polar polyphenolics in herbal extracts [57(a)]. Their formulation method is obtained by incorporating the standardized plant extract or the phyto-active compounds into phospholipids. The development of phytosome complexes dates to the late 80s at Indena (Milan, Italy), and its first applicability was in cosmetic delivery systems [57(b)]. Phytosomes maintain a hydrophilic-hydrophobic structure that makes them a great choice for wide-range phytochemical targeting, whereby inner aqueous core is the appropriate location for hydrophilic components and the phospholipid bilayer supplies the second lipophilic space for hydrophobic agents [57(b), 58]. Phytosomes have lots in common with liposomes like vesicular and phospholipid-based structure (Figure 8.3). Besides the similarities, the distinguished features are described in Table 8.6 [57(a), 70].

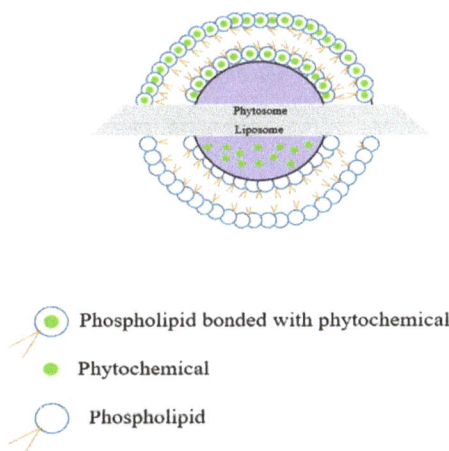

FIGURE 8.3   Schematic structure of phytosomes versus liposomes.

TABLE 8.6 Differences between Liposomes and Phytosomes

Characteristics	Phytosomes	Liposomes	References
Molecular linkages	Phytochemical is linked through chemical bonds to phospholipid molecules.	No chemical bond exists.	[55]
The molar ratio of phospholipid to phyto-active ingredient	1:1 or 2:1	Up to 10 times	[71]
The location of active phytochemical component	It is chemically bonded with the hydrophilic head of phospholipids and interacts via hydrogen bonds with the polar head of phospholipids (i.e., phosphate, and ammonium groups).	The hydrophilic compounds are located in the inner portion, while the lipophilic ones are inserted into the membranes	[55, 71, 72]
Entrapment efficacy (EE)	The drug itself is conjugated with lipids, forming the vesicles, as a result, the EE is higher.	In contrast, EE is lower.	[55]
The solvent in preparation method	In the past, just aprotic solvents (such as aromatic hydrocarbons, halogen derivatives, methylene chloride, ethyl acetate, or cyclic ethers, etc.), or solvents with dielectric constant such as acetone, dioxane, metyhlenechloride, hexane, and ethyl acetate, etc., were used; yet today they have been largely replaced by protic solvents like ethanol.	Usually, water or buffer solution is used.	[71, 72]

A major obstacle in the way of oral administration of water-soluble phytochemicals is their sensitivity to acidic conditions and bacterial degradation in the gut. Thus, a suitable formulation for most of the bioactive water-soluble phytochemicals like flavonoids, glycosides, tannins, terpenoids, etc., is phytosomal delivery systems. Hydrogen bonds are formed between the polar sections of the phytochemicals and the polar part of the phospholipids (i.e., phosphate groups) in phytosomes. By means of molecular complexation of water-soluble phytoconstituents with phospholipids, it is possible to gain the higher oral bioavailability. Stability enhancement of chemically bonded phytoconstituents with PC molecules is the most prominent advantage that have attracted the formulators' attentions towards phytosomal delivery systems [57(a)].

It is evident that by entrapping the phytochemical in phytosomal structure, which leads to enhancement of drug loading and the ability of phyto-components to cross cell membranes, the dosing intervals and bioavailability of active component will increase successfully. Furthermore, protection against bacterial and digestive secretions of GI-tract is another significant advantage of phytosomal delivery systems. Targeted delivery by means of surface modifications is also noticeable as it could provide efficient delivery systems [57(a)]. Methods of phytosomes preparation may include modified methanol reflux [73], polycarbonate membrane extrusion (film-dispersion method) [66], thin-film hydration [74], reverse evaporation technique [75] and solvent injection [76]. The practical use of nanotechnology has attracted attention in phytosomes preparations as it could provide a simple fabrication method promoting commercial-scale formulations [77, 78]. Some commercial herbal products encapsulated in liposomes or phytosomes have been introduced in Table 8.7.

TABLE 8.7   Commercially Available Liposomal or Phytosomal Herbal Extracts

Brand Name	Ingredients	References
Ginkgoselect®	Flavone glycosides, terpenoids, tannins, and lipids	[79]
Meriva®	Curcumin-soy phosphatidylcholine complex	[80]
Greenselect ® phytosome	Epigallocatechin 3-O-gallate from green tea	[72]
Silybin® phytosome	Silybin from silymarin	[72]
Glycyrrhiza® phytosome	18-beta glycyrrhetinic acid	[72]
Leucoselect®	Procyanidins from Vitis vinifera	[72]
Merinoselect®	Polyphenol from *Curcuma longa* (Curcumin)	[72]
Oleaselect®	Polyphenols from olive oil	[72]
Sabalselect®	An extract of saw palmet	[72]
PA2 phytosome horse®	Proanthocyanidin A2 from Chestnut bark	[72]
Zanthalene®	Zanthalene from *Zanthoxylum bungeanum*	[72]
Centella®	Terpenes	[72]
Hawthorn®	Flavonoids from *Crataegus* sp.	[72]

8.4.3   ESSENTIAL OILS (EOS)

Essential oils (EO) are natural, volatile compounds founded as secondary metabolites in aromatic plants. A wide range of biological activity is attributed to EOs in the literature such as antioxidant, antifungal, bactericidal, and anticancer activity [81]. However, essentials oils are prone to oxidation, temperature, and light, and upon exposure, irreversible chemical changes occur which affect their normal function. For example, b-caryophyllene in exposure to air is subject to oxidation turning into caryophyllene oxide. Additionally, EOs are poorly soluble in water. Thus, encapsulation of the EOs within lipid-based delivery systems might be beneficial to enhance the physicochemical properties of the EO compounds. EO-loaded liposomes might be prepared through widely-known liposome preparation techniques, including thin-film hydration method, sonication, freeze-thaw, supercritical solution technique and precipitation from gas saturated solution. Liposomal formulations of EO are reported to be stable for a long period of storage. EOs driven from *Origanum dictamnus* L. was encapsulated within liposomes of phosphatidylcholine and evaluated in terms of antibacterial activity against both Gram-positive and negative bacteria as well as human pathogenic fungi and antioxidant capability using differential scanning calorimetry. As per study results, higher antioxidant action was observed with liposomal formulations compared to free compounds in their pure form. Carvacrol, one of the main components of the oil, presented significantly higher antibacterial effect when encapsulated within the liposomes. Liposomal formulations containing as low as 25.0×10^{-8} g/ml carvacrol had equal antibacterial activity to 6.0×10^{-3} g/ml of the free compound [82]. In another study, essential oil E. camaldulensis encapsulated within soya lecithin and cholesterol (Chol)-based liposomes showed significantly improved antifungal activity *in vitro* [83]. Frankincense and myrrh oil (FMO) have multiple biological activities such as antimicrobial, anti-inflammatory, and antitumor properties; however, due to the poor water solubility and instability of FMO, oral route does not provide the sufficient bioavailability for FMO to be clinically effective and thus novel formulations possessing the desired pharmacokinetic properties would be useful. Shi et al. prepared solid lipid nanoparticles (SLN) containing FMO and assessed their antitumor efficacy in H22 tumor-bearing mice. Animals were subjected to oral administration of saline, FMO suspension, voided SLN, and FMO-SLN, for 10 consecutive days.

Results of the study revealed that neither saline nor voided SLN induced any meaningful inhibition in tumor growth, whereas all other treatment modalities demonstrated significant antitumor effects. At FMO concentration of 100 mg/kg, the inhibition rates of the FMO suspension, and FMO-SLNs were recorded 31.23% and 43.66%, respectively, indicating the role of liposomal encapsulation of FMO in significant improvement in antitumor efficacy *in vivo* [84]. Zhao et al. reported a self-nano-emulsified delivery system (SNEDDS) containing Zedoary turmeric oil with 1.7-fold and 2.5-fold increase in AUC and Cmax of the main metabolite of the oil, respectively in oral administration of SNEDDS in rats [85]. A similar observation was made with Zeodoray turmeric oil-loaded SNEDDS in another study conducted by Xi and coworkers [86].

8.5   ESSENTIAL FATTY ACIDS

Essential polyunsaturated fatty acid (PUFAs), like omega-3-fatty acids (ω-3 FAs) are nutrients that human cells cannot produce them in sufficient quantity, so diet and supplementation play a prominent role in providing human needs. DHA (Docosahexaenoic acid), EPA (eicosapentaenoic acid) and ALA (alpha-linolenic acid) are three major ω-3 FAs. Common food sources of ω-3 FAs are fatty fishes (salmon, sardines), flaxseed, microalgal oil, and walnuts [87]. Because of the presence of contaminants such as polychlorinated biphenyls (PCBs) and mercury in seafood compounds, the usage of omega 3 fatty acid supplements is highly recommended by FDA (Food and Drug Administration) [88]. The diversity of pharmacologic effects of DHA and EPA is evident in extensive studies. Some reported activities are anti-inflammatory, immunomodulatory, antioxidant, anti-obesity, anti-hyperlipidemia (hypertriglyceridemia) and anti-diabetic as well as cardioprotective and neuroprotective activities [89].

In order to elevate the ω-3 FA intake, supplementation of the diet with fish oil capsules seems to be the easiest way. Another way, which attracted the consumer's interest, is food fortification. The biggest challenge of fortification, nevertheless, is undesirable flavors and instability of fish oil during processing [90], which is the result of autoxidation of PUFAs leading to rancidity and shorter shelf life [91]. Today, fortification of ω-3 to foods (like chocolate, yogurt, bread, cheese, milk, pasta, bakery products, mayonnaise, margarines, spreads, egg, and egg products

[90, 92–94]) is carried out by means of liposomes, as an encapsulation technique. These liposomal functional foods have the ability to mask the undesirable sensory characteristics and protect the oils against oxidation. Using appropriate encapsulation process and materials have a crucial role in stability of ω-3 FA. Size reduction improves the oxidative stability of encapsulate ω3 PUFAs, especially in liposomal forms [95]. However, because of the low sensory threshold of off-flavor, the enriched products can be rejected by slight oxidation of ω3 PUFAs [92]. Table 8.8 provides encapsulation technologies of PUFAs along with their properties.

TABLE 8.8   Encapsulation Technologies of PUFAs

Method	Properties	References
Spray-drying (spray-dried emulsions)	• Most commonly used microencapsulation technique with the ability to control the mean particle size of the powders for spray-dried emulsions.	[96]
	• Low operational cost (economical) and easy to scale-up.	
	• The process is flexible, efficient, and uses easily available equipment.	
	• Oxidation of oil could be occurred due to high temperature used in the drying process.	
	• Only a limited number of materials are compatible with this technology.	
Freeze-dried emulsions	• The removal of oxygen and the application of low temperatures minimize product oxidation.	[97]
	• Suitable technology for the microencapsulation of highly sensitive ingredients, such as probiotics and PUFAs (Polyunsaturated fatty acids).	
	• Due to creation of highly porous, irregular, and flake-like structures, short shelf-life of freeze-dried microencapsulated fish oils was observed	
Spray granulation	Spray granulation is an efficient method to produce fish oil powders because of its low drying temperature (max. 70°C).	[98]
Carbohydrate/ sodium caseinate wall systems	It has been shown that HBCD (highly branched cyclic dextrin) is useful to increase the oxidative stability of lipids.	[99]

TABLE 8.8   *(Continued)*

Method	Properties	References
Nano-encapsulation	• Stability enhancement of reactive, sensitive, or volatile additives (vitamins, enzymes, antioxidants, slimming agents, etc.). • Significant reduction in acidity, syneresis, and proxide value while increasing DHA and EPA stability. • Sensory evaluation studies showed unpleasant fishy flavor in nano-liposomal formulations.	[90, 92, 100]
Mozafari method	• Non-toxicity of produced liposomes • Rapid production • Scalability	[95(b)]
Probe ultrasound of pre-formed liposomes	Improvement of loading capacity of DHA and EPA into the nanoliposomal membranes.	[101]
Simple and Complex coacervation	Simple coacervation: • In simple coacervation, the wall material is limited to gelatin. Complex coacervation: • In Complex coacervation, two oppositely charged polymers are used. • Complex coacervation is known as the most promising technologies for stabilization of omega-3 oils, and it is mainly owing to: o Highest entrapment efficacy o Small particle size o Low surface oil concentration, which is responsible for maintaining the sensory properties of the complex coacervate omega-3 powders during storage • Process-based parameters such as pH, temperature, cooling, and solidification rates play a significant role in size and the shape of the cells in the coacervation method.	[96(d), 102]
Inclusion complexation	The cyclodextrins are used as encapsulants, and provide protection against oxidation.	[103]
Liposome entrapment	High-cost low stability are two limiting factors in indication of this method for production of microencapsulated omega-3 fatty acids.	[104]

TABLE 8.8   *(Continued)*

Method	Properties	References
Electrospraying	A promising technology for food industry with the advantage of improvement of omega-3 oxidative stability	[105]
Spray granulation and fluid bed film coating	The coating agent was hydroxypropyl beta-cyclodextrin, which it could not provide the adequate prevention of lipid oxidation.	[106]
Encapsulation using ultrasonic atomizer	Emulsion preparation step is eliminated, and consequently, the oxidation rate will reduce. However, its low encapsulation efficiency clarifies the need for further studies.	[107]

The unsaturated nature of PUFAs results in susceptibility to oxidation. Two types of damage, which are common for lipid structures, are oxidative and hydrolytic degradation. The onset of degradation in oxidative reactions is much earlier than hydrolytic ones. Oxidative deterioration of omega-3 fatty acids will reduce their nutritional quality and sensory properties, yet encapsulation methods like liposomal forms of these fatty acids can protect them against oxidative factors by creating a barrier between PUFAs and the environment. Furthermore, the antioxidant materials like tocopherol (Vitamin E), Chol, and Tempol (a stable cyclic nitroxide free radical) preserve the PUFAs against oxidation. The protective role of Chol in liposomal forms is achievable by its hindering effect on hydration in the lipid bilayer [88]. Studies on PUFA's liposomal forms are summarized in Table 8.9.

TABLE 8.9   Liposomal Formulations of PUFAs

Liposomal Form	Properties	References
Liposome with high DHA percentage	A liposomal formulation with high DHA% was prepared by the means of a salmon egg phosphatidylcholine.	[108]
Liposomes and nanoliposomes containing fish oil, with 40% DHA and EPA	The stability studies showed that the surface charge and stability of liposomes decreases with increasing size.	[95(b)]
DHA/EPA encapsulated nanoliposomes	By means of thin-film hydration and then ultra-sonification of pre-formed liposomes, nano-liposomes with significant loading of DHA and EPA were prepared.	[101]

8.6   ANTIOXIDANT COMPOUNDS

Oxidative stress is a state of an imbalance between generation and accumulation of reactive oxygen species (ROS) such as hydrogen peroxide, superoxide radicals, hydroxyl radicals, and singlet oxygen within the cells and tissues. ROS production is an inevitable part of the natural biological process in the cells and are technically byproducts of routine metabolic reactions taken place during energy consumption. Moreover, ROS might be produced in various stages of different types of pathological conditions. A considerable number of vital functions in human cells are dependent on ROS generation, including protein phosphorylation and differentiation [109]. Nevertheless, levels of ROS inside the cells have to be precisely controlled in order to prevent unwanted damages caused by these materials to vital cell organelles or cellular structures such as cell genetic contents, proteins, and lipids. Massive data available on oxidative stress reveal an association between initiation and progression of several diseases and cellular ROS stress level. Clinical trials in human as well as *in vivo* studies demonstrated either direct or indirect link between cellular ROS level and cancer, diabetes, cardiovascular, and metabolic disorders [110]. Thus, explicit regulation of ROS generation in human cells is required for cell survival. There are a number of enzymatic and non-enzymatic strategies to neutralize the effects of ROS and oxidative stress. Among which, endogenous antioxidants such as glutathione (GSH), coenzyme Q10, lipoic acid, and enzymes such as GPx (glutathione peroxidase), CAT (catalase), SOD (Superoxide dismutase) are most widely known examples in this category. Nevertheless, the natural capacity of cell organelles, especially mitochondria scavenging machinery, is not sufficient to completely sequester the free radical components within the cell environment and thus exogenous antioxidants outside the body are necessary to clear the remaining ROS. There are several animal- or vegetal-based antioxidants available in diet or nutritional supplements possessing protective effects for human health, including vitamins, phytonutrients, proteins, minerals, hormones, coenzymes, and some other uncategorized molecules [111].

Liposomal delivery systems own favorable features making them promising carriers for oral application of antioxidants. Liposomal carriers can facilitate intracellular delivery of both lipid-soluble and water-soluble antioxidants as well as antioxidant enzymes through various proposed mechanisms, including receptor-mediated endocytosis, phagocytosis, and enhancement in endosomal escape via fusion of liposomes with endosomal membrane [58(a), 112]. The following sections provide a general picture of some of the widely-studied liposomal antioxidants as well as their clinical potentials in the treatment of various diseases.

8.6.1   SUPEROXIDE DISMUTASE (SOD) AND CATALASE (CAT)

SOD is an enzyme responsible for the conversion of superoxide radical-sradicals into molecular oxygen and hydrogen peroxide, a reaction that is also known as dismutation. Since superoxide radical is abundantly produced as a byproduct of many natural metabolic processes within the cell, SOD is a stunning antioxidant defense in approximately all cell types [113]. However, it should be noted that hydrogen peroxide is also a harming molecule that has to be converted into its degraded form by other enzymes such as CAT, another crucial enzyme that protects the cell from ROS-induced oxidative damage [114]. Nonetheless, poor physicochemical properties of both SOD and CAT limit their practical application in humans. SOD is a charged high molecular weight that is not able to readily cross the cell membrane. Additionally, SOD triggers immunological responses within the plasma and is readily trapped within the reticuloendothelial system (RES), which significantly reduces the effectiveness of antioxidant treatment [115]. Petelin et al. reported a liposomal formulation of SOD, CAT, and both, for the treatment of gingival inflammation in dogs. Treatment modalities were classified into (1) supragingival scaling only (2) supragingival scaling followed by subgingival use of enzyme-loaded liposomes (3) supragingival and subgingival scaling and root planning and the last group (4) received supragingival and subgingival scaling and root planing followed by subgingival application of enzyme-loaded liposomes. The maximum suppression was observed in the group 4 where scaling and root planning was followed by liposomal treatment [116]. Data upon oral delivery of SOD and/or CAT is limited, but promising advantages of oral lipid-based delivery systems with other molecules might be cautiously extended to oral liposomal formulations of either SOD or CAT. Nevertheless, firm conclusions can only be drawn after empirical and clinical use of the said delivery systems.

8.6.2   CO-ENZYME Q10 (COQ10)

Co-enzyme Q_{10} (CoQ_{10}), also known as ubiquinone, is a naturally occurring benzoquinone that is abundantly found in cellular and inner mitochondrial membrane. It is one of the components of the mitochondrial electron chain and is essential for ATP production [117]. As a dietary supplement, it occurs in a wide range of human food such as fruits (oranges and strawberries), vegetables (cauliflower and broccoli), fatty fish (trout, mackerel, and sardine), oils (soybean and canola oil), and nuts and seeds (sesame seeds and pistachios). As an antioxidant, CoQ_{10} has been shown to be either directly or indirectly connected with various diseases including cardiovascular, Parkinson's, and migraine disease [118]. Higher levels of CoQ_{10} in myocardial cells protect the tissue from ischemia-reperfusion injury [119]. Practical application of CoQ_{10} suffers from high molecular weight and low solubility of the compound, thus delivery approaches using liposomal carriers is of clinical value [120]. Xia et al. developed a liposomal formulation of CoQ_{10} using egg yolk phospholipid to evaluate drug absorption in Caco-2 cell monolayers as the model of human intestinal epithelium. Results of the study revealed a higher apparent permeability coefficient of CoQ_{10} for liposomal formulations (4.19×10^{-6} cm/s) compared to free CoQ_{10} (1×10^{-6} cm/s) indicating a significant enhancement in CoQ_{10} transport through the gastrointestinal cells. Moreover, Tween-80 presented in liposomal formulation significantly decrease the anisotropy of DPH fluorescent probe, implying the increase in fluidization of the lipid components of the Caco-2 cell membrane. Cellular membrane fluidization effect of liposomal formulation also increases the permeability of CoQ_{10} [121]. In another study conducted by Chee Ho Choi et al. neutral, anionic, and cationic lipid formulations of CoQ_{10} were prepared using via film hydration technique and evaluated in case of oral delivery of CoQ_{10} in male rats. It was observed that neutral lipid formulation was able to achieve the higher plasma CoQ_{10} concentrations compared to anionic and cationic possessing the second and third place, respectively. This observation is mainly due to the electrostatic interaction between the liposome and enterocyte cell membrane. However, all three formulations were able to achieve CoQ_{10} maximum plasma concentrations more than 287 ng/ml [122]. Considering the low solubility of CoQ_{10}, liposomal formulations of CoQ_{10} seem to be promising options in clinical practice. Currently, there are several liposomal formulations in the market containing CoQ_{10} such as Liquid Q® LiQsorb®, Coenzyme Q10 Liposomal from Vitaplex and Dr. Mercola.

8.6.3 GLUTATHIONE (GSH)

Glutathione (GSH), a non-protein thiol greatly found in animal cells, plays an important role in endogenous antioxidant system. As an antioxidant, it is involved in the reduction of hydrogen peroxidase, protecting cells from detrimental effects of ROS [123]. Hydrogen peroxidase is a byproduct of many biological reactions within the cells and is proved to damage the cellular proteins, membrane lipids and DNA through oxidation; hence, GSH action in neutralizing agents is indispensable to maintain the balance of antioxidants and oxidants. Low levels of GSH are clinically proven to be associated with the initiation and progression of several diseases, including cardiovascular, cancer, and diabetes [124]. Nevertheless, oral administration of GSH is practically hindered by the inappropriate pharmacokinetic characteristics such as poor solubility, significant first-pass metabolism, limited ability to transfer the cell membrane, and rapid clearance from blood circulation [125]. Therefore, liposomal formulation of GSH might be beneficial to improve the pharmacokinetic profile of the enzyme, enabling the drug to be administrated in oral route. A recent clinical trial showed oral supplementation of GSH is effective in increasing GSH levels in a variety of cellular blood compartments; however, large doses of the drug is needed to maintain the adequate serum concentration since GSH is an acid-liable molecule which is extensively destroyed within the acidic environment of the stomach. Liposomal GSH, on the other hand, is able to protect the entrapped cargo from chemical degradation and can be regarded as one of the promising means of GSH delivery *in vivo*. In a recent clinical study conducted by Sinha et al. a novel liposomal formulation of GSH was administrated at two doses (500 and 1000 mg per day) to 12 healthy adults for one month. GSH serum levels was increased more than 20% one week after treatment and over 40% at two weeks for the 500 mg/d GSH group. Further analysis on immune function revealed that following oral administration of liposomal GSH at both doses, 60% increase in lymphocyte proliferation in less than one week was recorded, and natural killer (NK) cells cytotoxicity was enhanced up to 4-times within the first two weeks. These two findings are also in agreement with the results of previous clinical trials using non-liposomal GSH, demonstrating a strong correlation between GSH levels and NK cells activity [126]. In another study, GSH was loaded into solid lipid microparticles composed of Dynasan 114 and Dynasan 118 and evaluated in terms of prolonged oral delivery. Three formulations were prepared using different ratios of Dynasan 114 to Dynasan 118. Maximum drug loading was observed with the formulation composed of 80% Dynasan 114 and 20% GSH with drug loading of roughly 22% compared to those of other formulations, both with 5%. The obtained microparticles were subjected to dissolution studies in a simulated gastrointestinal fluid with distinct pH values of 1.2 as the stomach environment and 6.8, simulating the intestinal conditions. Free GSH was dissolved in less than 5 min in both simulated fluids, whereas liposomal formulations, albeit with little difference, demonstrated a prolonged-release pattern with an initial burst release of around 30% during the first 30 minutes followed by a plateau in the concentration of GSH with the function of time. Moreover, the antioxidant activity of liposomal formulations was significantly higher compared to free GSH. These observations are in line with the results of the earlier studies evaluating the role of GSH encapsulation within lipid particles in enhancing overall efficiency of GSH therapy *in vivo* [127].

8.7 CHALLENGES AND OPPORTUNITIES

Liposomes involved in oral delivery systems such as nutritional supplements maintain several valuable advantages such as safety and biocompatibility, encapsulating both hydrophilic and hydrophobic compounds simultaneously, improving nutrient solubilization and protecting active ingredients from environmental issues, and increasing absorption; however, there are some drawbacks that have restricted the vast application of liposomes in pharmaceutical industry [1, 4, 6, 12]. Drug leakage and fast release, instability upon storage and under gastrointestinal conditions (low pH and enzymes), poor loading capacity for hydrophilic bioactive components, lack of understanding of the interaction between liposomes and food components and the high cost of purified phospholipids and food-grade raw materials for liposomes preparation are among the most important barriers of liposomes application [4, 6, 12]. Therefore, overcoming these problems is the current challenge in liposomal research.

Liposomes can aggregate, fuse, and eventually precipitate over time. As degradation of liposome carriers occurs, the encapsulated materials are released, and the shelf life of liposomes is decreased during storage. There are several factors, including composition of particles, pH, ionic strength, encapsulated agents, light, temperature, and dissolved oxygen, that can reduce the stability of liposomal carriers upon storage conditions [4].

Liposome behavior is highly dependent on the composition and physical properties of the bilayers that makes it difficult to evaluate the liposome system about function under different stresses [8]. Another related issue upon stability is during gastric ingestion that could be improved with well-organized assembly of phospholipids and Chol and surface coating in liposomes restricting enzymes to contact with liposomal phospholipids [6].

Many attempts for improving the stability of liposomes have been considered including using of Chol and neutral long-chain saturated phospholipids; however, phytosterols obtained from plants are suitable substitutes for Chol in liposomes as they are appropriate for people with hypercholesterolemia and they could also provide several possible benefits for human health such as lowering blood Chol levels, anti-inflammatory, antibacterial, anti-atherosclerotic, anti-oxidative, anti-ulcerative, anti-tumor, and anti-carcinogenic functions. Another successful improvement in stability of liposomes could be achieved by modification with several agents such as poly(ethylene glycol) (PEG), poloxamer, polysorbate 80, chitosan, and dextran derivatives [4]. Freeze-drying could also increase the stability of liposomes if appropriate cryoprotectants are added in the formulation [128]. Polymerizing liposomes or using lipid compositions having strong inter-bilayer cohesiveness could increase the stability of liposomes against detergents and proteins [1]. Coating liposome surface with polymers can provide steric stabilization and reduce absorption of macromolecules on the liposome surface. These liposomes are called stealth liposomes due to their invisibility to the immune system [129].

It is worth mentioning that the viability of the economics in preparation processes of liposomes should be considered as an important factor in the suitability of liposomal carriers [4]. Reducing various steps in the production of liposomes can prevent the high cost of manufacturing processes. As an example, remaining solvents such as propylene glycol or ethanol present at concentrations below 5–10%, when liposomes are manufactured from alcoholic solutions, can act as a preservative, and eliminate the solvent removal step [1]. Industry has determined its efforts to find a large-scale and continuous production method for liposomes. Using high-pressure homogenizers (e.g., microfluidization) as a recent development in manufacturing processes has provided several advantages such as continuous production of large volumes of liposomes and reproducible manner without the use of detergents and solvents as they could induce potential health and instability problems. The development of microfluidization along with use of cheaper commercial phospholipid fractions may lead to economical production in liposomes industry [9]. Considering all mentioned above, further studies are needed to pave the way for bottleneck points in using liposomes as a carrier for nutraceutical products such as supplements and focus on the production of liposomes from low-cost materials and on liposomes with long-term stability and controlled release properties (Figure 8.4).

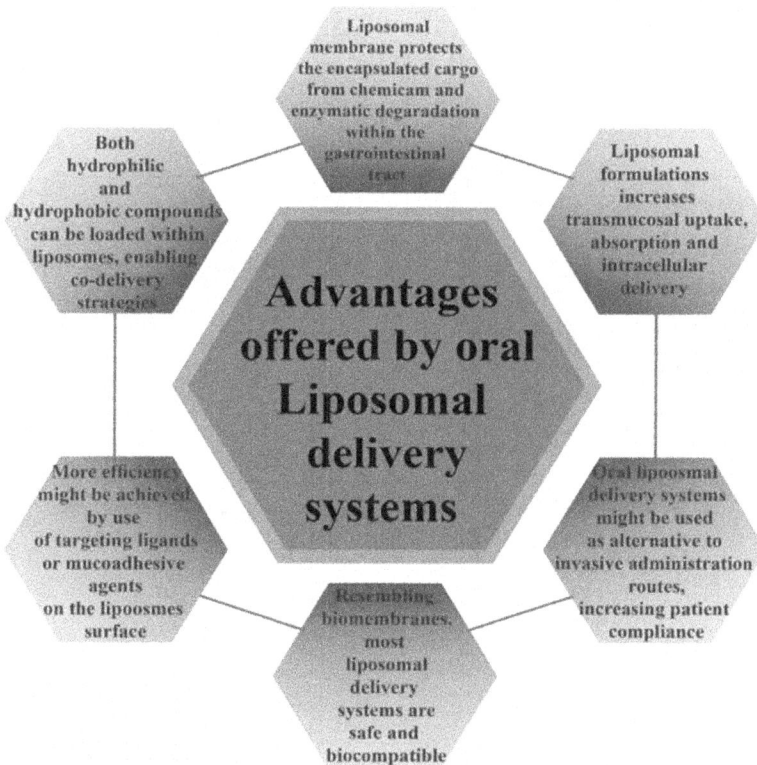

FIGURE 8.4   Advantages of oral liposomal delivery systems.

8.8 CONCLUSION

Due to several benefits including natural ingredients, entrapment efficiency, large-scale production, encapsulation ability of both hydrophilic and lipophilic agents, biocompatibility, and increasing absorption, liposome carriers could be widely used in oral delivery of nutritional supplements such as vitamins, minerals, herbal, and antioxidant compounds and essential fatty acids. However, there are some critical challenges in the application of liposomes, they can open a new door for researches in the supplement industry, and further studies have to focus on the production of lipid vesicles through safe, scalable methods by using low-cost ingredients.

ACKNOWLEDGMENT

The authors declare no conflict of interest.

KEYWORDS

- antioxidants
- essential fatty acids
- herbal extracts
- liposomal supplements
- minerals
- supplement industry
- vitamins

REFERENCES

1. Keller, B. C., (2001). Liposomes in nutrition. *Trends in Food Science & Technology, 12*(1), 25–31.
2. Deshpande, P. P., Biswas, S., & Torchilin, V. P., (2013). Current trends in the use of liposomes for tumor targeting. *Nanomedicine, 8*(9), 1509–1528.
3. Samimi, S., Maghsoudnia, N., Eftekhari, R. B., & Dorkoosh, F., (2019). Lipid-based nanoparticles for drug delivery systems. In: *Characterization and Biology of Nanomaterials for Drug Delivery* (pp. 47–76). Elsevier.

4. Emami, S., Azadmard-Damirchi, S., Peighambardoust, S. H., Valizadeh, H., & Hesari, J., (2016). Liposomes as carrier vehicles for functional compounds in food sector. *Journal of Experimental Nanoscience, 11*(9), 737–759.

5. Yuan, L., Geng, L., Ge, L., Yu, P., Duan, X., Chen, J., & Chang, Y., (2013). Effect of iron liposomes on anemia of inflammation. *International Journal of Pharmaceutics, 454*(1), 82–89.

6. Abd, E. A. H., Nafee, N., Ramadan, A., & Khalafallah, N., (2015). Liposomal buccal mucoadhesive film for improved delivery and permeation of water-soluble vitamins. *International Journal of Pharmaceutics, 488*(1, 2), 78–85.

7. Gibbs, F., S. K., Inteaz, A., Catherine, N. M., & Bernard, (1999). Encapsulation in the food industry: A review. *International Journal of Food Sciences and Nutrition, 50*(3), 213–224.

8. Singh, H., Thompson, A., Liu, W., & Corredig, M., (2012). *Liposomes as Food Ingredients and Nutraceutical Delivery Systems*, 287–318.

9. Ding, B., Zhang, X., Hayat, K., Xia, S., Jia, C., Xie, M., & Liu, C., (2011). Preparation, characterization, and the stability of ferrous glycinate nanoliposomes. *Journal of Food Engineering, 102*(2), 202–208.

10. Xu, Z., Liu, S., Wang, H., Gao, G., Yu, P., & Chang, Y., (2014). Encapsulation of iron in liposomes significantly improved the efficiency of iron supplementation in strenuously exercised rats. *Biological Trace Element Research, 162*(1–3), 181–188.

11. Khanniri, E., Bagheripoor-Fallah, N., Sohrabvandi, S., Mortazavian, A. M., Khosravi-Darani, K., & Mohammad, R., (2016). Application of liposomes in some dairy products. *Critical Reviews in Food Science and Nutrition, 56*(3), 484–493.

12. Liu, W., Ye, A., Liu, W., Liu, C., & Singh, H., (2013). Liposomes as food ingredients and nutraceutical delivery systems. *Agro Food Industry Hi-Tech, 24*(2), 68–71.

13. Schrooyen, P. M., Van, D. M. R., De Kruif, C. G., (2001). Microencapsulation: Its application in nutrition. *The Proceedings of the Nutrition Society, 60*(4), 475–479.

14. Baomiao, D., Xiangzhou, Y., Li, L., & Hualin, Y., (2017). Evaluation of iron transport from ferrous glycinate liposomes using Caco-2 cell model. *African Health Sciences, 17*(3), 933–941.

15. (a) Fairfield, K. M., & Fletcher, R. H., (2002). Vitamins for chronic disease prevention in adults: Scientific review. *JAMA, 287*(23), 3116–3126. (b) Henríquez-Sánchez, P., Sánchez-Villegas, A., Doreste-Alonso, J., Ortiz-Andrellucchi, A., Pfrimer, K., & Serra-Majem, L., (2009). Dietary assessment methods for micronutrient intake: A systematic review on vitamins. *British Journal of Nutrition, 102*(S1), S10–S37.

16. (a) Gregory, J. F., (2012). Accounting for differences in the bioactivity and bioavailability of vitamers. *Food & Nutrition Research, 56*(1), 5809. (b) Eitenmiller, R. R., Landen, Jr. W., & Ye, L., (2016). *Vitamin Analysis for the Health and Food Sciences*. CRC Press.

17. (a) Van, A. S. A., Koymans, L. M., & Bast, A., (1993). Molecular pharmacology of vitamin E: Structural aspects of antioxidant activity. *Free Radical Biology and Medicine, 15*(3), 311–328. (b) Dutta, A., & Dutta, S. K., (2003). Vitamin E and its role in the prevention of atherosclerosis and carcinogenesis: A review. *Journal of the American College of Nutrition, 22*(4), 258–268.

18. Jayaprakash, G., Sathiyabarathi, M., Robert, M. A., & Tamilmani, T., (2016). Rumen-protected choline: A significance effect on dairy cattle nutrition. *Veterinary World, 9*(8), 837.

19. Davis, J. L., Paris, H. L., Beals, J. W., Binns, S. E., Giordano, G. R., Scalzo, R. L., Schweder, M. M., et al., (2016). Liposomal-encapsulated ascorbic acid: Influence on vitamin C bioavailability and capacity to protect against ischemia-reperfusion injury. *Nutrition and Metabolic Insights, 9*, NMI. S39764.

20. (a) Depeint, F., Bruce, W. R., Shangari, N., Mehta, R., & O'Brien, P. J., (2006). Mitochondrial function and toxicity: Role of the B vitamin family on mitochondrial energy metabolism. *Chemico-Biological Interactions, 163*(1, 2), 94–112. (b) Afonsky, D., (1950). Oral aspect of vitamin B complex deficiency. *Oral Surgery, Oral Medicine, Oral Pathology, 3*(10), 1299–1330.

21. Lonsdale, D., (2006). A review of the biochemistry, metabolism and clinical benefits of thiamin(e) and its derivatives. *Evidence-Based Complementary and Alternative Medicine: ECAM, 3*(1), 49–59.

22. Fischer, M., & Bacher, A., (2008). Biosynthesis of vitamin B2: Structure and mechanism of riboflavin synthase. *Arch Biochem. Biophys., 474*(2), 252–265.

23. Gasperi, V., Sibilano, M., Savini, I., & Catani, M. V., (2019). Niacin in the central nervous system: An update of biological aspects and clinical applications. *International Journal of Molecular Sciences, 20*(4), 974.

24. Patassini, S., Begley, P., Xu, J., Church, S. J., Kureishy, N., Reid, S. J., Waldvogel, H. J., et al., (2019). Cerebral vitamin B5 (D-pantothenic acid) deficiency as a potential cause of metabolic perturbation and neurodegeneration in Huntington's disease. *Metabolites, 9*(6), 113.

25. Parra, M., Stahl, S., & Hellmann, H., (2018). Vitamin B$_6$ and its role in cell metabolism and physiology. *Cells, 7*(7), 84.

26. Zempleni, J., Hassan, Y. I., & Wijeratne, S. S., (2008). Biotin and biotinidase deficiency. *Expert Review of Endocrinology & Metabolism, 3*(6), 715–724.

27. Kunisawa, J., Hashimoto, E., Ishikawa, I., & Kiyono, H., (2012). A pivotal role of vitamin B9 in the maintenance of regulatory T cells *in vitro* and *in vivo*. *PloS One, 7*(2), e32094–e32094.

28. Giedyk, M., Goliszewska, K., & Gryko, D., (2015). Vitamin B12 catalyzed reactions. *Chemical Society Reviews, 44*(11), 3391–3404.

29. (a) Fernández-Bañares, F., Monzón, H., & Forné, M., (2009). A short review of malabsorption and anemia. *World Journal of Gastroenterology: WJG, 15*(37), 4644. (b) Moore, E., Mander, A., Ames, D., Carne, R., Sanders, K., & Watters, D., (2012). Cognitive impairment and vitamin B12: A review. *International Psychogeriatrics, 24*(4), 541–556. (c) Tucker, K. L., Hannan, M. T., Qiao, N., Jacques, P. F., Selhub, J., Cupples, L. A., & Kiel, D. P., (2005). Low plasma vitamin B12 is associated with lower BMD: The Framingham osteoporosis study. *Journal of Bone and Mineral Research, 20*(1), 152–158.

30. Tucker, K. L., Rich, S., Rosenberg, I., Jacques, P., Dallal, G., Wilson, P. W., & Selhub, J., (2000). Plasma vitamin B-12 concentrations relate to intake source in the Framingham Offspring Study. *The American Journal of Clinical Nutrition, 71*(2), 514–522.

31. (a) Doscherholmen, A., & Hagen, P. S., (1957).A dual mechanism of vitamin B_{12} plasma absorption. *The Journal of Clinical Investigation, 36*(11), 1551–1557. (b) Herbert, V., (1968). Absorption of vitamin B12 and folic acid. *Gastroenterology, 54*(1), 110–115.

32. Vitetta, L., Zhou, J., Manuel, R., Dal, F. S., Hall, S., & Rutolo, D., (2018). Route and type of formulation administered influences the absorption and disposition of vitamin B12 levels in serum. *Journal of Functional Biomaterials, 9*(1), 12.

33. Katsogiannis, I., Nikolaos, F., Christos, K., & Theodoros, C., (2018). Evaluation of liposomal B12 supplementation in a case series study. *Glob. Drugs Therap., 3*(5), 1–4.

34. (a) De Jager, J., Kooy, A., Lehert, P., Wulffelé, M. G., Van, D. K. J., Bets, D., Verburg, J., Donker, A. J., & Stehouwer, C. D., (2010). Long term treatment with metformin in patients with type 2 diabetes and risk of vitamin B-12 deficiency: Randomized placebo-controlled trial. *BMJ, 340*, c2181. (b) Buvat, D., (2004). Use of metformin is a cause of vitamin B12 deficiency. *American Family Physician, 69*(2), 264. (c) Andrès, E., Noel, E., & Goichot, B., (2002). Metformin-associated vitamin B12 deficiency. *Archives of Internal Medicine, 162*(19), 2251, 2252.

35. Sharabi, A., Cohen, E., Sulkes, J., & Garty, M., (2003). Replacement therapy for vitamin B12 deficiency: Comparison between the sublingual and oral route. *British Journal of Clinical Pharmacology, 56*(6), 635–638.

36. Duarte, T. L., & Lunec, J., (2005). When is an antioxidant not an antioxidant?: A review of novel actions and reactions of vitamin C. *Free Radical Research, 39*(7), 671–686.

37. Bendich, A., & Langseth, L., (1995). The health effects of vitamin C supplementation: A review. *Journal of the American College of Nutrition, 14*(2), 124–136.

38. (a) Fidge, N. H., Shiratori, T., Ganguly, J., & Goodman, D. S., (1968). Pathways of absorption of retinal and retinoic acid in the rat. *Journal of Lipid Research, 9*(1), 103–109. (b) Vogel, S., Gamble, M., & Blaner, W., (1999). Biosynthesis, absorption, metabolism, and transport of retinoids. In: *Retinoids* (pp. 31–95). Springer.

39. Fragoso, Y. D., Campos, N. S., Tenrreiro, B. F., & Guillen, F. J., (2012). Systematic review of the literature on vitamin A and memory. *Dementia & Neuropsychologia, 6*(4), 219–222.

40. Rando, R. R., (1990). The chemistry of vitamin A and vision. *Angewandte Chemie International Edition in English, 29*(5), 461–480.

41. (a) Griffiths, C., Russman, A. N., Majmudar, G., Singer, R. S., Hamilton, T. A., & Voorhees, J. J., (1993). Restoration of collagen formation in photodamaged human skin by tretinoin (retinoic acid). *New England Journal of Medicine, 329*(8), 530–535. (b) Niederreither, K., & Dollé, P., (2008). Retinoic acid in development: Towards an integrated view. *Nature Reviews Genetics, 9*(7), 541–553.

42. Sachaniya, J., Savaliya, R., Goyal, R., & Singh, S., (2018). Liposomal formulation of vitamin A for the potential treatment of osteoporosis. *International Journal of Nanomedicine, 13*(T-NANO 2014 Abstracts), 51.

43. Singh, A. K., & Das, J., (1998). Liposome encapsulated vitamin A compounds exhibit greater stability and diminished toxicity. *Biophysical Chemistry, 73*(1, 2), 155–162.

44. (a) DeLuca, H. F., (2004). Overview of general physiologic features and functions of vitamin D. *The American Journal of Clinical Nutrition, 80*(6), 1689S–1696S.

(b) DeLuca, H. F., & Zierold, C., (1998). Mechanisms and functions of vitamin D. *Nutrition Reviews, 56*(1), S4–S10.

45. Shearer, M., (1992). Vitamin K metabolism and nutriture. *Blood Reviews, 6*(2), 92–104.

46. Booth, S. L., (1997). Skeletal functions of vitamin K-dependent proteins: Not just for clotting anymore. *Nutrition Reviews, 55*(7), 282–284.

47. Dalmoro, A., Bochicchio, S., Lamberti, G., Bertoncin, P., Janssens, B., & Barba, A. A., (2019). Micronutrients encapsulation in enhanced nanoliposomal carriers by a novel preparative technology. *RSC Advances, 9*(34), 19800–19812.

48. Brigelius-Flohe, R., & Traber, M. G., (1999). Vitamin E: Function and metabolism. *The FASEB Journal, 13*(10), 1145–1155.

49. Saremi, A., & Arora, R., (2010). Vitamin E and cardiovascular disease. *American Journal of Therapeutics, 17*(3), e56–e65.

50. Suntres, Z. E., (2011). Liposomal antioxidants for protection against oxidant-induced damage. *Journal of Toxicology, 2011*.

51. Nacka, F., Cansell, M., Méléard, P., & Combe, N., (2001). Incorporation of α-tocopherol in marine lipid-based liposomes: *In vitro* and *in vivo* studies. *Lipids, 36*(12), 1313–1320.

52. Blusztajn, J. K., (1998). Choline, a vital amine. *Science, 281*(5378), 794–795.

53. Marwaha, S. S., Sethi, R. P., & Kennedy, J. F., (1983). Role of amino acids, betaine and choline in vitamin B12 biosynthesis by strains of Propionibacterium. *Enzyme and Microbial Technology, 5*(6), 454–456.

54. Mödinger, Y., Schön, C., Wilhelm, M., & Hals, P. A., (2019). Plasma kinetics of choline and choline metabolites after a single dose of superba Boost™ krill oil or choline bitartrate in healthy volunteers. *Nutrients, 11*(10), 2548.

55. Kumar, D., Vats, N., Saroha, K., & Rana, A. C., (2020). Phytosomes as emerging nanotechnology for herbal drug delivery. In: *Sustainable Agriculture Reviews* (Vol. 43, pp. 217–237). Springer.

56. Khursheed, R., Singh, S. K., Wadhwa, S., Gulati, M., & Awasthi, A., (2020). Enhancing the potential preclinical and clinical benefits of quercetin through novel drug delivery systems. *Drug Discovery Today, 25*(1), 209–222.

57. (a) Ajay, S., Vikas, P., Rajesh, S., Punit, B., & Suchit, J., (2012). Herbosomes: A current concept of herbal drug technology, an overview. *Journal of Medical Pharmaceutical and Allied Sciences, 1*. (b) Semalty, A., Semalty, M., Rawat, M. S. M., & Franceschi, F., (2010). Supramolecular phospholipids-polyphenolics interactions: The PHYTOSOME® strategy to improve the bioavailability of phytochemicals. *Fitoterapia, 81*(5), 306–314.

58. (a) Samad, A., Sultana, Y., & Aqil, M., (2007). Liposomal drug delivery systems: An updated review. *Current Drug Delivery, 4*(4), 297–305. (b) Sapra, P., & Allen, T., (2003). Ligand-targeted liposomal anticancer drugs. *Progress in Lipid Research, 42*(5), 439–462. (c) Huwyler, J., Drewe, J., & Krähenbühl, S., (2008). Tumor targeting using liposomal antineoplastic drugs. *International Journal of Nanomedicine, 3*(1), 21. (d) Muehlmann, L., Joanitti, G., Silva, J., Longo, J., & Azevedo, R., (2011). Liposomal photosensitizers: Potential platforms for anticancer photodynamic therapy. *Brazilian Journal of Medical and Biological Research, 44*(8), 729–737. (e) Nagar, G., (2019). *Phytosomes: A Novel Drug Delivery for Herbal Extracts.*

59. Valenti, D., De Logu, A., Loy, G., Sinico, C., Bonsignore, L., Cottiglia, F., Garau, D., & Fadda, A. M., (2001). Liposome-incorporated *Santolina insularis* essential oil: Preparation, characterization and *in vitro* antiviral activity. *Journal of Liposome Research, 11*(1), 73–90.

60. Ortan, A., Campeanu, G., Dinu-Pirvu, C., & Popescu, L., (2009). Studies concerning the entrapment of *Anethum graveolens* essential oil in liposomes. *Room. Biotechnol. Lett., 14*, 4411–4417.

61. Detoni, C. B., De Oliveira, D. M., Santo, I. E., Pedro, A. S., El-Bacha, R., Da Silva, V. E., Ferreira, D., et al., (2012). Evaluation of thermal-oxidative stability and anti-glioma activity of *Zanthoxylum tingoassuiba* essential oil entrapped into multi-and unilamellar liposomes. *Journal of Liposome Research, 22*(1), 1–7.

62. Fan, M., Xu, S., Xia, S., & Zhang, X., (2007). Effect of different preparation methods on physicochemical properties of salidroside liposomes. *Journal of Agricultural and Food Chemistry, 55*(8), 3089–3095.

63. Takahashi, M., Uechi, S., Takara, K., Asikin, Y., & Wada, K., (2009). Evaluation of an oral carrier system in rats: Bioavailability and antioxidant properties of liposome-encapsulated curcumin. *Journal of Agricultural and Food Chemistry, 57*(19), 9141–9146.

64. Isailović, B. D., Kostić, I. T., Zvonar, A., Đorđević, V. B., Gašperlin, M., Nedović, V. A., & Bugarski, B. M., (2013). Resveratrol loaded liposomes produced by different techniques. *Innovative Food Science & Emerging Technologies, 19*, 181–189.

65. Lin, Y. L., Liu, Y. K., Tsai, N. M., Hsieh, J. H., Chen, C. H., Lin, C. M., & Liao, K. W., (2012). A Lipo-PEG-PEI complex for encapsulating curcumin that enhances its antitumor effects on curcumin-sensitive and curcumin-resistance cells. *Nanomedicine: Nanotechnology, Biology and Medicine, 8*(3), 318–327.

66. Yu, H., Teng, L., Meng, Q., Li, Y., Sun, X., Lu, J., Lee, R. J., & Teng, L., (2013). Development of liposomal Ginsenoside Rg3: Formulation optimization and evaluation of its anti-cancer effects. *International Journal of Pharmaceutics, 450*(1, 2), 250–258.

67. Li, H., Li, S., & Duan, H., (2007). Preparation of liposomes containing extracts of *Tripterygium wilfordii* and evaluation of its stability. *Zhongguo Zhong Yao za zhi= Zhongguo Zhongyao Zazhi= China Journal of Chinese Materia Medica, 32*(20), 2128–2131.

68. Priprem, A., Watanatorn, J., Sutthiparinyanont, S., Phachonpai, W., & Muchimapura, S., (2008). Anxiety and cognitive effects of quercetin liposomes in rats. *Nanomedicine: Nanotechnology, Biology and Medicine, 4*(1), 70–78.

69. El-Samaligy, M. S., Afifi, N. N., & Mahmoud, E. A., (2006). Evaluation of hybrid liposomes-encapsulated silymarin regarding physical stability and *in vivo* performance. *International Journal of Pharmaceutics, 319*(1, 2), 121–129.

70. (a) Bombardelli, E., & Mustich, G., (1991). *Bilobalide Phospholipid Complex, Their Uses and Formulation Containing Them.* US Patent No. EPO-275005. (b) Amin, T., & Bhat, S. V., (2012). A review on phytosome technology as a novel approach to improve the bioavailability of nutraceuticals. *Int. J. Adv. Res. Technol., 1*(3), 1–5. (c) Tripathy, S., Patel, D. K., Barob, L., & Naira, S. K., (2013). A review on phytosomes, their characterization, advancement & potential for transdermal application. *Journal of Drug Delivery and Therapeutics, 3*(3), 147–152. (d) Dhase, A. S., & Saboo, S. S.,

(2015). Preparation and evaluation of phytosomes containing methanolic extract of leaves of Aegle marmelos (Bael). *Int. J. PharmTech Res., 8*(6), 231–240.

71. Amit, P., Tanwar, Y., Rakesh, S., & Poojan, P., (2013). Phytosome: Phytolipid drug delivery system for improving bioavailability of herbal drug. *J. Pharm. Sci. Biosci. Res., 3*(2), 51–57.

72. Lu, M., Qiu, Q., Luo, X., Liu, X., Sun, J., Wang, C., Lin, X., et al., (2019). Phyto-phospholipid complexes (phytosomes): A novel strategy to improve the bioavailability of active constituents. *Asian Journal of Pharmaceutical Sciences, 14*(3), 265–274.

73. Rahman, S., Cao, S., Steadman, K. J., Wei, M., & Parekh, H. S., (2012). Native and β-cyclodextrin-enclosed curcumin: Entrapment within liposomes and their *in vitro* cytotoxicity in lung and colon cancer. *Drug Delivery, 19*(7), 346–353.

74. Moghimi, H. R., Shirazi, F. H., Shafiee, A. M., Oghabian, M. A., Saffari, M., & Sojoudi, J., (2015). *In vitro* and *in vivo* enhancement of antitumoral activity of liposomal antisense oligonucleotides by cineole as a chemical penetration enhancer. *Journal of Nanomaterials, 2015.*

75. El-Samaligy, M., Afifi, N., & Mahmoud, E., (2006). Increasing bioavailability of silymarin using a buccal liposomal delivery system: Preparation and experimental design investigation. *International Journal of Pharmaceutics, 308*(1, 2), 140–148.

76. Shariare, M. H., Rahman, M., Lubna, S. R., Roy, R. S., Abedin, J., Marzan, A. L., Altamimi, M. A., et al., (2020). Liposomal drug delivery of *Aphanamixis polystachya* leaf extracts and its neurobehavioral activity in mice model. *Scientific Reports, 10*(1), 1–16.

77. Aw-Yong, P. Y., Gan, P. H., Sasmita, A. O., Mak, S. T., & Ling, A., (2018). Nanoparticles as carriers of phytochemicals: Recent applications against lung cancer. *Int. J. Res. Biomed. Biotechnol, 7*, 1–11.

78. Babazadeh, A., Zeinali, M., & Hamishehkar, H., (2018). Nano-phytosome: A developing platform for herbal anti-cancer agents in cancer therapy. *Current Drug Targets, 19*(2), 170–180.

79. Naik, S. R., & Panda, V. S., (2008). Hepatoprotective effect of Ginkgoselect Phytosome® in rifampicin induced liver injury in rats: Evidence of antioxidant activity. *Fitoterapia, 79*(6), 439–445.

80. (a) Marczylo, T. H., Verschoyle, R. D., Cooke, D. N., Morazzoni, P., Steward, W. P., & Gescher, A. J., (2007). Comparison of systemic availability of curcumin with that of curcumin formulated with phosphatidylcholine. *Cancer Chemotherapy and Pharmacology, 60*(2), 171–177. (b) Kidd, P. M., (2009). Bioavailability and activity of phytosome complexes from botanical polyphenols: The silymarin, curcumin, green tea, and grape seed extracts. *Altern Med. Rev., 14*(3), 226–246.

81. Bakkali, F., Averbeck, S., Averbeck, D., & Idaomar, M., (2008). Biological effects of essential oils-a review. *Food and Chemical Toxicology, 46*(2), 446–475.

82. Liolios, C., Gortzi, O., Lalas, S., Tsaknis, J., & Chinou, I., (2009). Liposomal incorporation of carvacrol and thymol isolated from the essential oil of *Origanum dictamnus* L. and *in vitro* antimicrobial activity. *Food Chemistry, 112*(1), 77–83.

83. Moghimipour, E., Aghel, N., Mahmoudabadi, A. Z., Ramezani, Z., & Handali, S., (2012). Preparation and characterization of liposomes containing essential oil of *Eucalyptus camaldulensis* leaf. *Jundishapur Journal of Natural Pharmaceutical Products, 7*(3), 117.

84. Shi, F., Zhao, J. H., Liu, Y., Wang, Z., Zhang, Y. T., & Feng, N. P., (2012). Preparation and characterization of solid lipid nanoparticles loaded with frankincense and myrrh oil. *International Journal of Nanomedicine, 7,* 2033.

85. Zhao, Y., Wang, C., Chow, A. H., Ren, K., Gong, T., Zhang, Z., & Zheng, Y., (2010). Self-nano emulsifying drug delivery system (SNEDDS) for oral delivery of zedoary essential oil: Formulation and bioavailability studies. *International Journal of Pharmaceutics, 383*(1, 2), 170–177.

86. Xi, J., Chang, Q., Chan, C. K., Meng, Z. Y., Wang, G. N., Sun, J. B., Wang, Y. T., et al., (2009). Formulation development and bioavailability evaluation of a self-nano emulsified drug delivery system of oleanolic acid. *AAPS Pharmscitech, 10*(1), 172–182.

87. Abedi, E., & Sahari, M. A., (2014). Long-chain polyunsaturated fatty acid sources and evaluation of their nutritional and functional properties. *Food Science & Nutrition, 2*(5), 443–463.

88. Hadian, Z., (2016). A review of nanoliposomal delivery system for stabilization of bioactive omega-3 fatty acids. *Electronic Physician, 8*(1), 1776.

89. (a) Saini, R. K., & Keum, Y. S., (2018). Omega-3 and omega-6 polyunsaturated fatty acids: Dietary sources, metabolism, and significance: A review. *Life Sciences, 203,* 255–267. (b) Belchior, T., Paschoal, V. A., Magdalon, J., Chimin, P., Farias, T. M., Chaves-Filho, A. B., Gorjão, R., et al., (2015). Omega-3 fatty acids protect from diet-induced obesity, glucose intolerance, and adipose tissue inflammation through PPARγ-dependent and PPARγ-independent actions. *Molecular Nutrition & Food Research, 59*(5), 957–967.

90. Ghorbanzade, T., Jafari, S. M., Akhavan, S., & Hadavi, R., (2017). Nano-encapsulation of fish oil in nano-liposomes and its application in fortification of yogurt. *Food Chemistry, 216,* 146–152.

91. Schreiner, M., & Windisch, W., (2006). Supplementation of cow diet with rapeseed and carrots: Influence on fatty acid composition and carotene content of the butterfat. *Journal of Food Lipids, 13*(4), 434–444.

92. Rasti, B., Erfanian, A., & Selamat, J., (2017). Novel nanoliposomal encapsulated omega-3 fatty acids and their applications in food. *Food Chemistry, 230,* 690–696.

93. Jacobsen, C., Let, M. B., Nielsen, N. S., & Meyer, A. S., (2008). Antioxidant strategies for preventing oxidative flavor deterioration of foods enriched with n-3 polyunsaturated lipids: A comparative evaluation. *Trends in Food Science & Technology, 19*(2), 76–93.

94. (a) Trautwein, E., (2001). n-3 Fatty acids—physiological and technical aspects for their use in food. *European Journal of Lipid Science and Technology, 103*(1), 45–55. (b) Martini, S., Thurgood, J., Brothersen, C., Ward, R., & McMahon, D. J., (2009). Fortification of reduced-fat Cheddar cheese with n-3 fatty acids: Effect on off-flavor generation. *Journal of Dairy Science, 92*(5), 1876–1884. (c) Henna, L. F., & Norziah, M., (2011). Contribution of microencapsulated n-3 PUFA powder toward sensory and oxidative stability of bread. *Journal of Food Processing and Preservation, 35*(5), 596–604. (d) Liu, Q., Wang, J., Bu, D., Liu, K., Wei, H., Zhou, L., & Beitz, D. C., (2010). Influence of linolenic acid content on the oxidation of milk fat. *Journal of Agricultural and Food Chemistry, 58*(6), 3741–3746.

95. (a) Araseki, M., Yamamoto, K., & Miyashita, K., (2002). Oxidative stability of polyunsaturated fatty acid in phosphatidylcholine liposomes. *Bioscience, Biotechnology, and Biochemistry, 66*(12), 2573–2577. (b) Rasti, B., Jinap, S., Mozafari, M., & Yazid, A., (2012). Comparative study of the oxidative and physical stability of liposomal and nanoliposomal polyunsaturated fatty acids prepared with conventional and Mozafari methods. *Food Chemistry, 135*(4), 2761–2770.

96. (a) Pourashouri, P., Shabanpour, B., Razavi, S. H., Jafari, S. M., Shabani, A., & Aubourg, S. P., (2014). Impact of wall materials on physicochemical properties of microencapsulated fish oil by spray drying. *Food and Bioprocess Technology, 7*(8), 2354–2365. (b) Rosenberg, M., Kopelman, I., & Talmon, Y., (1990). Factors affecting retention in spray-drying microencapsulation of volatile materials. *Journal of Agricultural and Food Chemistry, 38*(5), 1288–1294. (c) Kagami, Y., Sugimura, S., Fujishima, N., Matsuda, K., Kometani, T., & Matsumura, Y., (2003). Oxidative stability, structure, and physical characteristics of microcapsules formed by spray drying of fish oil with protein and dextrin wall materials. *Journal of Food Science, 68*(7), 2248–2255; (d) Kaushik, P., Dowling, K., Barrow, C. J., & Adhikari, B., (2015). Microencapsulation of omega-3 fatty acids: A review of microencapsulation and characterization methods. *Journal of Functional Foods, 19*, 868–881.

97. Heinzelmann, K., Franke, K., Jensen, B., & Haahr, A. M., (2000). Protection of fish oil from oxidation by microencapsulation using freeze-drying techniques. *European Journal of Lipid Science and Technology, 102*(2), 114–121.

98. Anwar, S. H., & Kunz, B., (2011). The influence of drying methods on the stabilization of fish oil microcapsules: Comparison of spray granulation, spray drying, and freeze-drying. *Journal of Food Engineering, 105*(2), 367–378.

99. Hannah, S., (2009). *Microencapsulation of an Omega-3 Polyunsaturated Fatty Acid Source with Polysaccharides for Food Applications.* Virginia Tech.

100. Shahidi, F., & Zhong, Y., (2010). Lipid oxidation and improving the oxidative stability. *Chemical Society Reviews, 39*(11), 4067–4079.

101. Hadian, Z., Sahari, M. A., Moghimi, H. R., & Barzegar, M., (2014). Formulation, characterization and optimization of liposomes containing eicosapentaenoic and docosahexaenoic acids; a methodology approach. *Iranian Journal of Pharmaceutical Research: IJPR, 13*(2), 393.

102. (a) Liu, S., Low, N., & Nickerson, M. T., (2010). Entrapment of flaxseed oil within gelatin-gum Arabic capsules. *Journal of the American Oil Chemists' Society, 87*(7), 809–815. (b) Barrow, C. J., Nolan, C., & Jin, Y., (2007). Stabilization of highly unsaturated fatty acids and delivery into foods. *Lipid Technology, 19*(5), 108–111.

103. Choi, M. J., Ruktanonchai, U., Min, S. G., Chun, J. Y., & Soottitantawat, A., (2010). Physical characteristics of fish oil encapsulated by β-cyclodextrin using an aggregation method or polycaprolactone using an emulsion-diffusion method. *Food Chemistry, 119*(4), 1694–1703.

104. Kubo, K., Sekine, S., & Saito, M., (2003). Docosahexaenoic acid-containing phosphatidylethanolamine in the external layer of liposomes protects docosahexaenoic acid from 2, 2'-azobis (2-aminopropane) dihydrochloride-mediated lipid peroxidation. *Archives of Biochemistry and Biophysics, 410*(1), 141–148.

105. Torres-Giner, S., Martinez-Abad, A., Ocio, M. J., & Lagaron, J. M., (2010). Stabilization of a nutraceutical omega-3 fatty acid by encapsulation in ultrathin electrosprayed zein prolamine. *Journal of Food Science, 75*(6), N69–N79.

106. Anwar, S. H., Weissbrodt, J., & Kunz, B., (2010). Microencapsulation of fish oil by spray granulation and fluid bed film coating. *Journal of Food Science, 75*(6), E359–E371.

107. Legako, J., & Dunford, N. T., (2010). Effect of spray nozzle design on fish oil-whey protein microcapsule properties. *Journal of Food Science, 75*(6), E394–E400.

108. Nara, E., Miyashita, K., Ota, T., & Nadachi, Y., (1998). The oxidative stabilities of polyunsaturated fatty acids in salmon egg phosphatidylcholine liposomes. *Fisheries Science, 64*(2), 282–286.

109. (a) Winterbourn, C. C., (2008). Reconciling the chemistry and biology of reactive oxygen species. *Nature Chemical Biology, 4*(5), 278. (b) Matés, J. M., & Sánchez-Jiménez, F. M., (2000). Role of reactive oxygen species in apoptosis: Implications for cancer therapy. *The International Journal of Biochemistry & Cell Biology, 32*(2), 157–170. (c) Dickinson, B. C., & Chang, C. J., (2011). Chemistry and biology of reactive oxygen species in signaling or stress responses. *Nature Chemical Biology, 7*(8), 504.

110. (a) Uttara, B., Singh, A. V., Zamboni, P., & Mahajan, R., (2009). Oxidative stress and neurodegenerative diseases: A review of upstream and downstream antioxidant therapeutic options. *Current Neuropharmacology, 7*(1), 65–74. (b) Rao, A., & Agarwal, S., (1999). Role of lycopene as antioxidant carotenoid in the prevention of chronic diseases: A review. *Nutrition Research, 19*(2), 305–323. (c) Alfadda, A. A., & Sallam, R. M., (2012). Reactive oxygen species in health and disease. *BioMed. Research International, 2012.*

111. Guerra-Araiza, C., Álvarez-Mejía, A. L., Sánchez-Torres, S., Farfan-García, E., Mondragón-Lozano, R., Pinto-Almazán, R., & Salgado-Ceballos, H., (2013). Effect of natural exogenous antioxidants on aging and on neurodegenerative diseases. *Free Radical Research, 47*(6, 7), 451–462.

112. Allen, T. M., & Cullis, P. R., (2013). Liposomal drug delivery systems: From concept to clinical applications. *Advanced Drug Delivery Reviews, 65*(1), 36–48.

113. Oberley, L. W., & Buettner, G. R., (1979). Role of superoxide dismutase in cancer: A review. *Cancer Research, 39*(4), 1141–1149.

114. Mahaseth, T., & Kuzminov, A., (2017). Potentiation of hydrogen peroxide toxicity: From catalase inhibition to stable DNA-iron complexes. *Mutation Research/Reviews in Mutation Research, 773*, 274–281.

115. (a) Reddy, M. K., & Labhasetwar, V., (2009). Nanoparticle-mediated delivery of superoxide dismutase to the brain: An effective strategy to reduce ischemia-reperfusion injury. *The FASEB Journal, 23*(5), 1384–1395. (b) Giovagnoli, S., Blasi, P., Ricci, M., & Rossi, C., (2004). Biodegradable microspheres as carriers for native superoxide dismutase and catalase delivery. *AAPS PharmSciTech, 5*(4), 1–9.

116. Petelin, M., Pavlica, Z., Ivanuša, T., Šentjurc, M., & Skalerič, U., (2000). Local delivery of liposome-encapsulated superoxide dismutase and catalase suppress periodontal inflammation in beagles. *Journal of Clinical Periodontology, 27*(12), 918–925.

117. Al-Hasso, S., (2001). Coenzyme Q10: A review. *Hospital Pharmacy, 36*(1), 51–66.
118. (a) Quinzii, C. M., Hirano, M., & DiMauro, S., (2007). CoQ_{10} deficiency diseases in adults. *Mitochondrion, 7*, S122–S126. (b) Spindler, M., Beal, M. F., & Henchcliffe, C., (2009). Coenzyme Q10 effects in neurodegenerative disease. *Neuropsychiatric Disease and Treatment, 5*, 597.
119. Yokoyama, H., Lingle, D. M., Crestanello, J. A., Kamelgard, J., Kott, B. R., Momeni, R., Millili, J., et al., (1996). Coenzyme Q10 protects coronary endothelial function from ischemia-reperfusion injury via an antioxidant effect. *Surgery, 120*(2), 189–196.
120. Balakrishnan, P., Lee, B. J., Oh, D. H., Kim, J. O., Lee, Y. I., Kim, D. D., Jee, J. P., et al., (2009). Enhanced oral bioavailability of coenzyme Q10 by self-emulsifying drug delivery systems. *International Journal of Pharmaceutics, 374*(1, 2), 66–72.
121. Xia, S., Xu, S., Zhang, X., Zhong, F., & Wang, Z., (2009). Nanoliposomes mediate coenzyme Q10 transport and accumulation across human intestinal Caco-2 cell monolayer. *Journal of Agricultural and Food Chemistry, 57*(17), 7989–7996.
122. Choi, C. H., Kim, S. H., Shanmugam, S., Baskaran, R., Park, J. S., Yong, C. S., Choi, H. G., et al., (2010). Relative bioavailability of coenzyme Q10 in emulsion and liposome formulations. *Biomolecules & Therapeutics, 18*(1), 99–105.
123. Salinas, A. E., & Wong, M. G., (1999). Glutathione S-transferases-a review. *Current Medicinal Chemistry, 6*(4), 279–310.
124. Franco, R., Schoneveld, O., Pappa, A., & Panayiotidis, M., (2007). The central role of glutathione in the pathophysiology of human diseases. *Archives of Physiology and Biochemistry, 113*(4, 5), 234–258.
125. Bernkop-Schnürch, A., Kast, C., & Guggi, D., (2003). Permeation enhancing polymers in oral delivery of hydrophilic macromolecules: Thiomer/GSH systems. *Journal of Controlled Release, 93*(2), 95–103.
126. Sinha, R., Sinha, I., Calcagnotto, A., Trushin, N., Haley, J. S., Schell, T. D., & Richie, J., (2018). Oral supplementation with liposomal glutathione elevates body stores of glutathione and markers of immune function. *European Journal of Clinical Nutrition, 72*(1), 105–111.
127. Bertoni, S., Albertini, B., Facchini, C., Prata, C., & Passerini, N., (2019). Glutathione-loaded solid lipid microparticles as innovative delivery system for oral antioxidant therapy. *Pharmaceutics, 11*(8).
128. Strauss, G., Schurtenberger, P., & Hauser, H., (1986). The interaction of saccharides with lipid bilayer vesicles: Stabilization during freeze-thawing and freeze-drying. *Biochimica et Biophysica Acta (BBA)-Biomembranes, 858*(1), 169–180.
129. Lasic, D. D., & Martin, F. J., (1995). *Stealth Liposomes* (Vol. 20). CRC Press.

Index

For Product Safety Concerns and Information please contact our EU
representative GPSR@taylorandfrancis.com
Taylor & Francis Verlag GmbH, Kaufingerstraße 24, 80331 München, Germany

www.ingramcontent.com/pod-product-compliance
Lightning Source LLC
Chambersburg PA
CBHW060335220326
41598CB00023B/2716

* 9 7 8 1 7 7 4 6 3 7 5 5 5 *